T0281005

Lecture Notes in Computer Science 14664

The series Lecture Notes in Computer Science (LNCS), including its subseries Lecture Notes in Artificial Intelligence (LNAI) and Lecture Notes in Bioinformatics (LNBI), has established itself as a medium for the publication of new developments in computer science and information technology research, teaching, and education.

LNCS enjoys close cooperation with the computer science R & D community, the series counts many renowned academics among its volume editors and paper authors, and collaborates with prestigious societies. Its mission is to serve this international community by providing an invaluable service, mainly focused on the publication of conference and workshop proceedings and postproceedings. LNCS commenced publication in 1973.

Albert Meroño Peñuela · Anastasia Dimou ·
Raphaël Troncy · Olaf Hartig · Maribel Acosta ·
Mehwish Alam · Heiko Paulheim ·
Pasquale Lisena
Editors

The Semantic Web

21st International Conference, ESWC 2024
Hersonissos, Crete, Greece, May 26–30, 2024
Proceedings, Part I

 Springer

Editors
Albert Meroño Peñuela 🆔
King's College London
London, UK

Anastasia Dimou 🆔
KU Leuven
Sint-Katelijne-Waver, Belgium

Raphaël Troncy 🆔
EURECOM
Biot, France

Olaf Hartig 🆔
Linköping University
Linköping, Sweden

Maribel Acosta 🆔
Technical University of Munich
Heilbronn, Germany

Mehwish Alam 🆔
Polytechnic Institute of Paris
Palaiseau, France

Heiko Paulheim 🆔
University of Mannheim
Mannheim, Germany

Pasquale Lisena 🆔
EURECOM
Biot, France

ISSN 0302-9743   ISSN 1611-3349 (electronic)
Lecture Notes in Computer Science
ISBN 978-3-031-60625-0   ISBN 978-3-031-60626-7 (eBook)
https://doi.org/10.1007/978-3-031-60626-7

Preface

This volume contains the main proceedings of the 21st edition of the European Semantic Web Conference (ESWC 2024). ESWC is a major venue for discussing the latest in scientific results and innovations related to the semantic web, knowledge graphs, and web data. This year we aimed at acknowledging recent developments in AI with a special tagline, "Fabrics of Knowledge: Knowledge Graphs and Generative AI". By doing so, and adapting the program as described below, we intended to open up the conference to fundamental questions about how we acquire, represent, use, and interact with knowledge in the advent of Generative AI and large language models.

This year ESWC's Research track addressed the theoretical, analytical, and empirical aspects of the Semantic Web, semantic technologies, knowledge graphs and semantics on the Web in general. The In-use track focused on contributions that reuse and apply state-of-the-art semantic technologies or resources to real-world settings. The Resource track welcomed resource contributions that are on the one hand innovative or novel and on the other hand sharable and reusable (e.g. datasets, knowledge graphs, ontologies, workflows, benchmarks, frameworks), and provide the necessary scaffolding to support the generation of scientific work and advance the state of the art.

The main scientific program of ESWC 2024 contained 32 papers selected out of 138 submissions (62 research, 19 in-use, 57 resource): 13 papers in the Research track, 5 in the In-Use track, and 14 in the Resource track. The overall acceptance rate was 23% (20% research, 26% in-use, 24% resource). Due to last year's success in innovating with the ESWC review process, this year we kept the approach of not including a final overall score before the rebuttal phase in the Research and Resource tracks. This enabled reviewers to focus on their reviews rather than scores, and on asking specific questions and to use the answers provided by authors in the rebuttal in their final assessment. The program chairs are grateful to the 40 senior PC members, the 240 PC members, and the 25 external reviewers for providing their feedback on the scientific program, and to all other community members who contributed in reviewing. Each paper received an average of 3.6 reviews, with both the Research and Resources being dual anonymous, and In-Use being single anonymous. We adopted ACM's terminology in all calls to improve ESWC's Diversity, Equity, and Inclusion principles in how we communicated review guidelines.[1]

We welcomed invited keynotes from three world renowned speakers, spanning industry and academia and in keeping with our theme of building understanding between the knowledge graph and generative AI communities: Elena Simperl (King's College London & Open Data Institute), Peter Clark (Allen Institute for Artificial Intelligence, AI2), and Katariina Kari (IKEA Systems B.V.).

As part of the conference's special topic, ESWC 2024 featured a Special Track on Large Language Models for Knowledge Engineering, providing a venue for scientific discussion and community building for early work on this exciting new area of research.

---

[1] See "Words Matter" https://www.acm.org/diversity-inclusion/words-matter.

The track had 52 submissions, demonstrating the importance of research in this direction. The number of selected papers for presentation and publication for this track is reported in the Satellite Volume.

ESWC 2024 also had a record-breaking number of workshop and tutorial proposals, of which 16 workshops and 2 tutorials were accepted. Their large variety of topics included data management, natural language processing, sustainability, and generative neuro-symbolic AI, aligning with this year's conference topic.

The conference also offered other opportunities to discuss the latest research and innovation work, including a poster and demo session, workshops and tutorials, a PhD symposium, an EU project networking session, and an industry track. We thank Joe Raad and Bruno Sartini for organising the Workshop and Tutorials track, which hosted 16 workshops and 2 tutorials covering topics ranging from knowledge graph construction to deep learning with knowledge graphs. We are also thankful to María Poveda-Villalón and Andrea Nuzzolese for successfully running the Posters and Demos track. We are grateful to Marta Sabou and Valentina Presutti for coordinating a very special PhD Symposium aiming at bridging the ESWC PhD and the International Semantic Web Summer School (ISWS) alumni, which welcomed 14 PhD students who had the opportunity to present their work and receive feedback in a constructive environment. Thanks go to Irene Celino and Artem Revenko for their management of the Industry Track which welcomed submissions from several large industry players. We also thank Cassia Trojahn and Sabrina Kirrane for increasing the networking potential of ESWC by running the Project Networking Session. A special thanks to Tabea Tietz and Stefano De Giorgis for their amazing job as Web and Publicity chairs, and to Pasquale Lisena for preparing this volume with Springer and the conference metadata. We thank STI International for supporting the conference organization, and in particular Umut Serles and Juliette Opdenplatz for their invaluable support. We thank our sponsors for supporting ESWC 2024 and also our sponsorship chairs Nitisha Jain and Jan-Christoph Kalo for securing them. Finally, we are also grateful to Dieter Fensel, John Domingue, Elena Simperl, Paul Groth, Catia Pesquita and the ESWC 2023 organising committee for their invaluable support and advice.

As we reflect on both the past and the future, and our role as researchers and technologists, our thoughts go out to all those impacted by war around the world.

April 2024

Albert Meroño Peñuela
Anastasia Dimou
Raphaël Troncy
Olaf Hartig
Maribel Acosta
Mehwish Alam
Heiko Paulheim
Pasquale Lisena

Organization

General Chair

Albert Meroño Peñuela   King's College London, UK

Research Track Program Chairs

Anastasia Dimou   KU Leuven, Belgium
Raphaël Troncy   EURECOM, France

Resource Track Program Chairs

Mehwish Alam   Télécom Paris, Institut Polytechnique de Paris,
     France
Heiko Paulheim   University of Mannheim, Germany

In-Use Track Program Chairs

Olaf Hartig   Linköpings Universitet, Sweden
Maribel Acosta   Ruhr University Bochum, Germany

Special Track on Large Language Models for Knowledge Engineering

Oscar Corcho   Universidad Politécnica de Madrid, Spain
Paul Groth   University of Amsterdam, Netherlands
Elena Simperl   King's College London, UK
Valentina Tamma   University of Liverpool, UK

Workshops and Tutorials Chairs

Joe Raad   University of Paris-Saclay, France
Bruno Sartini   Ludwig-Maximilians University of Munich,
     Germany

Poster and Demo Chairs

Andrea Nuzzolese   ISTC-CNR, Italy
María Poveda-Villalón   Universidad Politécnica de Madrid, Spain

PhD Symposium Chairs

Marta Sabou   Vienna University of Economics and Business,
                                   Austria
Valentina Presutti   University of Bologna, Italy

Industry Track Program Chairs

Irene Celino   Cefriel, Italy
Artem Revenko   Semantic Web Company GmbH, Austria

Sponsorship

Nitisha Jain   King's College London, UK
Jan-Christoph Kalo   University of Amsterdam, Netherlands

Project Networking

Cassia Trojahn   IRIT, France
Sabrina Kirrane   WU Vienna, Austria

Web and Publicity

Stefano De Giorgis   University of Bologna, Italy
Tabea Tietz   FIZ Karlsruhe, Germany

Proceedings and Conference Metadata

Pasquale Lisena   EURECOM, France

Program Committee

Nora Abdelmageed	Friedrich-Schiller-Universität Jena, Germany
Ghadeer Abuoda	Aalborg University, Denmark
Maribel Acosta	TU Munich, Germany
Shqiponja Ahmetaj	TU Wien, Austria
Aljbin Ahmeti	Semantic Web Company GmbH and TU Wien, Austria
Mehwish Alam	Télécom Paris, France
Céline Alec	Université de Caen-Normandie, France
Vladimir Alexiev	Ontotext Corp., Bulgaria
Panos Alexopoulos	Textkernel B.V., Netherlands
Alsayed Algergawy	University of Jena, Germany
Reham Alharbi	University of Liverpool, UK
Bradley Allen	University of Amsterdam, Netherlands
Doerthe Arndt	TU Dresden, Germany
Natanael Arndt	eccenca GmbH, Germany
Luigi Asprino	University of Bologna, Italy
Ghislain Auguste Atemezing	ERA, France
Maurizio Atzori	University of Cagliari, Italy
Sören Auer	TIB Leibniz Information Center Science, Germany & Technology and University of Hannover, Germany
Carlos Badenes-Olmedo	Universidad Politécnica de Madrid, Spain
Booma Sowkarthiga Balasubramani	University of Illinois at Chicago, USA
Konstantina Bereta	National and Kapodistrian University of Athens, Greece
Abraham Bernstein	University of Zurich, Switzerland
Russa Biswas	Hasso Plattner Institute, Germany
Christian Bizer	University of Mannheim, Germany
Peter Bloem	Vrije Universiteit Amsterdam, Netherlands
Martin Blum	University of Trier, Germany
Carlos Bobed	University of Zaragoza, Spain
Fernando Bobillo	University of Zaragoza, Spain
Pieter Bonte	Ghent University, Belgium
Andreas Both	DATEV eG, Germany
Alexandros Bousdekis	Institute of Communication and Computer Systems-National Technical University of Athens, Greece
Loris Bozzato	Fondazione Bruno Kessler, Italy
Janez Brank	Jozef Stefan Institute, Slovenia

Danilo Dessì   GESIS – Leibniz Institute for the Social Sciences, Germany

Gayo Diallo   University of Bordeaux, France

Stefan Dietze   GESIS – Leibniz Institute for the Social Sciences, Germany

Dimitar Dimitrov   GESIS – Leibniz Institute for the Social Sciences, Germany

Anastasia Dimou   KU Leuven, Belgium

Christian Dirschl   Wolters Kluwer Germany, Germany

Daniil Dobriy   Vienna University of Economics and Business, Austria

Milan Dojchinovski   Czech Technical University in Prague, Czech Republic

Ivan Donadello   Free University of Bozen-Bolzano, Italy

Mauro Dragoni   Fondazione Bruno Kessler, Italy

Kai Eckert   Mannheim University of Applied Sciences, Germany

Vasilis Efthymiou   Harokopio University of Athens, Greece

Shusaku Egami   National Institute of Advanced Industrial Science and Technology, Japan

Fajar J. Ekaputra   Vienna University of Economics and Business (WU), Austria

Vadim Ermolayev   Ukrainian Catholic University, Ukraine

Paola Espinoza Arias Universidad Politécnica de Madrid, Spain

Lorena Etcheverry   Universidad de la República, Uruguay

Pavlos Fafalios   Technical University of Crete and FORTH-ICS, Greece

Alessandro Faraotti   IBM, Italy

Catherine Faron   Université Côte d'Azur, France

Anna Fensel   Wageningen University and Research, Netherlands

Javier D. Fernández   F. Hoffmann-La Roche AG, Switzerland

Mariano Fernández López Universidad San Pablo CEU, Spain

Jesualdo Tomás Fernández-Breis Universidad de Murcia, Spain

Sebastián Ferrada   Universidad de Chile, Chile

Agata Filipowska   Poznan University of Economics, Poland

Erwin Filtz   Siemens AG Österreich, Austria

Giorgos Flouris   FORTH-ICS, Greece

Flavius Frasincar   Erasmus University Rotterdam, Netherlands

Naoki Fukuta   Shizuoka University, Japan

Michael Färber   Karlsruhe Institute of Technology, Germany

Mohamed H. Gad-Elrab Bosch Center for Artificial Intelligence, Germany

Alban Gaignard   CNRS, France

Luis Galárraga	Inria, France
Fabien Gandon	Inria, France
Aldo Gangemi	Università di Bologna and ISTC-CNR, Italy
Raúl García-Castro	Universidad Politécnica de Madrid, Spain
Andrés García-Silva	Expert.ai, Spain
Daniel Garijo	Universidad Politécnica de Madrid, Spain
Manas Gaur	Wright State University, USA
Yuxia Geng	Zhejiang University, China
Genet Asefa Gesese	FIZ Karlsruhe – Leibniz-Institut für Informationsinfrastruktur, Germany
Pouya Ghiasnezhad Omran	Australian National University, Australia
Martin Giese	University of Oslo, Norway
Jose M. Gimenez-Garcia	Universidad de Valladolid, Spain
Francois Goasdoue	Université de Rennes 1, France
Jose Manuel Gomez-Perez	Expert.ai, Spain
Simon Gottschalk	Leibniz Universität Hannover, Germany
Floriana Grasso	University of Liverpool, UK
Damien Graux	Huawei Research Ltd., UK
Paul Groth	University of Amsterdam, Netherlands
Claudio Gutierrez	Universidad de Chile, Chile
Peter Haase	metaphacts GmbH, Germany
Mohad-Saïd Hacid	Université Lyon 1, France
Torsten Hahmann	University of Maine, USA
George Hannah	University of Liverpool, UK
Andreas Harth	Friedrich-Alexander-Universität Erlangen-Nürnberg and Fraunhofer IIS, Germany
Olaf Hartig	Linköping University, Sweden
Oktie Hassanzadeh	IBM, USA
Ivan Heibi	University of Bologna, Italy
Veronika Heimsbakk	Capgemini, Norway
Nicolas Heist	University of Mannheim, Germany
Lars Heling	Stardog Union, Germany
Nathalie Hernandez	IRIT, France
Daniel Herzig	metaphacts GmbH, Germany
Ryohei Hisano	ETH Zurich, Switzerland
Pascal Hitzler	Kansas State University, USA
Rinke Hoekstra	Elsevier, Netherlands
Aidan Hogan	Universidad de Chile, Chile
Andreas Hotho	University of Wuerzburg, Germany
Wei Hu	Nanjing University, China
Thomas Hubauer	Siemens AG Corporate Technology, Germany

Andreea Iana	University of Mannheim, Germany
Luis Ibanez-Gonzalez	University of Southampton, UK
Ana Iglesias-Molina	Universidad Politécnica de Madrid, Spain
Filip Ilievski	Vrije Universiteit Amsterdam, Netherlands
Antoine Isaac	Europeana & Vrije Universiteit Amsterdam, Netherlands
Hajira Jabeen	GESIS – Leibniz Institute for the Social Sciences, Germany
Nitisha Jain	King's College London, London, UK
Mustafa Jarrar	Birzeit University, Palestine
Ernesto Jimenez-Ruiz	City, University of London, UK
Milos Jovanovik	Ss. Cyril and Methodius University in Skopje, North Macedonia
Simon Jupp	Elsevier, Netherlands
Jan-Christoph Kalo	University of Amsterdam, Netherlands
Eduard Kamburjan	University of Oslo, Norway
Maulik R. Kamdar	Optum Health, USA
Katariina Kari	Inter IKEA Systems B.V., Finland
Tomi Kauppinen	Aalto University, Finland
Mayank Kejriwal	University of Southern California, USA
Natthawut Kertkeidkachorn	Japan Advanced Institute of Science and Technology, Japan
Ali Khalili	Deloitte, Netherlands
Sabrina Kirrane	Vienna University of Economics and Business, Austria
Tomas Kliegr	Prague University of Economics and Business, Czech Republic
Matthias Klusch	DFKI, Germany
Haridimos Kondylakis	Institute of Computer Science, FORTH, Greece
George Konstantinidis	University of Southampton, UK
Stasinos Konstantopoulos	NCSR Demokritos, Greece
Roman Kontchakov	Birkbeck, University of London, UK
Manolis Koubarakis	National and Kapodistrian University of Athens, Greece
Kozaki Kouji	Osaka Electro-Communication University, Japan
Maria Koutraki	Leibniz Universität Hannover, Germany
Anelia Kurteva	TU Delft, Netherlands
Tobias Käfer	Karlsruhe Institute of Technology, Germany
Birgitta König-Ries	Friedrich Schiller University of Jena, German
Jose Emilio Labra Gayo	Universidad de Oviedo, Spain
Frederique Laforest	INSA Lyon, France
Sarasi Lalithsena	IBM Watson, USA

Andre Lamurias	NOVA School of Science and Technology, Portugal
Davide Lanti	Free University of Bozen-Bolzano, Italy
Danh Le Phuoc	TU Berlin, Germany
Maxime Lefrançois	École des Mines de Saint-Étienne, France
Huanyu Li	Linköping University, Sweden
Sven Lieber	Royal Library of Belgium (KBR), Belgium
Stephan Linzbach	GESIS – Leibniz Institute for the Social Sciences, Germany
Anna-Sofia Lippolis	University of Bologna and ISTC-CNR, Italy
Pasquale Lisena	EURECOM, France
Wenqiang Liu	Xi'an Jiaotong University, China
Giorgia Lodi	Istituto di Scienze e Tecnologie della Cognizione (CNR), Italy
Vanessa Lopez	IBM, Ireland
Pierre Maillot	Inria, France
Maria Maleshkova	Helmut-Schmidt-Universität/Universität der Bundeswehr Hamburg, Germany
Maria Vanina Martinez	IIIA-CSIC, Spain
Miguel A. Martinez-Prieto	University of Valladolid, Spain
Jose L. Martinez-Rodriguez	Autonomous University of Tamaulipas, Mexico
Patricia Martín-Chozas	Universidad Politécnica de Madrid, Spain
Edgard Marx	Leipzig University of Applied Sciences (HTWK), Germany
Philipp Mayr	GESIS – Leibniz Institute for the Social Sciences, Germany
Jamie McCusker	Rensselaer Polytechnic Institute, USA
Lionel Medini	CNRS, France
Albert Meroño-Peñuela	King's College London, UK
Franck Michel	Université Côte d'Azur, CNRS, I3S, France
Nandana Mihindukulasooriya	IBM Research AI, USA
Nada Mimouni	CEDRIC lab - CNAM Conservatoire National des Arts et Métiers Pari, France
Daniel Miranker	The University of Texas at Austin, USA
Victor Mireles	Semantic Web Company, Austria
Pascal Molli	University of Nantes, France
Pierre Monnin	Université Côte d'Azur, Inria, CNRS, I3S, Sophia Antipolis, France
Boris Motik	University of Oxford, UK
Enrico Motta	The Open University, UK
Diego Moussallem	Paderborn University, German
Paul Mulholland	The Open University, UK
Varish Mulwad	GE Research, India

Raghava Mutharaju	IIIT-Delhi, India
María Navas-Loro	Universidad Politécnica de Madrid, Spain
Fabian Neuhaus	University of Magdeburg, Germany
Vinh Nguyen	National Library of Medicine, NIH, USA
Andriy Nikolov	AstraZeneca, UK
Nikolay Nikolov	SINTEF, Norway
Andrea Nuzzolese	ISTC-CNR, Italy
Cliff O'Reilly	City London University, UK
Femke Ongenae	Ghent University, Belgium
Francesco Osborne	The Open University, UK
Ankur Padia	UBMC, USA
George Papadakis	National Technical University of Athens, Greece
Pierre-Henri Paris	Telecom Paris, France
Heiko Paulheim	University of Mannheim, German
Terry Payne	University of Liverpool, UK
Tassilo Pellegrini	University of Applied Sciences St. Pölten, Austria
Maria Angela Pellegrino	Università degli Studi di Salerno, Italy
Bernardo Pereira Nunes	Australian National University, Australia
Romana Pernisch	Vrije Universiteit Amsterdam, Netherlands
Catia Pesquita	Universidade de Lisboa, Portugal
Rafael Peñaloza	University of Milano-Bicocca, Italy
Guangyuan Piao	National University of Ireland, Ireland
Francesco Piccialli	University of Naples Federico II, Italy
Lydia Pintscher	Wikimedia Deutschland, Germany
Dimitris Plexousakis	Institute of Computer Science, FORTH, Greece
Axel Polleres	Vienna University of Economics and Business, Austria
Livio Pompianu	University of Cagliari, Italy
María Poveda-Villalón	Universidad Politécnica de Madrid, Spain
Nicoleta Preda	University of Versailles Saint-Quentin-en-Yvelines, France
Valentina Presutti	University of Bologna, Italy
Joe Raad	University of Paris-Saclay, France
Alexandre Rademaker	IBM Research and EMAp/FGV, Brazil
Helen Mair Rawsthorne	École des Mines de Saint-Étienne, France
Simon Razniewski	Bosch Center for Artificial Intelligence, Germany
Diego Reforgiato	Università degli studi di Cagliari, Italy
Artem Revenko	Semantic Web Company GmbH, Austria
Mariano Rico	Universidad Politécnica de Madrid, Spain
Célian Ringwald	Université Côte d'Azur, Inria, CNRS, I3S, France
Giuseppe Rizzo	LINKS Foundation, Italy
Mariano Rodríguez Muro	Google, USA

Steffen Staab	UNI Stuttgart, Germany
Bram Steenwinckel	Ghent University, Belgium
Kostas Stefanidis	Tampere University, Finland
Nadine Steinmetz	University of Applied Sciences Erfurt, Germany
Armando Stellato	University of Rome Tor Vergata, Italy
Simon Steyskal	Siemens AG Österreich, Austria
Lise Stork	Vrije Universiteit Amsterdam, Netherlands
Umberto Straccia	ISTI-CNR, Italy
Chang Sun	Institute of Data Science at Maastricht University, Netherlands
Zequn Sun	Nanjing University, China
Vojtěch Svátek	Prague University of Economics and Business, Czech Republic
Ruben Taelman	Ghent University, Belgium
Yousouf Taghzouti	École des Mines de Saint-Étienne, France
Valentina Tamma	University of Liverpool, UK
Olivier Teste	IRIT, France
Krishnaprasad Thirunarayan	Wright State University, USA
Elodie Thiéblin	Logilab, France
Ilaria Tiddi	Vrije Universiteit Amsterdam, Netherlands
Tabea Tietz	FIZ Karlsruhe, Germany
Konstantin Todorov	University of Montpellier, France
Ioan Toma	STI Innsbruck, Austria
Riccardo Tommasini	INSA Lyon, France
Sebastian Tramp	eccenca GmbH, Germany
Trung-Kien Tran	Bosch Center for Artificial Intelligence, Germany
Cassia Trojahn	IRIT, France
Raphaël Troncy	EURECOM, France
Yannis Tzitzikas	University of Crete and FORTH-ICS, Greece
Jürgen Umbrich	Vienna University of Economy and Business (WU), Austria
Ricardo Usbeck	Leuphana University Lüneburg, Germany
Marieke van Erp	KNAW Humanities Cluster, Netherlands
Frank Van Harmelen	Vrije Universiteit Amsterdam, Netherlands
Miel Vander Sande	Meemoo, Belgium
Guillermo Vega-Gorgojo	Universidad de Valladolid, Spain
Ruben Verborgh	Ghent University, Belgium
Maria-Esther Vidal	TIB, Germany
Serena Villata	CNRS, France
Fabio Vitali	University of Bologna, Italy
Domagoj Vrgoc	Pontificia Universidad Católica de Chile, Chile
Kewen Wang	Griffith University, Australia

Ruijie Wang	University of Zurich, Switzerland
Xander Wilcke	Vrije Universiteit Amsterdam, Netherlands
Honghan Wu	King's College London, UK
Zhe Wu	eBay, USA
Josiane Xavier Parreira	Siemens AG Österreich, Austria
Guohui Xiao	University of Bergen, Norway
Nadia Yacoubi Ayadi	Université Claude Bernard Lyon 1, France
Fouad Zablith	American University of Beirut, Lebanon
Hamada Zahera	Paderborn University, Germany
Ondřej Zamazal	Prague University of Economics and Business, Czech Republic
Xiaowang Zhang	Tianjin University, China
Ziqi Zhang	Accessible Intelligence, UK
Yihang Zhao	King's College London, UK
Antoine Zimmermann	École des Mines de Saint-Étienne, France
Sara Zuppiroli	ISTC-CNR, Italy
Hanna Ćwiek-Kupczyńska	University of Luxembourg, Luxembourg
Kārlis Čerāns	University of Latvia, Latvia

Additional Reviewers

Akaichi, Ines
Antakli, Andre
Bruns, Oleksandra
Cardellino, Cristian
Cintra, Paul
Djeddi, Warith Eddine
Fanourakis, Nikolaos
Fischer, Elisabeth
Gaur, Manas
Gautam, Nikita
Gui, Zhou
Martín Chozas, Patricia
Montiel-Ponsoda, Elena
Morales Tirado, Alba Catalina
Nayyeri, Mojtaba
Olivier, Inizan
Omeliyanenko, Janna

Ondraszek, Sarah Rebecca
Patkos, Theodore
Peng, Yiwen
Pons, Gerard
Qu, Yuanwei
Ragazzi, Luca
Raoufi, Ensiyeh
Ratta, Marco
Ringwald, Célian
Schlör, Daniel
Schraudner, Daniel
Shao, Chen
van der Weijden, Daniel
Viviurka Do Carmo, Paulo Ricardo
Xiong, Bo
Yumusak, Semih

Sponsors

Platinum Sponsors

VideoLectures.NET is an award-winning free and open access educational video lectures repository. The lectures are given by distinguished scholars and scientists at the most important and prominent events like conferences, summer schools, workshops and science promotional events from many fields of science. The portal is aimed at promoting science, exchanging ideas and fostering knowledge sharing by providing high quality didactic contents not only to the scientific community but also to the general public. All lectures, accompanying documents, information and links are systematically selected and classified through the editorial process taking into account also users' comments.

Gold Sponsors

Ontotext is a global leader in enterprise knowledge graph technology and semantic database engines. Ontotext employs big knowledge graphs to enable unified data access and cognitive analytics via text mining and integration of data across multiple sources. Ontotext™ engine and Ontotext Platform power business critical systems in the biggest banks, media, market intelligence agencies, car and aerospace manufacturers. Ontotext technology and solutions are spread wide across the value chain of the most knowledge intensive enterprises in financial services, publishing, healthcare, pharma, manufacturing and public sectors. Leveraging AI and cognitive technologies, Ontotext helps enterprises get competitive advantage, by connecting the dots of their proprietary knowledge and putting in the context of global intelligence.

Silver Sponsors

Springer is part of Springer Nature, a leading global research, educational and professional publisher, home to an array of respected and trusted brands providing quality content through a range of innovative products and services. Springer Nature is the world's largest academic book publisher, publisher of the world's most influential journals and a pioneer in the field of open research. The company numbers almost 13,000 staff in over 50 countries and has a turnover of approximately €1.5 billion. Springer Nature was formed in 2015 through the merger of Nature Publishing Group, Palgrave Macmillan, Macmillan Education and Springer Science+Business Media.

Founded over 350 years ago, the **University of Innsbruck** today is the most important research and educational institution in western Austria, offering a wide range of programmes across all disciplines. Located in the heart of the Alps, it offers 28,000 students and 5,500 employees the best conditions.

Bronze Sponsors

metaphacts is a German software company that empowers customers to drive knowledge democratization and decision intelligence using knowledge graphs. Built entirely on open standards and technologies, our product metaphactory delivers a low-code, FAIR Data platform that supports collaborative knowledge modeling and knowledge generation and enables on-demand citizen access to consumable, contextual and actionable knowledge. metaphacts serves customers in areas such as life sciences and pharma, engineering and manufacturing, finance and insurance, retail, cultural heritage, and more. For more information about metaphacts and its products and solutions please visit www.metaphacts.com.

eccenca Corporate Memory is cutting-edge Knowledge Graph technology. It digitally captures the expertise of knowledge workers so that it can be accessed and processed by machines. The fusion of human knowledge with large amounts of data, coupled with the computing power of machines, results in powerful artificial intelligence that enables companies to execute existing processes as well as innovation projects of all kinds at high speed and low cost. And it creates an impressive competitive advantage.

Through eccenca.my you can register and create an eccenca Corporate Memory Community Edition Sandbox for evaluation. Join pioneers like Bosch, Siemens, AstraZeneca and many other global market leaders – our world-class team of Linked Data Experts is ready when you are.

Contents – Part I

In-Use

Contents – Part II

Research

Do Similar Entities Have Similar Embeddings?

Nicolas Hubert[1,2]([✉]) [iD], Heiko Paulheim[3] [iD], Armelle Brun[2] [iD],
and Davy Monticolo[1] [iD]

[1] Université de Lorraine, ERPI, Nancy, France
`davy.monticolo@univ-lorraine.fr`
[2] Université de Lorraine, CNRS, LORIA, Nancy, France
`{nicolas.hubert,armelle.brun}@univ-lorraine.fr`
[3] University of Mannheim, Data and Web Science Group, Mannheim, Germany
`heiko@informatik.uni-mannheim.de`

Abstract. Knowledge graph embedding models (KGEMs) developed
for link prediction learn vector representations for entities in a knowledge graph, known as embeddings. A common tacit assumption is the
KGE entity similarity assumption, which states that these KGEMs retain
the graph's structure within their embedding space, *i.e.*, position similar
entities within the graph close to one another. This desirable property
make KGEMs widely used in downstream tasks such as recommender
systems or drug repurposing. Yet, the relation of entity similarity and
similarity in the embedding space has rarely been formally evaluated.
Typically, KGEMs are assessed based on their sole link prediction capabilities, using ranked-based metrics such as Hits@K or Mean Rank. This
paper challenges the prevailing assumption that entity similarity in the
graph is inherently mirrored in the embedding space. Therefore, we conduct extensive experiments to measure the capability of KGEMs to cluster similar entities together, and investigate the nature of the underlying
factors. Moreover, we study if different KGEMs expose a different notion
of similarity. Datasets, pre-trained embeddings and code are available at:
https://github.com/nicolas-hbt/similar-embeddings/.

Keywords: Knowledge Graph · Embedding · Representation
Learning · Entity Similarity

1 Introduction

Knowledge Graphs (KGs) such as DBpedia [3] and YAGO [31] represent facts as
triples (s, p, o) consisting of a subject s and an object o connected by a predicate
p defining their relationship. Common learning tasks with KGs include entity
clustering, node classification, and link prediction.

These tasks are predominantly tackled using Knowledge Graph Embedding
Models (KGEMs), which generate dense vector representations for entities and
relations of a KG, a.k.a. Knowledge Graph Embeddings (KGEs). The dense
numerical vectors that are learnt for entities and relations are expected to preserve the intrinsic semantics of the KG [37].

A. Meroño Peñuela et al. (Eds.): ESWC 2024, LNCS 14664, pp. 3–21, 2024.
https://doi.org/10.1007/978-3-031-60626-7_1

As KGEMs take into account the semantic relationship between two entities to learn embeddings, it is often taken for granted that the resulting embeddings capture both the semantics and attributes of entities and their relationships in the KG. Embeddings are thus widely used to measure the semantic similarity between entities and relations, facilitating data integration through entity or relation alignments [7,17]. They are also used in various similarity-based tasks including entity similarity [33] and conceptual clustering [9].

However, the widespread assumption that KGEMs create semantically meaningful representations of the underlying entities (*i.e.*, project similar entities closer than dissimilar ones) has been challenged recently [14]. Jain *et al.* [14] demonstrate that entity embeddings learnt with KGEMs are not well-suited to identify the concepts or classes for a vast majority of KG entities, while simple statistical approaches provide comparable or better performance. Concerned with word embeddings, Ilievski *et al.* [13] also point out that KGEMs are consistently outperformed by simpler heuristics for similarity-based tasks. The authors argue that many properties on which KGEMs heavily rely on are not useful for determining similarity, which introduces noise and subsequently decreases performance. Our work falls within this line of thought. More specifically, we formulate the following research question:

RQ1. To what extent does proximity in the embedding space align with the notion of entity similarity in the KG?

We call this the *KGE entity similarity assumption*. Notably, there is no universally accepted definition for entity similarity. In this work, we follow a straightforward approach: two entities in a KG are similar if we make similar statements about them. This aligns with the assumption of distributional semantics: words appearing in similar contexts are semantically similar. Answering **RQ1** requires remembering that most embedding-based models are trained to maximize rank-based metrics for link prediction, which disregards semantics. One could then argue that maximizing such metrics is at least partially decoupled from the task of learning similar vectors for similar entities. However, as related entities are more likely to appear in similar triples (*e.g.*, featuring the same predicate), it is reasonable to believe that semantically close entities – especially those connected to other entities through a shared set of predicates – are also more likely to be assigned similar vectors [23]. This raises our second research question:

RQ2. How do traditional rank-based metrics correlate with entity similarity?

In other words, we ask whether KGEMs with good link prediction performance w.r.t. metrics such as Mean Reciprocal Rank (MRR) and Hits@K necessarily group similar entities close in the embedding space. If so, it would be sensible to study which of the two is most likely to influence the other, and whether a form of causality (rather than just correlation) between these two aspects exists.

To delve further, it is worth noting that many link prediction train sets have been shown to suffer from extremely skewed distributions, especially in the occurrence of a subset of predicates [26]. Rossi *et al.* [26] demonstrate that relying on global metrics (*e.g.* Hits@K and MRR) over such heavily skewed distributions hinders our understanding of KGEMs. Consequently, we ultimately distance ourselves from analyzing rank-based metrics, and dive deeper into entity embeddings with the sole consideration of studying how and why they may differ between models. In line with Rossi *et al.* [26] findings that a given subset of predicates is often likely to heavily influence entity representations – we formulate our last research question as follows:

RQ3. Do different KGEMs focus on different predicates to capture the notion of similarity in the embedding space?

The top-K neighbors in the embedding space for a given entity may differ between KGEMs. However, this does not tell us much about why this is the case. We posit that studying the distribution of predicates in the K-hop subgraph centered around each neighboring entity can provide insights into the relevance of certain predicates for particular KGEMs. In other words, the subset of predicates that Rossi *et al.* [26] found out to influence entity representations and KGEM performance w.r.t. rank-based metrics might not just be datasetdependent. Different KGEMs might also implicitly overweigh different subsets of predicates.

The main contributions of our work are summarized as follows.

- We show that different KGEMs fulfill the *KGE entity similarity assumption* only to a limited extent. Notably, even for a given KGEM, results can vary substantially on a per-class basis. Moreover, the semantics of classes is inequally captured by different KGEMs, thereby highlighting that different KGEMs expose different notions of similarity.
- We show that in most cases, performance in link prediction does not correlate with a model's adherence to the KGE entity similarity assumption. This demonstrates that rank-based metrics cannot be used as a reliable proxy for assessing the semantic consistency of the embedding space.
- We show that different KGEMs turn their attention to different predicate subsets for learning similar embeddings for related entities. This suggests that the notion of similarity in the embedding space is partially influenced by the predicate distribution in the close neighborhood around KG entities.

The remainder of the paper is structured as follows. Related work about KGEMs and their use in semantic-related tasks is presented in Sect. 2. Section 3 elaborates on our approach for measuring similarity. Section 4 details our experimental setting. Results are provided and discussed in Sect. 5. Lastly, Sect. 6 summarizes the key findings and outlines directions for future research.

2 Related Work

Knowledge Graph Embeddings. KGEMs have garnered significant attention in recent years due to their capacity to represent structured knowledge within a continuous vector space. The seminal translational model TransE [6] represents entities and relations as low-dimensional vectors and establishes the relationship between a head, a relation, and a tail through a translation operation in the embedding space. Subsequent KGEMs have primarily aimed to address its limitations and enhance the representational expressiveness of knowledge graph embeddings [37]. Representative models are DistMult [39], ComplEx [36], ConvE [8], and TuckER [4]. The embeddings learnt by these KGEMs have shown potential applications in tasks such as like link prediction [37], entity clustering [12], and node classification [12].

Using KGEs for Semantic-Related Tasks. As relational models, KGEMs are widely used for predicting links in KGs [25,37]. However, the vector representations learnt by these models can also be used for other tasks [16]. For example, pretrained language models with knowledge graphs have been used for Named Entity Recognition (NER) [19,32]. Other tasks aiming at discovering rich information about entities through their embeddings include entity typing [20] and entity alignment [33]. Many works also explore the use of KGEs for drug repurposing [30] and recommender systems [10]. These works are based on the premise that the distance between entities in the embedding space should reflect their intrinsic similarity, and can be leveraged, *e.g.* for recommending a common set of items to similar users [10].

Analzying the KGE Entity Similarity Assumption. While many of the approaches *implicity* rely on the KGE entity similarity assumption to hold, there are only few works actually *explicitly* validating this assumption. Portisch *et al.* [23] provide anecdotic examples for a few embedding approaches, also suggesting that the underlying notions of similarity might differ, but do not conduct any formal evaluation. Jain *et al.* [14] analyze KGEMs on the basis of class assignments, showing that the original class assignments can only be reconstructed to a limited extent with classification and clustering methods, which rather questions the assumption of similar entities being close in the vector space. A similar study is conducted by Alshargi *et al.* [2], also concluding that "the current quality of the embeddings for ontological concepts is not in a satisfactory state". While Portisch *et al.* provide anecdotic examples, and Jain *et al.* and Alshargi *et al.* only look at class assignments, this work is the first one to empirically study the relation of entity similarity and similarity in the KGE space. Moreover, our study is more fine-grained than the previous ones, which end at the class level: while those only inspect whether entities of the same class have similar embeddings (*e.g.*, two movies are embedded closer than a movie and a person), we also consider similarity *within* a class (*e.g.*, analyze whether similar movies are embedded closer than less similar ones).

3　Approach

In this section, we detail our proposal for quantifying the KGE entity similarity assumption. While similarity in embedding spaces is usually measured using cosine similarity, there is no uniform definition of entity similarity in KGs. In Sect. 3.1, we elaborate on the metrics used to capture the notion of similarity between entities in the KG. Section 3.2 discusses how the aforementioned notions of similarity can be compared. It is worth noting that the line of research closest to ours is the one of Jain *et al.* [14]. Both [14] and our work question common assumptions that are taken for granted in the KG community. However, [14] looks at the capability of KGEMs to learn the semantics of classes, and considers the specific tasks of entity classification and clustering. In contrast, we are concerned with similarity measures of different embedding models and how they reflect entity similarity. As such, even though our work fits within an existing literature on the concept of semantic capture in embeddings, to the best of our knowledge this paper is the first one to present an approach to thoroughly investigate the extent to which entity similarity in KGs is mirrored in the embedding space.

3.1 Towards a Graph-Based Notion of Similarity

In this section, we detail our attempt at capturing the notion of similarity in the original KG. Unlike the notion of similarity in the embedding space, which is typically measured as the cosine similarity of two vectors, it should be noted that graph-based similarity cannot be measured in a single, uniform way. As previosuly said, there is no universal definition for entity similarity in KGs. Multiple metrics and approaches can be used, and the choice between one or another largely depends on the particular aspect to be measured. In what follows, we intend to explain the rationale behind our modeling choices, and briefly mention alternatives that we ultimately discarded.

To measure the similarity between two entities e_1 and e_2 in one KG, we (1) determine the set of common statements about e_1 and e_2, *i.e.*, relations to other entities that e_1 and e_2 have in common. While many entities will have such relations in common with central entities (*i.e.*, those with many ingoing and outgoing edges), we also (2) need to make sure that centrality does not skew our similarity measure. Finally, (3) since an entity cannot be fully described by its immediate neighbors, indirect dependencies also need to be captured.

To satisfy our first desideratum (1), we compare the subgraphs around e_1 and e_2. We experiment with the 1-hop only vs. 2-hop subgraph neighborhood. The latter option addresses desideratum (3), as we also consider indirect dependencies. In Fig. 1, we give a concrete example of the 1-hop and 2-hop subgraphs for entities Bob (e_1) and Julie (e_2).[1] The similarity of e_1 and e_2 can now be measured by the similarity of their respective subgraphs. Graph Edit Distance (GED) [28] has been used for this intent. However, GED is NP-hard and thus computationally demanding, it comes with the need for arbitrarily defining weights for vertex and edge insertion/deletion/substitution, and we experimentally found it was sensitive to subgraphs' sizes, which is detrimental to desideratum (2). We considered other metrics such as Katz centrality and the common-neighbors metric. However, as the name suggests, Katz is a measure

---

[1] We consider both ingoing and outgoing edges to define those neighborhoods.

of centrality, not similarity. Besides, it takes into account all the paths between two entities and is therefore sensitive to the absolute number of paths. Common neighbors does not consider predicates and is only suited for unlabeled graphs.[2]

In our experiments, we use the Jaccard coefficient is used to measure the overlap in the 1-hop and 2-hop subgraphs of entities. In particular, its value denotes how similar the respective subgraphs of two entities are, which gives us insight into how much two entities are related based on graph-based information. The central entity (e_1 or e_2) is replaced by a unique token, e.g. :dummy. Triples forming the 1-hop neighborhood are extracted as is, while for the 2-hop paths the intermediate entity is ignored (written as $\square$ below).

More formally: let $T_1(A)$ be the set of all triples $\{(A, p, o) \cup (s, p, A)\}$, where A is replaced by a dummy identifier. Then, $J_1(e_1, e_2)$ is the Jaccard overlap of $T_1(e_1)$ and $T_1(e_2)$, i.e., $\frac{|T_1(e_1) \cap T_1(e_2)|}{|T_1(e_1) \cup T_1(e_2)|}$.

Let $T_2(A)$ be the set of all paths $\{(A, p1, \square, p2, o) \cup (s, p1, \square, p2, A)\}$, where A is replaced by a dummy identifier. Then $J_2(e_1, e_2)$ is the Jaccard overlap of $T_2(e_1)$ and $T_2(e_2)$, i.e., $\frac{|T_2(e_1) \cap T_2(e_2)|}{|T_2(e_1) \cup T_2(e_2)|}$.

Following Fig. 1, triples in the two 1-hop subgraphs are:

Bob	Julie
(:dummy, plays, Guitar)	(:dummy, plays, Guitar)
(:dummy, livesIn, Mannheim)	(:dummy, livesIn, Karlsruhe)
(:dummy, friend, Anna)	(:dummy, friend, Roger)

When considering 2-hop information, we also consider the following paths which are shared between Bob's and Julie's subgraphs (reverse edges included):

$$\text{(:dummy, livesIn, } \square \text{, inCountry, Germany)}$$

$$\text{(:dummy, livesIn, } \square \text{, inRegion, Baden-Württemberg)}$$

$$\text{(:dummy, friend, } \square \text{, staffMember}^{-1} \text{, Vodafone)}$$

The similarity of Bob and Julie considering the 1-hop subgraphs only is $J_1(Bob, Julie) = 0.2$ (since they share one out of five triples). Considering the 2-hop neighborhood graphs: $J_2(Bob, Julie) = 0.5$ (since they share three out of six two-hop paths). For instance, we consider (:dummy, livesIn, $\square$, inCountry, Germany) as the full path. This allows for capturing indirect dependencies, such as *both Julie and Bob live in a city in Germany*, or *both Julie and Bob know someone who works for Vodafone*. This approach better fits the general and commonsense idea that two entities should already be similar if they share relations to entities which are similar themselves according to their respecitve 1-hop subgraphs. Moreover, keeping the intermediate entity in the path would overweigh information already counted as part of the 1-hop subgraph, especially if the intermediate entity were a high degree entity (desideratum (3)). For any entity in the 1-hop neighborhood of a central entity, its own 1-hop

---

[2] Given, e.g., (SAP, headquarter, Germany), (BOSCH, headquarter, Germany), (Berlin, capitalOf, Germany), with common neighbors, SAP, BOSCH, and Berlin would be equally similar.

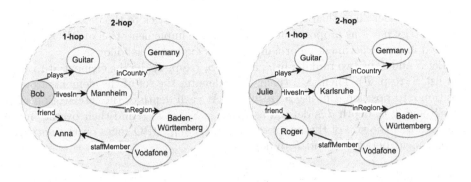

Fig. 1. 1-hop and 2-hop subgraphs for Bob and Julie.

neighborhood is known and fixed. Consequently, whenever there is a match between triples in the 1-hop neighborhood of two central entities being compared, there would be as many matches in their respective 2-hop subgraphs as there are relations connecting their shared 1-hop neighbors. For example, if Bob and Julie were living in the same city, *e.g.* Mannheim, this information would be counted multiple times – in (:dummy, livesIn, Mannheim, inCountry, Germany), (:dummy, livesIn, Mannheim, inRegion, Baden-Württemberg), etc. The resulting relatedness between Bob and Julie would thus be overly influenced by that single fact. Similarly, if they were living in distinct cities – which is the case in our example – their relatedness score would be unduly penalized.

3.2   Embedding vs. Graph: A Different Notion of Similarity?

Based on a given entity e, it is possible to get its N closest neighbors in the KG (using the J_1 and J_2 metrics defined above) and in the embedding space (using cosine similarity). To assess if the two notions of similarity are different, we need to measure how much the two lists of closests neighbors between these two approaches actually overlap.

A common measure to compare two ranked lists is Kendall's Tau [18]. However, it suffers from two important caveats: the two lists should be of the same size and they should contain the same set of items. Considering our experimental purposes, the latter limitation is undesirable, as there is no reason why these two lists (of the top N similar entities according to J_1 or J_2, and in the embedding space) should share the same set of entities.

Consequently, we use the Rank-Biased Overlap (RBO) [38] to compare the similarity of two ranked lists. Unlike Kendall's Tau which is correlation-based, RBO is intersection-based. Most importantly, RBO can handle lists containing different items. It also allows for the weighting of rankings, giving higher importance to items at the top of the lists through tweaking the *persistence* parameter p. A small value for p will only consider the first few items, whereas a larger value will encompass more items. However, we experimentally found that results are very sensitive to the choice of p. Besides, when comparing two ranked lists, we want to consider all of their items. Consequently, we stick to the default parameter strategy $p = 1$ as proposed in an open-source implementation of RBO[3] that we used in our experiments. Formally, RBO is expressed as follows.

Let S and T be two ranking lists, and let S_i (resp. T_i) be the element at rank i in list S (resp. T). Then, $S_{c:d}$ (resp. $T_{c:d}$) denotes the set of the elements from position c to position d in list S (resp. T). At each depth d, the intersection of lists S and T to depth d is defined as $I_{S,T,d} = S_{1:d} \cap T_{1:d}$, and their overlap up to depth d is the size of the intersection, i.e. $|I_{S,T,d}|$. RBO relies on the notion of *agreement* between S and T to depth d:

$$A_{S,T,d} = \frac{|I_{S,T,d}|}{d} \tag{1}$$

In the case whre the persistence parameter $p = 1$, the RBO formula is expressed as follows:

$$RBO(S,T,k) = \frac{1}{k} \sum_{d=1}^{k} A_{S,T,d} \tag{2}$$

where k is the size of the two lists S and T being compared. For the case where $p \neq 1$, we refer the reader to [38].

In Fig. 2, we illustrate how RBO@5 is calculated for two lists of top-5 neighbors, where one relies on similarity in the embedding space whereas the other one depends on the Jaccard overlap between entities' subgraphs. We see that (i) RBO can be calculated on lists with different sets of entities, (ii) RBO weighs more matches that occur at the top of the lists, and (iii) RBO is sensitive to the order of matching entities in the respective lists.

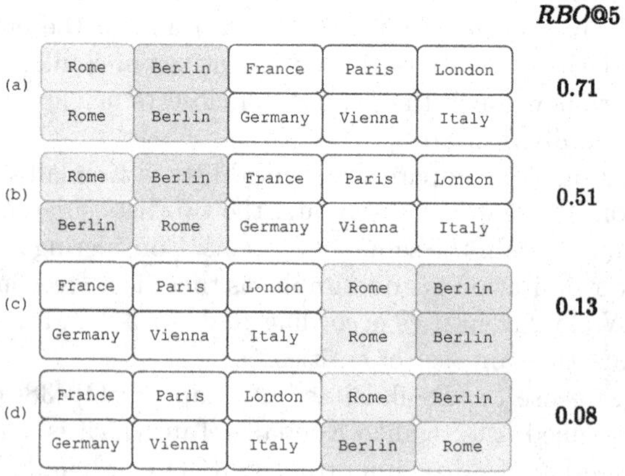

Fig. 2. Illustration of RBO@5 values in four cases.

---

[3] https://github.com/changyaochen/rbo/.

4   Experiments

In this section, we detail our experimental setting, *i.e.* the KGEMs (Sect. 4.1) and datasets (Sect. 4.2) used in the experiments.

4.1   Knowledge Graph Embedding Models

In our experiments, we use seven popular KGEMs from different families of models: geometric-based (TransE [6], TransD [15], BoxE [1]), multiplicative (RESCAL [22], DistMult [39], TuckER [4]), and convolutional-based (ConvE [8]) models. Other models have been considered, *e.g.* ComplEx [36] and RotatE [34]. However, these models generate complex-valued embeddings, which requires a different approach than using simple cosine similarity. We trained these KGEMs using PyKEEN[4] with the provided configuration files, when available. For those datasets with no reported best hyperparameters, we used the hyperparameters reported in the original paper (*e.g.* for YAGO4-19K, with best hyperparameters found in [11]) or performed manual hyperparameter search and kept the sets of hyperparameters leading to the best results on the validation sets (*e.g.* for AIFB). We also trained RDF2Vec [24], which is a versatile embedding approach that can be adapted for different downstream applications and that relies on path-based information to encode entities (only). Even though RDF2Vec is not specifically designed to handle the link prediction task and is not evaluated w.r.t. it, its entity embeddings are expected to reflect their similarity or relatedness in the KG to some extent [23]. We used the implementation provided by pyRDF2Vec[5] with the default hyperparameters reported in [23]: embeddings of dimension 200, with 2,000 walks maximum, a depth of 4, a window size of 5, and 25 epochs for training word2vec [21] with the continuous skip-gram architecture.

4.2   Datasets

Since we also want to analyze results on a per-class basis, we consider only benchmark datasets which also come with a schema of multiple classes and relations. We use the following datasets in our evaluation: AIFB [5], Codex-S and Codex-M [27], DBpedia50 [29], FB15K-237 [35], and YAGO4-19K [11]. Entity types are not directly available in the original repository of DBpedia50[6] and FB15k-237[7]. However, we ran SPARQL queries against DBpedia to get entity types for DBpedia50. The resulting class hierarchy is a subset of the DBpedia ontology, with a maximum depth of 8, an average fan-out (branching factor) of 3.40, and each entity being typed with an average of 4.55 classes. For FB15K-237, we reused the entity-typed version presented in [11]. Table 1 provides finer-grained statistics for these datasets.

---

[4] https://github.com/pykeen/pykeen/.
[5] https://github.com/IBCNServices/pyRDF2Vec/.
[6] https://github.com/bxshi/ConMask/.
[7] https://www.microsoft.com/en-us/download/details.aspx?id=52312/.

Table 1. Datasets used in the experiments. Column header from left to right: number of entities, relations (predicates), classes, train triples, validation triples, and test triples.

Dataset	$|\mathcal{E}|$	$|\mathcal{R}|$	$|\mathcal{C}|$	$|\mathcal{T}_{train}|$	$|\mathcal{T}_{valid}|$	$|\mathcal{T}_{test}|$
AIFB	2,389	16	18	14,170	745	785
Codex-S	2,034	42	502	32,888	1,827	1,827
Codex-M	17,050	51	1,503	185,584	10,310	10,311
DBpedia50	24,624	351	285	32,388	123	2,098
FB15k237	14,541	237	643	272,115	17,535	20,466
YAGO4-19k	18,960	74	1,232	27,447	485	463

5   Results and Discussion

5.1   Different Notions of Similarity (RQ1)

Table 2 reports results w.r.t. rank-based metrics (MRR, Hits@K) and RBO@K for all the KGEMs and datasets considered in this work. Similar to MRR and Hits@K, RBO is computed for all entities in a KG and the average is reported.

When retaining 1-hop subgraphs (from R1@3 to R1@100 in Table 2), in most cases TuckER has the highest RBO values. In other words, it appears to fulfill the KGE entity similarity assumption best, having the highest tendency to position similar entities closer in the vector space. When moving to 2-hop subgraphs, Dist-Mult and TuckER are the models with the best alignment capabilities (Table 2). BoxE, RDF2Vec, RESCAL, and TransD fare worse than other models. The substantially lower RBO values for RDF2Vec can be related to the fact that the vector distance in RDF2vec space mixes similarity and relatedness, where the latter does not show a strong overlap in common paths [23]. For example, it will position *USA* close to *Washington D.C.*, but they have only few common paths starting/ending in the respective entities.

A crucial observation from our study is that the behavior of KGEMs varies not only across datasets but also across different classes within these datasets. Table 3 illustrates this insight on classes that were selected to provide a more intuitive understanding of how different KGEMs are differently able to capture similarities between instances of the same class. In particular, we clearly see that on YAGO4-19K, TransE captures the notion of SpanishMunicipality quite well compared to other models, but does not do well for MusicPlaylist entities (highlighted in blue). This means that different models are better suited for capturing the nuances of specific classes. Jain *et al.* [14] demonstrated that semantic representation in embeddings is not consistent across all KG entities, but is restricted to a small subset of them. Our results point in the same direction and suggest that KGEMs are not equipped with consistent capabilities to learn similar embeddings for entities within the same class. This problem appears at two different levels: First, we noted a general misalignment in the representation of certain classes across various datasets, regardless of the KGEM used. Second, there is noticeable inconsistency within specific classes, where entities might align well with their graph-based proximity in one KGEM but not in another.

Table 2. Rank-based and RBO results. As RDF2Vec is not designed for link prediction, it is not evaluated w.r.t. rank-based metrics. H@3 and H@10 stand for Hits@3 and Hits@10, respectively. RBO is abbreviated as R, while the number directly following it denotes whether 1-hop subgraphs or 2-hop subgraphs are considered. Bold fonts indicate which KGEM performs best for a given configuration (dataset-metric). Underlined results denote the second-best performing model.

FB15K-237 / YAGO4-19K

	MRR	H@3	H@10	R1@3	R1@10	R1@100	R2@3	R2@10	R2@100	MRR	H@3	H@10	R1@3	R1@10	R1@100	R2@3	R2@10	R2@100
RDF2Vec	–	–	–	0.023	0.033	0.063	0.014	0.021	0.054	–	–	–	0.126	0.154	0.177	0.164	0.170	0.183
TransE	0.240	0.264	0.404	0.132	0.178	0.228	0.035	0.050	0.108	0.762	0.836	0.895	0.247	0.301	0.360	0.368	0.362	**0.260**
TransD	0.184	0.205	0.378	0.176	0.220	0.259	0.065	0.083	0.135	0.763	0.895	0.915	0.197	0.252	0.365	0.181	0.237	0.137
DistMult	0.226	0.247	0.392	0.194	0.245	0.343	**0.103**	**0.131**	**0.222**	0.809	0.838	0.870	0.310	0.350	_0.407_	_0.378_	**0.372**	0.225
RESCAL	0.279	0.305	0.447	0.239	0.269	0.284	0.064	0.081	0.138	0.676	0.692	0.754	0.176	0.215	0.176	0.298	0.271	0.103
TuckER	**0.341**	**0.373**	**0.516**	_0.321_	_0.364_	**0.411**	_0.099_	_0.121_	_0.205_	0.897	0.901	0.910	**0.362**	**0.461**	**0.498**	**0.385**	_0.362_	_0.228_
ConvE	0.300	0.327	0.474	**0.341**	**0.370**	_0.357_	0.083	0.102	0.153	**0.905**	**0.907**	**0.916**	_0.328_	_0.368_	0.365	0.336	0.313	0.168
BoxE	0.299	0.326	0.477	0.221	0.261	0.313	0.060	0.081	0.166	0.895	0.902	0.914	0.136	0.146	0.113	0.244	0.231	0.096

Codex-S / Codex-M

	MRR	H@3	H@10	R1@3	R1@10	R1@100	R2@3	R2@10	R2@100	MRR	H@3	H@10	R1@3	R1@10	R1@100	R2@3	R2@10	R2@100
RDF2Vec	–	–	–	0.030	0.043	0.096	0.021	0.036	0.099	–	–	–	0.013	0.019	0.042	0.011	0.017	0.041
TransE	0.293	0.331	0.526	0.201	0.266	0.451	_0.154_	**0.210**	**0.345**	0.227	0.260	0.371	0.082	0.101	0.153	0.058	0.076	0.105
TransD	0.226	0.294	0.515	0.252	0.322	0.462	0.136	0.178	0.297	0.215	0.281	0.420	0.155	0.188	0.263	0.055	0.070	0.117
DistMult	0.261	0.298	0.409	_0.452_	**0.511**	**0.616**	**0.163**	0.205	_0.339_	0.209	0.233	0.356	0.173	0.212	0.312	**0.144**	**0.178**	**0.256**
RESCAL	0.283	0.307	0.475	0.135	0.193	0.336	0.132	0.176	0.274	0.149	0.159	0.253	0.067	0.084	0.115	0.025	0.036	0.057
TuckER	0.393	0.442	0.618	**0.461**	**0.511**	_0.615_	_0.154_	_0.209_	_0.339_	0.282	0.310	0.424	0.198	0.239	0.328	_0.120_	_0.153_	_0.222_
ConvE	**0.414**	**0.466**	0.611	0.450	0.495	0.574	0.146	0.195	0.306	**0.290**	**0.319**	0.425	**0.315**	**0.343**	**0.384**	0.098	0.119	0.158
BoxE	0.398	0.453	**0.622**	0.312	0.366	0.486	0.118	0.179	0.291	0.290	0.318	**0.431**	0.241	_0.272_	_0.334_	0.084	0.106	0.152

AIFB / DBpedia50

	MRR	H@3	H@10	R1@3	R1@10	R1@100	R2@3	R2@10	R2@100	MRR	H@3	H@10	R1@3	R1@10	R1@100	R2@3	R2@10	R2@100
RDF2Vec	–	–	–	0.089	0.131	0.223	0.137	0.174	0.210	–	–	–	0.04	0.053	0.179	0.059	0.082	0.189
TransE	0.469	0.524	0.699	0.413	0.456	0.571	0.304	0.349	_0.455_	0.397	**0.490**	**0.567**	0.109	0.123	0.297	_0.214_	0.246	0.282
TransD	0.472	0.689	0.817	0.472	0.507	_0.617_	0.305	0.330	0.382	0.260	0.376	0.442	0.111	0.136	0.318	0.126	0.187	0.244
DistMult	0.500	0.568	0.763	0.423	0.463	0.588	_0.358_	**0.406**	**0.495**	0.384	0.423	0.478	0.131	0.144	0.348	**0.241**	**0.275**	_0.299_
RESCAL	0.440	0.486	0.628	0.345	0.387	0.454	0.245	0.273	0.322	0.225	0.240	0.295	0.063	0.079	0.276	0.142	0.171	0.197
TuckER	0.797	0.824	0.888	**0.693**	**0.712**	**0.726**	**0.375**	0.380	0.394	0.424	0.442	0.502	_0.151_	**0.166**	**0.418**	0.235	_0.271_	**0.305**
ConvE	0.779	0.799	0.870	0.454	0.505	0.567	0.263	0.292	0.321	0.434	0.456	0.536	_0.137_	0.146	0.324	0.201	0.228	0.253
BoxE	**0.806**	**0.841**	**0.904**	_0.570_	_0.587_	0.614	0.343	_0.374_	0.437	**0.456**	0.480	0.556	0.137	_0.147_	_0.351_	_0.214_	0.247	0.278

To generalize this observation, we computed correlation matrices for RBO values across models and datasets, focusing on class-specific performance. Our analysis, averaging these matrices, gives us a broader view of how different KGEMs perceive similarity on a per-class basis. Figure 4 shows the rank correlations between models when considering 2-hop subgraphs, averaged over all datasets and RBO values. Notably, RDF2Vec stands out as it seems to capture a distinct notion of similarity compared to other KGEMs (cf. [23]). It demonstrates that RDF2Vec fares better on different classes than other KGEMs. We also observe a high degree of variation between models of the same family: TransE and TransD expose a different notion of similarity (0.64), as RESCAL and TuckER do (0.6). Strong rank correlations are observed between DistMult and BoxE (0.85), DistMult and TransE (0.84), and BoxE and TransE (0.84) (Fig. 4). This demonstrates that while some KGEMs exhibit a close conceptualization of entity similarity, this is not universally the case. Therefore, **RQ1** can be answered as follows: proximity in the embedding space does *not* consistently align with the notion of entity similarity in the KG, since this property substantially differs between models and datasets. Consequently, careful consideration is necessary when using KGEs for drawing conclusions about similar entities (*e.g.* for recommending items to similar users).

Table 3. RBO@10 values per model for selected classes. Numbers in parenthesis indicate the rank of a class for a given model.

Dataset	Class	RDF2Vec	TransE	TransD	BoxE	RESCAL	DistMult	Tucker	ConvE
AIFB	Book	0.383 (1)	0.560 (1)	0.314 (7)	0.551 (1)	0.613 (1)	0.621 (1)	0.374 (5)	0.214 (13)
	inProceedings	0.048 (13)	0.419 (3)	0.227 (11)	0.461 (3)	0.412 (2)	0.453 (3)	0.332 (11)	0.326 (3)
Codex-S	Republic	0.000 (26)	0.556 (3)	0.236 (12)	0.463 (5)	0.363 (3)	0.339 (10)	0.399 (9)	0.311 (12)
	NationalAcademy	0.000 (27)	0.558 (2)	0.217 (14)	0.446 (8)	0.383 (2)	0.323 (11)	0.366 (12)	0.301 (14)
Codex-M	UrbanDistrict	0.002 (66)	0.538 (9)	0.178 (40)	0.482 (11)	0.057 (66)	0.494 (14)	0.504 (8)	0.327 (12)
	CollegeTown	0.000 (88)	0.615 (2)	0.205 (30)	0.653 (2)	0.133 (30)	0.670 (1)	0.640 (2)	0.337 (27)
DBpedia50	Album	0.134 (52)	0.336 (33)	0.270 (23)	0.351 (30)	0.300 (9)	0.310 (42)	0.387 (27)	0.315 (29)
	EthnicGroup	0.028 (136)	0.469 (9)	0.227 (36)	0.452 (9)	0.400 (1)	0.458 (9)	0.434 (16)	0.388 (12)
FB15k237	Periodicals	0.002 (414)	0.115 (109)	0.255 (64)	0.150 (163)	0.124 (181)	0.139 (280)	0.590 (11)	0.638 (3)
	BaseballPlayer	0.001 (420)	0.146 (78)	0.336 (31)	0.236 (93)	0.143 (160)	0.277 (104)	0.631 (4)	0.598 (12)
YAGO-19k	MusicPlaylist	0.233 (35)	**0.325 (94)**	0.322 (37)	0.411 (10)	0.350 (41)	0.366 (83)	0.627 (4)	0.554 (7)
	SpanishMunicipality	0.226 (40)	**0.636 (1)**	0.460 (4)	0.327 (33)	0.236 (95)	0.531 (20)	0.551 (20)	0.505 (17)

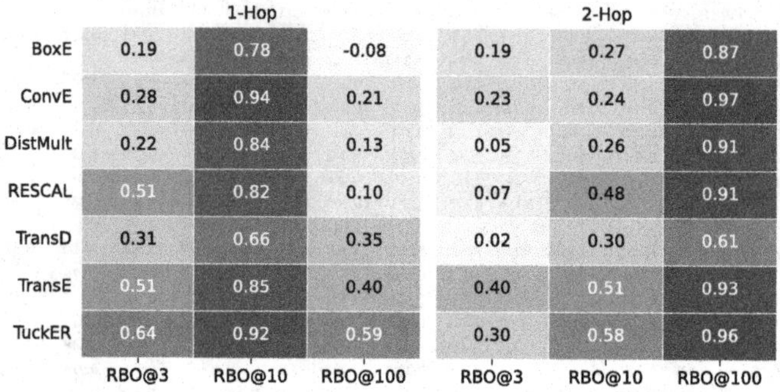

Fig. 3. Correlations between MRR and RBO values with 1-hop and 2-hop subgraphs.

5.2 Correlation Between Rank-Based Metrics and RBO (RQ2)

As previously mentioned, some KGEMs fare better than others in terms of RBO values, *e.g.* DistMult and TuckER. It is important to note that RBO values are designed to measure how well a KGEM entity embedding aligns with the concept of proximity within the original KG. This goal is distinct from the aim of rank-based metrics, which assess how effectively a KGEM assigns higher plausibility scores to ground-truth triples. In what follows, we aim study the relationship between these two types of metrics: how does KGEM performance in rank-based metrics correlate with RBO values?

We analyzed this using Pearson's correlation coefficient, comparing MRR and RBO values for 1-hop and 2-hop subgraphs across our 6 datasets, as depicted in Fig. 3. The results revealed notable disparities among the models. For instance, TransE and TuckER consistently show moderate to high correlation between MRR and RBO values. This implies that for these models, higher plausibility scoring of triples often aligns with better representation of entity proximity in the KG. However, this trend is not uniform across all models, and we do not observe general trends across families of models: while TransE and TransD are both translational models, correlation results are quite different.

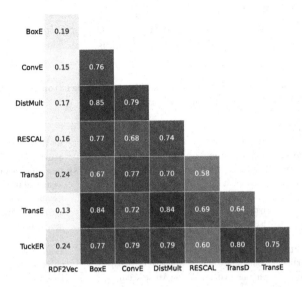

Fig. 4. Rank correlations between KGEMs averaged over all datasets and all RBO metrics ($K = 3, 10, 100$).

From a coarse-grained viewpoint, we note that MRR *generally* does not correlate well with RBO values. Two specific cases are that MRR is heavily (and positively) correlated with RBO@10 (1-hop) and RBO@100 (2-hop), which means that MRR can be seen as a reliable proxy measure for those two metrics (and vice versa). However, it still remains to be explained why the correlation between MRR and RBO@10 (1-hop) is so high, while a substantial drop in correlation between MRR and RBO@100 (1-hop) is observed. Another result to be assessed is why MRR correlates more with RBO@K with lower K values when considering 1-hop subgraph and higher K values when considering 2-hop subgraphs. Therefore, **RQ2** cannot be answered in a conclusive way: while in the general case, the correlation between performance in grouping similar entities and link prediction performance is moderate at best, some metrics show a higher correlation. This has a severe practical implication: the commonly observed practice to pick a KGE model which is good at link prediction for a task heavily relying on entity similarity has to be considered a suboptimal strategy.

5.3 Analyzing Predicate Importance (RQ3)

We previously highlighted that different KGEMs are equipped with different notions of similarity in their respective embedding spaces. We also demonstrated that the adherence to the KGE entity similarity assumption can vary a lot between classes: for a given combination of dataset and model, how much the top-

K neighbors of an entity in the embedding space align with its top-K neighbors in the graph largely depends on which class this entity belongs to.

To examine this behavior further, we pick random classes and count the predicates in the common triples of the top-K neighbors (in the embedding space) for entities of this class differs between models.

Table 4. Top 5 predicates for entities of classes `MusicComposition` and `MusicGroup` on YAGO4-19K, along with their importance score.

	MusicComposition		MusicGroup	
	K=3	K=10	K=3	K=10
RDF2Vec	isPartOf (0.52)	isPartOf (0.46)	**genre (0.53)**	**genre (0.60)**
	inLanguage (0.18)	inLanguage (0.30)	**foundingLocation (0.38)**	**foundingLocation (0.32)**
	hasPart (0.17)	hasPart (0.15)	**award (0.10)**	**award (0.07)**
	composer (0.09)	composer (0.07)	memberOf (0.00)	memberOf (0.01)
	lyricist (0.02)	lyricist (0.02)	byArtist (0.00)	byArtist (0.00)
TransE	isPartOf (0.40)	isPartOf (0.36)	genre (0.79)	genre (0.85)
	hasPart (0.28)	hasPart (0.27)	foundingLocation (0.17)	foundingLocation (0.12)
	inLanguage (0.16)	inLanguage (0.26)	award (0.04)	award (0.02)
	composer (0.09)	composer (0.06)	memberOf (0.00)	memberOf (0.00)
	citation (0.05)	citation (0.04)	byArtist (0.00)	knowsLanguage (0.00)
TransD	isPartOf (0.40)	inLanguage (0.37)	genre (0.84)	genre (0.87)
	inLanguage (0.24)	isPartOf (0.31)	foundingLocation (0.13)	foundingLocation (0.11)
	hasPart (0.23)	hasPart (0.23)	award (0.03)	award (0.02)
	composer (0.08)	composer (0.06)	memberOf (0.00)	memberOf (0.00)
	citation (0.03)	citation (0.03)	byArtist (0.00)	knowsLanguage (0.00)
RESCAL	isPartOf (0.42)	inLanguage (0.33)	genre (0.74)	genre (0.81)
	inLanguage (0.23)	isPartOf (0.33)	foundingLocation (0.22)	foundingLocation (0.16)
	hasPart (0.21)	hasPart (0.25)	award (0.03)	award (0.02)
	composer (0.09)	composer (0.06)	memberOf (0.01)	memberOf (0.00)
	citation (0.04)	citation (0.02)	knowsLanguage (0.00)	knowsLanguage (0.00)
DistMult	isPartOf (0.38)	inLanguage (0.39)	genre (0.81)	genre (0.86)
	hasPart (0.26)	isPartOf (0.29)	foundingLocation (0.14)	foundingLocation (0.11)
	inLanguage (0.22)	hasPart (0.22)	award (0.04)	award (0.02)
	composer (0.08)	composer (0.05)	memberOf (0.00)	memberOf (0.00)
	citation (0.05)	citation (0.03)	knowsLanguage (0.00)	knowsLanguage (0.00)
TuckER	isPartOf (0.38)	inLanguage (0.37)	genre (0.85)	genre (0.90)
	hasPart (0.25)	isPartOf (0.30)	foundingLocation (0.11)	foundingLocation (0.08)
	inLanguage (0.24)	hasPart (0.23)	award (0.04)	award (0.02)
	composer (0.08)	composer (0.05)	memberOf (0.01)	memberOf (0.00)
	citation (0.04)	citation (0.03)	byArtist (0.00)	byArtist (0.00)
BoxE	isPartOf (0.48)	isPartOf (0.40)	genre (0.73)	genre (0.83)
	hasPart (0.29)	hasPart (0.30)	foundingLocation (0.23)	foundingLocation (0.15)
	inLanguage (0.11)	**inLanguage (0.19)**	award (0.04)	award (0.02)
	composer (0.10)	composer (0.07)	byArtist (0.00)	byArtist (0.00)
	lyricist (0.02)	citation (0.02)	memberOf (0.00)	memberOf (0.00)
ConvE	isPartOf (0.39)	isPartOf (0.33)	genre (0.83)	genre (0.86)
	hasPart (0.25)	inLanguage (0.31)	foundingLocation (0.14)	foundingLocation (0.12)
	inLanguage (0.22)	hasPart (0.25)	award (0.02)	award (0.02)
	composer (0.08)	composer (0.06)	memberOf (0.01)	memberOf (0.00)
	citation (0.04)	citation (0.04)	byArtist (0.00)	byArtist (0.00)

Results for two picked classes are reported in Table 4, where each predicate is weighted in accordance to its frequency in the set of common triples between the 2-hop subgraphs of top-K neighbors in the embedding space (for $K = 3$ and $K = 10$) and a given entity of class MusicComposition and MusicGroup (YAGO4-19K). A class-based analysis reveals that predicates are differently ranked, *i.e.* different KGEMs *may* rely on different sets of predicates. For example, for $K = 10$ and entities of the class MusicComposition, either isPartOf or inLanguage appears as the most frequent predicate depending on the KGEM considered.

We additionally observe that even when the ranking is the same between KGEMs – especially since the predicate distribution can be severely skewed towards one or few predicates – the relative importance of predicates can still differ. For instance, although genre is consistently ranked as the most frequent predicate for MusicGroup entities, results for RDF2Vec suggest a more balanced predicate distribution and a lesser emphasis put on this single predicate (highlighted in blue in Table 4). It is the only model to put a signficant emphasis on award (*i.e.*, considers two MusicGroups as similar if they have won the same (1-hop) or similar (2-hop) awards). In some cases, a few KGEMs seem to assign a lower relevance to specific predicates. For instance, under BoxE the top-3 and top-10 neighbors of entities that are MusicCompositions contain the predicate inLanguage at a much lesser frequency in their subgraphs' intersection (highlighted in brown in Table 4), compared to other models. Answering **RQ3**, we observe that the different notions of entity similarity reflected by different KGEMs are comparable, as most of them put a focus on the same set of predicates when determining entity similarity.

6   Conclusion and Outlook

This work delved into the intricate relationship between entity similarity in KGs and their respective embeddings in KGEMs, questioning the widespread *KGE entity similarity assumption*. Contrary to this belief, we showed that the choice of KGEM significantly influences the notion of entity similarity encoded in the resulting vector space. This finding has profound implications for a variety of downstream tasks where accurate entity similarity is crucial, *e.g.* recommender systems and semantic searches. For instance, if the proximity of graph embeddings does not align with our proposed metric for measuring entity similarity, the tacit assumption of downstream systems exploiting embeddings for recommender systems does not hold and caution is needed when deploying such systems.

A common practice is to evaluate KGEMs by their performance in link prediction, then picking the one with the best performance for a downstream task relying on entity similarty. A critical takeaway from our study is that this practice might not yield the best results in terms of capturing true entity similarity. Instead, our results advocate for cautiousness. Moreover, in scenarios where the similarity of specific classes of entities is of paramount importance, a per-class analysis becomes essential. This approach allows for a tailored selection of

KGEMs that are more adept at capturing the nuances and semantics of particular classes, thereby ensuring a more accurate and meaningful representation of entity similarity.

References

1. Abboud, R., Ceylan, İ.İ., Lukasiewicz, T., Salvatori, T.: Boxe: a box embedding model for knowledge base completion. In: Advances in Neural Information Processing Systems 33: Annual Conference on Neural Information Processing Systems 2020, NeurIPS 2020, December 6-12, 2020, virtual (2020)
2. Alshargi, F., Shekarpour, S., Soru, T., Sheth, A.: Concept2vec: metrics for evaluating quality of embeddings for ontological concepts. arXiv preprint: arXiv:1803.04488 (2018)
3. Auer, S., Bizer, C., Kobilarov, G., Lehmann, J., Cyganiak, R., Ives, Z.G.: Dbpedia: a nucleus for a web of open data. In: Aberer, K., et al. (eds.) The Semantic Web. Lecture Notes in Computer Science, vol. 4825, pp. 722–735. Springer, Berlin (2007). https://doi.org/10.1007/978-3-540-76298-0_52
4. Balazevic, I., Allen, C., Hospedales, T.M.: TuckER: tensor factorization for knowledge graph completion. In: Proceedings of the 2019 Conference on Empirical Methods in Natural Language Processing and the 9th International Joint Conference on Natural Language Processing, EMNLP-IJCNLP 2019, Hong Kong, China, November 3-7, 2019, pp. 5184–5193. Association for Computational Linguistics (2019). https://doi.org/10.18653/v1/D19-1522
5. Bloehdorn, S., Sure, Y.: Kernel methods for mining instance data in ontologies. In: Aberer, K., et al. (eds.) The Semantic Web. Lecture Notes in Computer Science, vol. 4825, pp. 58–71. Springer, Berlin (2007). https://doi.org/10.1007/978-3-540-76298-0_5
6. Bordes, A., Usunier, N., García-Durán, A., Weston, J., Yakhnenko, O.: Translating embeddings for modeling multi-relational data. In: Conference on Neural Information Processing Systems (NeurIPS), pp. 2787–2795 (2013)
7. Chen, W., Zhu, H., Han, X., Liu, Z., Sun, M.: Quantifying similarity between relations with fact distribution. In: Proceedings of the 57th Conference of the Association for Computational Linguistics, ACL 2019, Florence, Italy, July 28- August 2, 2019, Volume 1: Long Papers, pp. 2882–2894. Association for Computational Linguistics (2019). https://doi.org/10.18653/V1/P19-1278
8. Dettmers, T., Minervini, P., Stenetorp, P., Riedel, S.: Convolutional 2D knowledge graph embeddings. In: Proceedings of the Thirty-Second AAAI Conference on Artificial Intelligence, (AAAI-18), the 30th innovative Applications of Artificial Intelligence (IAAI-18), and the 8th AAAI Symposium on Educational Advances in Artificial Intelligence (EAAI-18), New Orleans, Louisiana, USA, February 2-7, 2018, pp. 1811–1818. AAAI Press (2018)
9. Gad-Elrab, M.H., Stepanova, D., Tran, T., Adel, H., Weikum, G.: ExCut: explainable embedding-based clustering over knowledge graphs. In: Pan, J.Z., et al. (eds.) The Semantic Web - ISWC 2020. Lecture Notes in Computer Science(), vol. 12506, pp. 218–237. Springer, Cham (2020). https://doi.org/10.1007/978-3-030-62419-4_13
10. Guo, Q., et al.: A survey on knowledge graph-based recommender systems. IEEE Trans. Knowl. Data Eng. **34**(8), 3549–3568 (2022). https://doi.org/10.1109/TKDE.2020.3028705

11. Hubert, N., Monnin, P., Brun, A., Monticolo, D.: Sem@k: is my knowledge graph embedding model semantic-aware? Seman. Web **14**, 1–37 (2023). https://doi.org/10.3233/SW-233508

12. Hubert, N., Paulheim, H., Monnin, P., Brun, A., Monticolo, D.: Schema first! learn versatile knowledge graph embeddings by capturing semantics with MASCHInE. In: K-CAP '23: Knowledge Capture Conference, Pensacola, Florida, USA, December 5-7, 2023. ACM (2023). https://doi.org/10.48550/ARXIV.2306.03659

13. Ilievski, F., Shenoy, K., Chalupsky, H., Klein, N., Szekely, P.: A study of concept similarity in Wikidata. Seman. Web J. (2023)

14. Jain, N., Kalo, J.C., Balke, W.T., Krestel, R.: Do embeddings actually capture knowledge graph semantics? In: Verborgh, R., et al. (eds.) The Semantic Web. Lecture Notes in Computer Science(), vol. 12731, pp. 143–159. Springer, Cham (2021). https://doi.org/10.1007/978-3-030-77385-4_9

15. Ji, G., He, S., Xu, L., Liu, K., Zhao, J.: Knowledge graph embedding via dynamic mapping matrix. In: Proceedings of the 53rd Annual Meeting of the Association for Computational Linguistics and the 7th International Joint Conference on Natural Language Processing of the Asian Federation of Natural Language Processing, ACL 2015, July 26-31, 2015, Beijing, China, Volume 1: Long Papers, pp. 687–696. The Association for Computer Linguistics (2015).https://doi.org/10.3115/v1/p15-1067

16. Ji, S., Pan, S., Cambria, E., Marttinen, P., Yu, P.S.: A survey on knowledge graphs: representation, acquisition, and applications. IEEE Trans. Neural Networks Learn. Syst. **33**(2), 494–514 (2022). https://doi.org/10.1109/TNNLS.2021.3070843

17. Kalo, J., Ehler, P., Balke, W.: Knowledge graph consolidation by unifying synonymous relationships. In: Ghidini, C., et al. (eds.) The Semantic Web - ISWC 2019. Lecture Notes in Computer Science(), vol. 11778, pp. 276–292. Springer, Cham (2019). https://doi.org/10.1007/978-3-030-30793-6_16

18. Kendall, M.G.: A new measure of rank correlation. Biometrika **30**(1/2), 81–93 (1938)

19. Liu, W., et al.: K-BERT: enabling language representation with knowledge graph. In: The Thirty-Fourth AAAI Conference on Artificial Intelligence, AAAI 2020, The Thirty-Second Innovative Applications of Artificial Intelligence Conference, IAAI 2020, The Tenth AAAI Symposium on Educational Advances in Artificial Intelligence, EAAI 2020, New York, NY, USA, February 7-12, 2020, pp. 2901–2908. AAAI Press (2020). https://doi.org/10.1609/AAAI.V34I03.5681

20. Ma, Y., Cambria, E., Gao, S.: Label embedding for zero-shot fine-grained named entity typing. In: Calzolari, N., Matsumoto, Y., Prasad, R. (eds.) COLING 2016, 26th International Conference on Computational Linguistics, Proceedings of the Conference: Technical Papers, December 11-16, 2016, Osaka, Japan, pp. 171–180. ACL (2016). https://aclanthology.org/C16-1017/

21. Mikolov, T., Chen, K., Corrado, G., Dean, J.: Efficient estimation of word representations in vector space. In: 1st International Conference on Learning Representations, ICLR 2013, Scottsdale, Arizona, USA, May 2-4, 2013, Workshop Track Proceedings (2013)

22. Nickel, M., Tresp, V., Kriegel, H.: A three-way model for collective learning on multi-relational data. In: Proceedings of the 28th International Conference on Machine Learning, ICML, pp. 809–816 (2011)

23. Portisch, J., Heist, N., Paulheim, H.: Knowledge graph embedding for data mining vs. knowledge graph embedding for link prediction-two sides of the same coin? Seman. Web **13**(3), 399–422 (2022)

24. Ristoski, P., Paulheim, H.: Rdf2vec: RDF graph embeddings for data mining. In: Groth, P., et al. (eds.) The Semantic Web - ISWC 2016. Lecture Notes in Computer Science(), vol. 9981, pp. 498–514. Springer, Cham (2016). https://doi.org/10.1007/978-3-319-46523-4_30

25. Rossi, A., Barbosa, D., Firmani, D., Matinata, A., Merialdo, P.: Knowledge graph embedding for link prediction: a comparative analysis. ACM Trans. Knowl. Discovery Data **15**(2), 1–49 (2021)

26. Rossi, A., Matinata, A.: Knowledge graph embeddings: are relation-learning models learning relations? In: Proceedings of the Workshops of the EDBT/ICDT 2020 Joint Conference, Copenhagen, Denmark, March 30, 2020. CEUR Workshop Proceedings, vol. 2578. CEUR-WS.org (2020)

27. Safavi, T., Koutra, D.: Codex: A comprehensive knowledge graph completion benchmark. In: Proceedings of the 2020 Conference on Empirical Methods in Natural Language Processing, EMNLP 2020, Online, November 16-20, 2020, pp. 8328–8350. Association for Computational Linguistics (2020).https://doi.org/10.18653/v1/2020.emnlp-main.669

28. Sanfeliu, A., Fu, K.: A distance measure between attributed relational graphs for pattern recognition. IEEE Trans. Syst. Man Cybern. **13**(3), 353–362 (1983). https://doi.org/10.1109/TSMC.1983.6313167

29. Shi, B., Weninger, T.: Open-world knowledge graph completion. In: Proceedings of the Thirty-Second AAAI Conference on Artificial Intelligence, (AAAI-18), the 30th innovative Applications of Artificial Intelligence (IAAI-18), and the 8th AAAI Symposium on Educational Advances in Artificial Intelligence (EAAI-18), New Orleans, Louisiana, USA, February 2-7, 2018, pp. 1957–1964. AAAI Press (2018). https://doi.org/10.1609/AAAI.V32I1.11535

30. Sosa, D.N., Derry, A., Guo, M.G., Wei, E., Brinton, C., Altman, R.B.: A literature-based knowledge graph embedding method for identifying drug repurposing opportunities in rare diseases. In: Pacific Symposium on Biocomputing 2020, Fairmont Orchid, Hawaii, USA, January 3-7, 2020, pp. 463–474 (2020)

31. Suchanek, F.M., Kasneci, G., Weikum, G.: Yago: a core of semantic knowledge. In: Proceedings of the 16th International Conference on World Wide Web, WWW, pp. 697–706. ACM (2007)

32. Sun, Y., et al.: ERNIE 2.0: a continual pre-training framework for language understanding. In: The Thirty-Fourth AAAI Conference on Artificial Intelligence, AAAI 2020, The Thirty-Second Innovative Applications of Artificial Intelligence Conference, IAAI 2020, The Tenth AAAI Symposium on Educational Advances in Artificial Intelligence, EAAI 2020, New York, NY, USA, February 7-12, 2020, pp. 8968–8975. AAAI Press (2020). https://doi.org/10.1609/AAAI.V34I05.6428

33. Sun, Z., et al.: A benchmarking study of embedding-based entity alignment for knowledge graphs. Proc. VLDB Endow. **13**(11), 2326–2340 (2020)

34. Sun, Z., Deng, Z., Nie, J., Tang, J.: Rotate: Knowledge graph embedding by relational rotation in complex space. In: 7th International Conference on Learning Representations, ICLR (2019)

35. Toutanova, K., Chen, D.: Observed versus latent features for knowledge base and text inference. In: Proceedings of the 3rd Workshop on Continuous Vector Space Models and their Compositionality, pp. 57–66. Association for Computational Linguistics (2015)

36. Trouillon, T., Welbl, J., Riedel, S., Gaussier, É., Bouchard, G.: Complex embeddings for simple link prediction. In: Proceedings of the 33rd International Conference on Machine Learning, ICML, vol. 48, pp. 2071–2080 (2016)

37. Wang, M., Qiu, L., Wang, X.: A survey on knowledge graph embeddings for link prediction. Symmetry **13**(3), 485 (2021)
38. Webber, W., Moffat, A., Zobel, J.: A similarity measure for indefinite rankings. ACM Trans. Inf. Syst. **28**(4), 1–38 (2010). https://doi.org/10.1145/1852102.1852106
39. Yang, B., Yih, W., He, X., Gao, J., Deng, L.: Embedding entities and relations for learning and inference in knowledge bases. In: 3rd International Conference on Learning Representations, ICLR (2015)

Treat Different Negatives Differently: Enriching Loss Functions with Domain and Range Constraints for Link Prediction

Nicolas Hubert[1,2]([✉])[iD], Pierre Monnin[3][iD], Armelle Brun[2][iD],
and Davy Monticolo[1][iD]

[1] Université de Lorraine, ERPI, Nancy, France
davy.monticolo@univ-lorraine.fr
[2] Université de Lorraine, CNRS, LORIA, Nancy, France
{nicolas.hubert,armelle.brun}@univ-lorraine.fr
[3] Université Côte d'Azur, Inria, CNRS, I3S, Sophia-Antipolis, France
pierre.monnin@inria.fr

Abstract. Knowledge graph embedding models (KGEMs) are used for various tasks related to knowledge graphs (KGs), including link prediction. They are trained with loss functions that consider batches of true and false triples. However, different kinds of false triples exist and recent works suggest that they should not be valued equally, leading to specific negative sampling procedures. In line with this recent assumption, we posit that negative triples that are semantically valid w.r.t. signatures of relations (domain and range) are high-quality negatives. Hence, we enrich the three main loss functions for link prediction such that all kinds of negatives are sampled but treated differently based on their semantic validity. In an extensive and controlled experimental setting, we show that the proposed loss functions systematically provide satisfying results which demonstrates both the generality and superiority of our proposed approach. In fact, the proposed loss functions (1) lead to better MRR and Hits@10 values, and (2) drive KGEMs towards better semantic correctness as measured by the Sem@K metric. This highlights that relation signatures globally improve KGEMs, and thus should be incorporated into loss functions. Domains and ranges of relations being largely available in schema-defined KGs, this makes our approach both beneficial and widely usable in practice.

Keywords: Knowledge Graph Embeddings · Link Prediction · Schema-based Learning · Loss Functions

1 Introduction

A knowledge graph (KG) is a collection of triples (h, r, t) where h (head) and t (tail) are two entities of the graph, and r is a predicate that qualifies the nature of the relation holding between them. In this work, we do not consider literals. KGs are inherently incomplete, incorrect, or overlapping and thus major refinement tasks include entity matching and link prediction [29]. The latter is the focus of this paper. Link prediction (LP) aims at completing KGs by leveraging existing facts to infer missing ones. In the LP task, one is provided with a set of incomplete triples, where the missing head (resp. tail) needs to be predicted. This amounts to holding a set of triples where, for each triple, either the head h or the tail t is missing.

The LP task is often addressed using knowledge graph embedding models (KGEMs). They represent entities and relations as low-dimensional vectors in a latent embedding space that preserves as much as possible graph structural information and graph properties [15]. A plethora of KGEMs has been proposed in the literature. They usually differ w.r.t. their scoring function, *i.e.* how they model interactions between entities. Entities and relations embeddings are learned throughout several epochs, in an optimization process relying on loss functions. Loss functions aims at maximizing the score output by KGEMs for *positive* triples, *i.e.*, triples that exist in the graph, and minimizing the score of *negative* triples, *i.e.*, triples that are absent from the graph. A recent segment of the literature started investigating the influence of negative triples in the training of KGEMs [17]. Indeed, their generation – called negative sampling – is usually performed by corrupting true triples, *i.e.*, replacing their head or tail with another entity randomly chosen in the graph. Enhanced sampling mechanisms were then envisioned and, for example, involve selecting entities of the same type as the entity to replace [17].

The latter proposal comes within the scope of works considering the underpinning semantics of KGs as additional information to improve results w.r.t. the LP task. Over the past few years, semantic information has been incorporated in various parts of KGEMs, *e.g.*, as mentioned, the negative sampling procedure [14,18], but also the model itself [6,19,23,28,32], or the loss function [5,7,10,20]. Existing works proposing to include semantic information into loss functions showcase promising results. However, they are restricted to specific loss functions [7,10], or consider ontological axioms [7,20] that do not include domain and range axioms (or relation signatures) which are widely available in KGs. Interestingly, such axioms could be leveraged to generate different kinds of negative triples, *i.e.*, negative triples that respect relation signatures (called semantically valid) and those that do not (called semantically invalid). Additionally, to the best of our knowledge, negative triples were only studied in the sampling procedure, where specific ones are chosen to train on and the others are discarded. The possibility to sample all kinds of negatives but differently consider them within loss functions was left unassessed. This twofold observation motivates our first research question:

RQ1 how main loss functions used in LP can incorporate domain and range constraints to differently consider negative triples based on their semantic validity?

Precedent work also pointed out the performance gain of incorporating ontological information as measured by rank-based metrics such as MRR and

Hits@K [5,7,10,20]. However, while such approaches include semantic information as KGEM inputs, the semantic capabilities of the resulting KGEM are left unassessed, even though this would provide a fuller picture of its performance [11,13]. Hence, our second research question:

RQ2 what is the impact of incorporating relation signatures into loss functions on the overall KGEM performance?

To address both questions, we propose signature-driven loss functions, *i.e.* loss functions containing terms that depend on some background knowledge (BK) about types of entities and domains and ranges of relations. To broaden the impact of our approach, our work is concerned with the three most encountered loss functions in the literature: the pairwise hinge loss (PHL) [3], the 1-N binary cross-entropy loss (BCEL) [8], and the pointwise logistic loss (PLL) [27] (further detailed in Sect. 3). For each of them, we propose a tailored signature-driven version. The considered BK is available in many schema-defined KGs [9], which makes these newly introduced loss functions widely usable in practice. Furthermore, the impact of loss functions is evaluated using both rank-based metrics and Sem@K [11,13] – a metric that measures the consistency of KGEMs predictions for the LP task with relation signatures, *i.e.*, the semantic correctness of predictions.

To summarize, the main contributions of this work are:

- We propose signature-driven versions for the three mostly used loss functions for the LP task, leveraging BK about relation domains and ranges.
- We evaluate our approach in terms of traditional rank-based metrics, and also w.r.t. Sem@K, which gives more insight into the benefits of our proposal.
- We show that the designed signature-driven loss functions provide, in most cases, better performance w.r.t. both rank-based metrics and Sem@K. Consequently, our findings strongly indicate that signature information should be systematically incorporated into loss functions.

The remainder of the paper is structured as follows. Related work is presented in Sect. 2. In Sect. 3, we detail the signature-driven loss functions proposed in this work. Dataset descriptions and experimental settings are provided in Sect. 4. Key findings are presented in Sect. 5 and are further discussed in Sect. 6. Lastly, Sect. 7 sums up the main findings and outlines future research directions.

2   Related Work

This section firstly relates to former contributions that make use of semantic information to enhance model results regarding the LP task. Emphasis is placed on how semantic information can be incorporated in the loss functions (Sect. 2.1) – a research avenue which remains relatively unexplored compared to incorporating semantic information in other parts of the learning process, *e.g.* in the negative sampling or in the interaction function. Secondly, a brief background on the mainstream loss functions is provided (Sect. 2.2). This is to help position our contributions w.r.t. the vanilla loss functions used in practice.

2.1   Semantic-Enhanced Approaches

A significant body of the literature proposes approaches that incorporate semantic information for performing LP with KGEMs, with the purpose of improving KGEM performance w.r.t. traditional rank-based metrics.

The most straightforward way to do so is to embed semantic information in the model itself. For instance, AutoETER [23] is an automated type representation learning mechanism that can be used with any KGEM and that learns the latent type embedding of each entity. In TaRP [6], type information and instance-level information are simultaneously considered and encoded as prior probabilities and likelihoods of relations, respectively. TKRL [32] bridges type information with hierarchical information: while type information is utilized as relation-specific constraints, hierarchical types are encoded as projection matrices for entities. TKRL allows entities to have different representations in different types. Similarly, in [28], the proposed KGEM allows entities to have different vector representations depending on their respective types. TransC [19] encodes each concept of a KG as a sphere and each instance as a vector in the same semantic space. The relations between concepts and instances (`rdf:type`), and the relations between concepts and sub-concepts (`rdfs:subClassOf`) are based on the relative distance within this shared semantic space.

A few works incorporate semantic information to constrain the negative sampling (NS) procedure and generate meaningful negative triples [14,18,31]. For instance, type-constrained negative sampling (TCNS) [18] replaces the head or the tail of a triple with a random entity belonging to the same type (`rdf:type`) as the ground-truth entity. Jain *et al.* [14] go a step further and use ontological reasoning to iteratively improve KGEM performance by retraining the model on inconsistent predictions. It is noteworthy that our approach adopts an orthogonal direction of such works by proposing to sample all kinds of negatives but to treat them differently in the loss function when training.

A few work actually propose to include semantic information in the learning and optimization process. In [10], entities embeddings of the same semantic category are enforced to lie in a close neighborhood of the embedding space. However, their approach only fits single-type KGs. In addition, the only mainstream model benchmarked in this work is TransE [3], and only the pairwise hinge loss is used. Likewise, d'Amato *et al.* [7] solely consider the pairwise hinge loss, and their approach is benchmarked w.r.t. to translational models only. Moreover, BK is injected in the form of `equivalentClass`, `equivalentProperty`, `inverseOf`, and `subClassOf` axioms, similarly to [20] who incorporate `equivalentProperty` and `inverseOf` axioms as regularization terms in the loss function. However, the aforementioned axioms are rarely provided in KGs [9]. Cao *et al.* [5] propose a new regularizer called Equivariance Regularizer, which limits overfitting by using semantic information. However, their approach is data-driven and does not rely on a schema. In contrast, the approach presented in Sect. 3 leverages domain and range constraints which are available in most schema-defined KGs.

Finally, all the aforementioned semantic-driven approaches are only evaluated w.r.t.. rank-based metrics. However, semantic-driven approaches would benefit from a semantic-oriented evaluation. To the best of our knowledge, the work around Sem@K [11–13] is the only one to provide appropriate tools for measuring KGEM semantic correctness. Hence, our experiments will also be evaluated with this metric.

2.2   Loss Functions for the Link Prediction Task

Few works revolve around the influence of loss functions on KGEM performance [1,21,22]. Mohamed *et al.* [22] point out the lack of consideration regarding the impact of loss functions on KGEM performance. Experimental results
provided in [1] indicate that no loss function consistently provides the best
results, and that it is rather the combination between the scoring and loss
functions that impacts KGEM performance. In particular, some scoring functions better match with specific loss functions. For instance, Ali *et al.* show that
TransE can outperform state-of-the-art KGEMs when configured with an appropriate loss function. Likewise, Mohamed *et al.* [22] show that the choice of the loss
function significantly influence KGEM performance. Consequently, they provide
an extensive benchmark study of the main loss functions used in the literature.
Namely, their analysis relies on a commonly accepted categorization between
pointwise and pairwise loss functions. The main difference between pointwise
and pairwise loss functions lies in the way the scoring function, the triples, and
their respective labels are considered all together. Under the pointwise approach,
the loss function relies on the predicted scores for triples and their actual label
values, which is usually 1 for positive triple and 0 (or −1) for negative triples.
In contrast, pairwise loss functions are defined in terms of differences between
the predicted score of a true triple and the score of a negative counterpart.

In our approach (Sect. 3), we consider the three most commonly used loss
functions for performing LP [24]: the pairwise hinge loss (PHL) [3], the 1-N
binary cross-entropy loss (BCEL) [8], and the pointwise logistic loss (PLL) [27].
Their vanilla formulas are recalled in Eqs. (1), (2), and (3).

$$\mathcal{L}_{PHL} = \sum_{t\in\mathcal{T}^+}\sum_{t'\in\mathcal{T}^-} [\gamma + f(t') - f(t)]_+ \tag{1}$$

where $\mathcal{T}$, f, and $[x]_+$ denote a batch of triples, the scoring function, and the
positive part of x, respectively. $\mathcal{T}$ is further split into a batch of positive triples
$\mathcal{T}^+$ and a batch of negative triples $\mathcal{T}^-$. γ is a configurable margin hyperparameter specifying how much the scores of positive triples should be separated from
the scores of corresponding negative triples.

$$\mathcal{L}_{BCEL} = -\frac{1}{|\mathcal{E}|}\sum_{t\in\mathcal{T}} \ell(t)\log(f(t)) + (1 - \ell(t))\log(1 - f(t)) \tag{2}$$

where $\ell(t) \in \{0,1\}$ denotes the true label of t and $\mathcal{T}$ is a batch with all possible
$(h, r, *)$. $|\mathcal{E}|$ is the number of entities in the KG.

$$\mathcal{L}_{PLL} = \sum_{t\in\mathcal{T}} \log(1 + \exp^{-\ell(t)\cdot f(t)}) \tag{3}$$

where $\ell(t) \in \{-1,1\}$ denotes the true label of t.

3   Signature-Driven Loss Functions

Building on the limits of previous work (Sect. 2.1), we propose signature-driven loss functions that extend the three most frequently used loss functions [24] and leverage BK about domains and ranges of relations, which are provided in many KGs used in the literature.

The purpose of the proposed loss functions is to distinguish *semantically valid negatives* from *semantically invalid ones*. The former are defined as triples (h, r, t) respecting both the domain and range of the relation r, *i.e.*,

$$\text{type}(h) \cap \text{domain}(r) \neq \emptyset \wedge \text{type}(t) \cap \text{range}(r) \neq \emptyset$$

whereas the latter violate at least one of the constraints, *i.e.*

$$\text{type}(h) \cap \text{domain}(r) = \emptyset \vee \text{type}(t) \cap \text{range}(r) = \emptyset.$$

$\text{domain}(r)$ (resp. $\text{range}(r)$) is defined as the expected type as head (resp. tail) for the relation r. For example, the relation presidentOf expects a Person as head and a Country as tail. Starting from a positive triple (EmmanuelMacron, presidentOf, France) which represents a true fact, (BarackObama, presidentOf, France) and (EmmanuelMacron, presidentOf, Germany) are examples of semantically valid negative triples, whereas (Adidas, presidentOf, France) and (EmmanuelMacron, presidentOf, Christmas) are examples of semantically invalid negative triples. In this work, entities are multi-typed. Therefore, $\text{type}(e)$ returns the set of types (a.k.a. classes) that the entity e belongs to.

We introduce a loss-independent ϵ factor, which is dubbed as the *semantic factor* and aims at bringing the scores of semantically valid negative triples closer to the positive ones. This common semantic factor fitting into different loss functions shows the generality of our approach that can possibly be extended to other loss functions. Interestingly, this factor also allows to take into account to some extent the Open World Assumption (OWA) under which KGs are represented. Under the OWA, triples that are not represented in a KG are either false or missing positive triples. In traditional training procedures, these triples are indiscriminately considered negative, which corresponds to the Closed World Assumption. On the contrary, our proposal considers semantically invalid triples as true negative while semantically valid triples (and possibly missing positive or false negative under the OWA) are closer to true positive triples. This assumes that entity types are complete and correct.

$\mathcal{L}_{\mathbf{PHL}}$ defined in Eq. (1) relies on the margin hyperparameter γ. Increasing (resp. descreasing) the value of γ will increase (resp. descrease) the margin that will be set between the scores of positive and negative triples. However, this unique γ treats all negative triples indifferently: the same margin will separate the scores of semantically valid and semantically invalid negative triples from the score of the positive triple they both originate from. We suggest that the scores of these two kinds of negative triples should be treated differently. Hence, our approach redefines $\mathcal{L}_{PHL}$ as follows (Eq. (4)):

$$\mathcal{L}_{PHL}^S = \sum_{t \in \mathcal{T}^+} \sum_{t' \in \mathcal{T}^-} [\gamma \cdot \ell(t') + f(t') - f(t)]_+$$

(4)

$$\text{where } \ell(t') = \begin{cases} 1 \text{ if } t' \text{ is semantically invalid} \\ \epsilon \text{ otherwise} \end{cases}$$

The loss function in Eq. (4) now has a superscripted S to make it clear this is the signature-driven version of the vanilla $\mathcal{L}_{PHL}$ as defined in Eq. (1). A choice of $\epsilon < 1$ leads the KGEM to apply a higher margin between scores of positive and semantically invalid triples than between positive and semantically valid ones. For a given positive triple, this allows to keep the scores of its semantically valid negative counterparts relatively closer compared to the scores of its semantically invalid counterparts. Intuitively, when the KGEM outputs wrong predictions, more of them are still expected to meet the domain and range constraints imposed by relations. Thus, wrong predictions are assumed to be more meaningful, and, in a sense, semantically closer to the ground-truth triple.

$\mathcal{L}_{\mathbf{BCEL}}$ defined in Eq. (2) is adapted to $\mathcal{L}_{BCEL}^S$ by redefining the labelling function ℓ. In particular, when dealing with a KG featuring typed entities and providing information about domains and ranges of relations, the labelling function ℓ is no longer binary. Instead, the labels of semantically valid negative triples can be fixed to some intermediate value between the label value of positive triples and of semantically invalid negative triples, which leads to the labelling function ℓ defined in Eq. (5):

$$\ell(t') = \begin{cases} 1 \text{ if } t' \in \mathcal{T}^+ \\ \epsilon \text{ if } t' \in \mathcal{T}^- \text{ and } t' \text{ is semantically valid} \\ 0 \text{ if } t' \in \mathcal{T}^- \text{ and } t' \text{ is semantically invalid} \end{cases}$$

(5)

where the semantic factor ϵ is a tunable hyperparameter denoting the label value of semantically valid negative triples. The intuition underlying the refinement of the labelling function ℓ is to voluntarily cause some confusion between semantically valid negative triples and positive triples. By bridging their respective label values, it is expected that the KGEM will somehow consider the former as "less negative triples" and assign them a higher score compared to positive triples.

$\mathcal{L}_{\mathbf{PLL}}$ defined in Eq. (3) could be adapted to $\mathcal{L}_{PLL}^S$ similarly to $\mathcal{L}_{BCEL}^S$. In other words, the labelling function ℓ could also output an intermediate label value ϵ for semantically valid negative triples. Although this approach provides very good results in terms of Sem@K values, it does not provide consistently good results across datasets. Furthermore, obtained results in terms of MRR and Hits@K can be far below the ones obtained with the vanilla model (see supplementary materials for further details). That is why, here, to treat semantically valid and invalid negative triples differently, instead of modifying the labelling function ℓ, the semantic factor ϵ for $\mathcal{L}_{PLL}^S$ defines the probability with which semantically valid negative triples are considered as positive triples and there-fore are labelled the same way. For example, with $\epsilon = 0.05$, at each training epoch and for each batch, the semantically valid negative triples of the given training batch will be considered positive with a probability of 5%.

It is noteworthy that our approach can be applied in practice to KGs with or without types, domains and ranges. Indeed, in the absence of such background knowledge, our signature-driven loss functions reduce to their respective vanilla counterparts. Recall that our approach does not focus on negative sampling or complex negative sample generators such as KBGAN [4], NSCaching [34], and self-adversarial negative sampling [26]. Although these works are related to ours, they constrain negative sampling upstream. On the contrary, our approach does not constrain the sampling of negative triples but rather dynamically distributes the negative triples into different parts of the loss functions, based on their semantic validity.

4 Experimental Setting

4.1 Evaluation Metrics

In our experiments, KGEM performance is assessed w.r.t. MRR, Hits@K and Sem@K, with $K = 10$. **Mean Reciprocal Rank (MRR)** corresponds to the arithmetic mean over the reciprocals of ranks of the ground-truth triples. MRR is bounded in the $[0, 1]$ interval, where the higher the better. **Hits@K** accounts for the proportion of ground-truth triples appearing in the first K top-scored triples. This metric is bounded in the $[0, 1]$ interval and its values increases with K, where the higher the better. **Sem@K** [11,13] accounts for the proportion of triples that are semantically valid in the first K top-scored triples:

$$\text{Sem@}K = \frac{1}{|\mathcal{B}|} \sum_{q \in \mathcal{B}} \frac{1}{K} \sum_{q' \in \mathcal{S}_q^K} \text{compatibility}(q, q') \tag{6}$$

where, given a ground-truth triple $q = (h, r, t)$, $\mathcal{S}_q^K$ is the list of the top-K candidate triples scored by a KGEM (*i.e.*, by predicting the tail for $(h, r, ?)$ or the head for $(?, r, t)$). A candidate triple q' is assessed w.r.t. q by the compatibility function that checks whether the predicted head (resp. tail) belongs to the domain (resp. range) of the relation. In this work, class hierarchies are considered: if a relation has a given class as domain (resp. range), entities from its subclasses are considered semantically valid. Sem@K is bounded in the $[0, 1]$ interval.

4.2 Datasets and Models

Even though our approach could be applied in KGs with or without relation signatures, we evaluate our proposal in an ideal experimental setting to pre-cisely qualify the interest of considering domains and ranges. Firstly, all relations appearing in the training set have a defined domain and range. Secondly, both the head h and tail t of train triples have another semantically valid counterpart for negative sampling. These two conditions guarantee that each positive train triple can be paired with at least one semantically valid triple. Finally, validation and test sets contain triples whose relation have a well-defined domain (resp. range), as well as more than 10 semantically valid candidates as head (resp. tail). This ensures Sem@K is not unduly penalized and can be calculated on the same set of entities as Hits@K and MRR until $K = 10$.

To ensure these requirements, we filtered FB15k237-ET, DBpedia93k, and YAGO4-19k [13] so that they comply with the aforementioned criteria. In the following, their filtered versions are referred to as FB15k187, DBpedia77k, and Yago14k, respectively. Table 1 provides statistics for these datasets and reflects the diversity of their characteristics. In this work, several KGEMs are considered: TransE [3], TransH [30], and DistMult [33] using $\mathcal{L}_{PHL}$; ComplEx [27] and SimplE [16] using $\mathcal{L}_{PLL}$; ConvE [8], TuckER [2], and RGCN [25] using $\mathcal{L}_{BCEL}$. Datasets and codes are available in our GitHub repository.[1]

Table 1. Datasets used in the experiments. These are filtered versions of the standard FB15k237-ET, DBpedia93k, and YAGO4-19k.

Dataset	$|\mathcal{E}|$	$|\mathcal{R}|$	$|\mathcal{T}_{train}|$	$|\mathcal{T}_{valid}|$	$|\mathcal{T}_{test}|$	Split ratios		
FB15k187	14,305	187	245,350	15,256	17,830	88%	5.5 %	6.5%
DBpedia77k	76,651	150	140,760	16,334	32,934	74%	9 %	17%
Yago14k	14,178	37	18,263	472	448	95%	2.5%	2.5%

4.3 Implementation Details

For the sake of comparison, MRR, Hits@K and Sem@K are computed after training models from scratch. KGEMs used in the experiments were implemented in PyTorch. After training KGEMs for a large number of epochs, we noticed the best achieved results were found around epoch 400 or below. Consequently, a maximum of 400 epochs of training was set, as in LibKGE[2] (except RGCN which is trained during 4,000 epochs due to lower convergence to the best achieved results). Except when using the BCEL which does not require negative sampling, Uniform Random Negative Sampling [3] was used to pair each train triple with two corresponding negative triples: one which is semantically invalid, and one which is semantically valid. Regarding BCEL, each positive triple is scored against negative triples formed with all other entities in the graph. Hence, negative triples comprise both semantically valid and invalid triples. It should be noted that the best epsilon values for each combination of model and dataset were found on the validation sets. Once the best epsilon values are found, they remain fixed for all triples. In order to ensure fair comparisons between models, embeddings are initialized with the same seed and each model is fed with

---

[1] https://github.com/nicolas-hbt/semantic-lossfunc.
[2] https://github.com/uma-pi1/kge.

exactly the same set of negative triples at each epoch. Grid-search based on predefined hyperparameters was performed. Best hyperparameters and the full hyperparameter space are reported on GitHub (see footnote 1).

5   Results

5.1   Global Performance

Table 2. Rank-based and semantic-based results. Bold fonts indicate which model performs best w.r.t. a given metric. Suffixes V and S indicate whether the model is trained under the vanilla or signature-driven version of the loss function, respectively. Hits@10 and Sem@10 are abbreviated to H@10 and S@10. Underlined cells indicate results that are more specifically referred to in Sect. 5.1. We use the symbol † when the comparison of results in a duel (*e.g.* TransE-V vs. TransE-S) is statistically significant ($\alpha = .05$) for a given metric and dataset.

	FB15k187			DBpedia77k			Yago14k		
	MRR	H@10	S@10	MRR	H@10	S@10	MRR	H@10	S@10
TransE-V	.260	.446	.842	.274	.438	.936	.868	**.945**	.795
TransE-S	**.315**†	**.497**†	**.973**†	.275	.440	**.985**†	.876	.944	**.968**†
TransH-V	.266	.450	.855	.270	.437	.907	.836	.944	.581
TransH-S	**.319**†	**.501**†	**.973**†	**.274**†	**.442**†	**.980**†	**.857**†	**.945**	<u>**.831**†</u>
DistMult-V	.291	.457	.824	.295	.405	.784	.904	**.930**	.409
DistMult-S	**.332**†	**.504**†	**.971**†	**.300**†	**.416**†	**.901**†	**.912**†	.929	**.449**†
ComplEx-V	.280	.416	.472	**.309**†	**.415**†	.769	**.925**	**.932**	.333
ComplEx-S	**.316**†	**.476**†	<u>**.796**†</u>	.297	.409	**.897**	.923	.931	<u>**.667**†</u>
SimplE-V	.261	.387	.462	**.259**†	**.346**†	**.883**†	.926	**.931**	.355
SimplE-S	**.268**†	**.409**†	<u>**.759**†</u>	.230	.302	.850	.924	.927	<u>**.769**†</u>
ConvE-V	.273	.470	.973	.273	.382	.935	**.934**	**.942**	.904
ConvE-S	**.283**†	**.476**†	**.996**†	**.283**†	**.405**†	**.985**†	.933	.940	**.997**†
TuckER-V	.316	.516	.985	.311	.410	.912	.923	.927	.781
TuckER-S	**.320**†	**.522**†	**.996**†	**.312**	**.421**†	**.969**†	**.931**†	**.943**†	**.929**†
RGCN-V	.241	.386	.775	.194	.297	.872	.911	.923	.349
RGCN-S	**.260**†	**.415**†	**.860**†	**.197**†	**.320**†	**.957**†	**.927**†	**.934**†	<u>**.828**†</u>

Table 2 displays KGEM performance, datasets and evaluation metrics of interest. We performed t-tests (when prediction-related data follow a normal distribution according to Shapiro test) and Wilcoxon tests (when they do not) at the significance level $\alpha = .05$. Table 2 clearly shows that, with the sole exception of SimplE on DBpedia77k, the signature-driven loss functions $\mathcal{L}^S_{PHL}$, $\mathcal{L}^S_{BCEL}$, and $\mathcal{L}^S_{PLL}$ all lead to significant improvement in terms of Sem@10. Importantly, in some cases the relative gain in Sem@10 compared to the corresponding vanilla loss function is huge (underlined in Table 2): +137%, +117%, and +100% for RGCN, SimplE, and ComplEx on the smaller Yago14k dataset, respectively, and +69% and +64% for ComplEx and SimplE on FB15k187, respectively. These gains in terms of Sem@10 are observed regardless of the loss function at hand, which demonstrates that our designed signature-driven loss functions can all drive KGEM semantic correctness towards more satisfying results. It is worth noting that, in most cases, they also lead to better KGEM performance as measured by MRR and Hits@10. We observe that in 19 out of 24 ($\approx$ 79%) one-to-one comparisons between the same KGEM trained with vanilla vs. signature-driven loss functions, better MRR values are reported for the model trained under the signature-driven loss function. On the remaining comparisons ($\approx$ 21%), only two highlight statistically significant losses in terms of MRR. However, they are minimal and often for the benefit of significantly better Sem@10 values. This observation raises the question whether better MRR and Hits@K should be pursued at any cost, or whether a small drop w.r.t. to these metrics is acceptable if this leads to a significantly better KGEM semantic correctness (see Sect. 6.3). Plus, these promising results imply that even if the intended purpose is to only maximize MRR and Hits@K values, taking the available signature information into consideration is strongly advised: this does not only improve KGEM semantic correctness, but also provide performance gains in terms of MRR and Hits@K.

In the following, we provide a finer-grained results' analysis, which focuses on the different loss functions and datasets used in the experiments. Although we previously showed the effectiveness of our approach, the benefits brought from considering BK in the form of relation domains and ranges differ across loss functions. In particular, the gains achieved using $\mathcal{L}_{PHL}^{S}$ and $\mathcal{L}_{BCEL}^{S}$ are substantial. With these loss functions, we observe a systematic improvement w.r.t. Sem@10. Gains are also reported w.r.t. rank-based metrics in the vast majority of cases. The only exception is on Yago14k where semantic losses are sometimes slightly outperformed but still competitive w.r.t. their vanilla counterparts. However, the difference in terms of MRR and Hits@10 values is negligible and not statistically significant. This may come from the reduced number of triples in Yago14k, which results in the additional signature information improving Sem@K but not helping discriminate gold entities from others. Therefore, incorporating BK about relation domains and ranges into $\mathcal{L}_{PHL}^{S}$ and $\mathcal{L}_{BCEL}^{S}$ is a viable approach that provides consistent gains both in terms of MRR, Hits@10 and Sem@10. Regarding the benefits from doing so under $\mathcal{L}_{PLL}^{S}$, we can notice a slight decline in MRR and Hits@10 values in some cases. However, the other side of the coin is that achieved Sem@10 values are substantially higher: except for SimplE on DBpedia77k, Sem@10 values increase in a range from +17% to +117% for the remaining one-to-one comparisons. As such, our approach using $\mathcal{L}_{PLL}^{S}$ also provides satisfactory results, as long as a slight drop in rank-based metrics is acceptable if it comes with the benefit of significantly better KGEM semantic correctness. Besides, it is worth noting the following points: our hyperparameter tuning strategy relied on the choice of the best ϵ value on Yago14k

– for computational limitations. The ϵ value which was found to perform the best on Yago14k was subsequently used in all the remaining scenarios. A more thorough tuning of ϵ on the other datasets would have potentially provided even more satisfying results under the $\mathcal{L}^S_{PLL}$, thus strengthening the value of our approach.

5.2 Ablation Study

In this section, KGEMs are tested on three buckets of relations that feature narrow (B1), intermediate (B2), and large (B3) sets of semantically valid heads or tails, respectively. The cut-offs have been manually defined and are provided in supplementary materials (see footnote 1). The analysis of B1 allows us to gauge the impact of signature-driven loss functions on relations for which it is harder to predict semantically valid entities. Results reported in Table 3 for B1 clearly demonstrate that the impact of injecting BK into loss functions is exacerbated for them, thus supporting the value of our approach in a sparse and difficult setting. One might think that the better MRR values achieved with $\mathcal{L}^S_{PHL}$, $\mathcal{L}^S_{BCEL}$, and $\mathcal{L}^S_{PLL}$ are highly correlated to the better Sem@10 values. This is partially true, as for a relation in B1, placing all semantically valid candidates at the top of the ranking list is likely to uplift the rank of the ground-truth itself. However, we can see that RGCN-V and RGCN-S have almost equal MRR values on Yago14k, while Sem@10 values of RGCN-S are much higher than RGCN-V (+254%). Similar findings hold for ComplEx on Yago14k, ConvE on DBpedia77k, TransE on DBpedia77k and Yago14k. This shows that in a number of cases, signature-driven loss functions improve the semantic correctness of KGEM for small relations while leaving its performance untouched in terms of rank-based metrics. Results on B2 and B3 are provided in supplementary materials (see footnote 1). In particular, it can be noted that the relative benefit of our approach w.r.t. MRR and Hits@10 is more limited on such buckets. Regarding Sem@K, results achieved with the vanilla loss functions are already high, hence the relatively lower gain brought by injecting ontological BK. These already high Sem@K values may be explained by a higher number of semantically valid candidates.

6 Discussion

6.1 Treating Different Negatives Differently (RQ1)

The proposed loss functions $\mathcal{L}^S_{PHL}$, $\mathcal{L}^S_{BCEL}$, and $\mathcal{L}^S_{PLL}$ provide adequate training objective for KGEMs, as evidenced in Table 2. Most importantly, in Sect. 3 we clearly show how the inclusion of signature information into $\mathcal{L}^S_{PHL}$, $\mathcal{L}^S_{BCEL}$, and $\mathcal{L}^S_{PLL}$ can be brought under one roof thanks to a commonly defined semantic factor. Although this semantic factor operates at different levels depending on the loss function, its common purpose is to differentiate how semantically valid and semantically invalid negative triples should be considered compared to positive

Table 3. Rank-based and semantic-based results on the bucket of relations that feature a narrow set of semantically valid heads or tails (B1). The cut-offs have been manually defined and are provided in supplementary materials.

	FB15k187			DBpedia77k			Yago14k		
	MRR	H@10	S@10	MRR	H@10	S@10	MRR	H@10	S@10
TransE-V	.535	.727	.647	.498	.578	.460	.914	**.979**	.620
TransE-S	**.646**	**.805**	**.937**	**.539**	**.626**	**.910**	**.922**	.975	**.924**
TransH-V	.541	.734	.661	.475	.555	.425	.897	.972	.436
TransH-S	**.655**	**.814**	**.936**	**.541**	**.643**	**.828**	**.923**	**.981**	**.684**
DistMult-V	.589	.735	.628	.466	.462	.244	.949	**.959**	.304
DistMult-S	**.667**	**.802**	**.929**	**.498**	**.547**	**.451**	**.965**	.958	**.372**
ComplEx-V	.530	.567	.116	**.424**	**.425**	.161	.955	.956	.133
ComplEx-S	**.637**	**.723**	**.537**	.421	.399	**.198**	**.961**	.956	**.423**
SimplE-V	.507	.553	.136	**.396**	**.370**	**.273**	**.959**	.882	**.932**
SimplE-S	**.576**	**.671**	**.505**	.324	.259	.206	.958	**.883**	.930
ConvE-V	.549	.779	.973	.518	**.569**	.789	.969	**.972**	.915
ConvE-S	**.562**	**.783**	**.986**	.518	.566	**.927**	.969	.965	**.960**
TuckER-V	.597	.811	.969	.519	.568	.740	.949	.970	.846
TuckER-S	**.598**	**.815**	**.973**	**.526**	**.582**	**.797**	**.964**	.970	**.892**
RGCN-V	.510	.629	.468	.386	.387	.254	.963	.959	.141
RGCN-S	**.549**	**.705**	**.682**	**.396**	**.415**	**.398**	**.966**	**.967**	**.499**

triples, whereas traditional approaches treat all negative triples indifferently. Besides, our approach that transforms vanilla loss functions into signature-driven ones can also work for other loss functions such as the pointwise hinge loss or the pairwise logistic loss, as presented in [22], by including a semantic factor ϵ as well. The tailoring of these losses and the experiments are left for future work.

Recall that compared to complex negative sampler such as KBGAN [4], NSCaching [34], and self-adversarial NS [26], we do not introduce any potential overhead due to the need for maintaining a cache [34] or training an intermediate adversarial learning framework [4,26] for generating high-quality negatives. Instead, negative triples dynamically enter a different part of the loss function depending on their semantic validity. In future work, we will compare the performance and algorithmic complexity of our approach w.r.t. NS procedures, thereby highlighting the potential cost savings in computational resources and execution time compared to sophisticated NS. Our approach is also agnostic to the underlying NS procedure, and can work along with simple uniform random NS [3] as well as more complex procedures [4,26,34]. Besides, our approach can be applied even in the absence of BK. In this case, signature-driven loss functions reduce to their vanilla version.

6.2 Impact of Signature-Driven Losses on Performance (RQ2)

Mohamed *et al.* [21] investigate the effects of specific choices of loss functions on the scalability and performance of KGEMs w.r.t. rank-based metrics. In this present work, we also assess the semantic capabilities of such models. Based on the results provided and analyzed in Sect. 3, incorporating BK about relation domains and ranges into the loss functions clearly contribute to a better KGEM semantic correctness, as evidenced by the huge increase frequently observed w.r.t. Sem@K. Although it seems intuitive that our proposed approach will result in better predicted semantics, it should not be taken for granted. Hubert *et al.* demonstrate that in the special case of LP on a single target relation (*e.g.*, recommender systems), incorporating schema-based information during training (in their case, during negative sampling) actually decreases the semantic correctness of KGEMs [12]. This result leads us to posit that injecting ontological information during training does not necessarily lead to a better semantic correctness. The improvement might depend on (1) where this information is consumed (*e.g.*, in the loss function, during negative sampling) and (2) what the task at hand is.

It should also be noted that this increase is not homogeneously distributed across relations: relations with a smaller set of semantically valid entities as heads or tails are more challenging w.r.t. Sem@K (see Table 3 and results on B2 and B3 in supplementary materials(see footnote 1)). For such relations, the relative gain from signature-driven loss functions is more acute. In addition, signature-driven loss functions also drive most of the KGEMs towards better MRR and Hits@K values. As shown in Table 2, this is particularly the case when using the $\mathcal{L}^S_{PHL}$ and $\mathcal{L}^S_{BCEL}$. This result is particularly interesting, as it suggests that when semantic information about entities and relations is available, there are benefits in using it, even if the intended goal remains to enhance KGEM performance w.r.t. rank-based metrics only.

In addition, it has been noted that including signature information in loss functions has the highest impact on small relations (B1) (Table 3), both in terms of rank-based metrics (MRR, Hits@10) and semantic correctness (Sem@10). For relations with a larger pool of semantically valid entities, the impact is still positive w.r.t. Sem@10, but sometimes at the expense of a small drop in terms of rank-based metrics. If the latter metrics are the sole optimization objective, it would be reasonable to design an adaptive training strategy in which vanilla and signature-driven loss functions are alternatively used depending on the current relation and the number of semantically valid candidates as head or tail.

6.3 On the Evaluation of KGEM Predictions for LP

It should be noted that Sem@K measures the capability of models to predict semantically correct triples w.r.t. relation signatures and but does not measure falsehood, contrary to MRR or Hits@K. It can be argued that in specific use-cases, such as e-commerce recommender systems, models should predict the expected item (high MRR/Hits@K) but also predict semantically valid items (high Sem@K) for a good user experience. In other applications such as healthcare, aerospace, or security, users expect to notice when models are failing in order to be able to take over. This corresponds to high MRR/Hits@K and low Sem@K so that errors are clearly noticeable.

Our experiments and these use cases illustrate the need of both Hits@K and Sem@K metrics to precisely qualify model performance w.r.t. the needs of the considered application (*e.g.*, e-commerce, healthcare). This opens perspectives on sampling and differently considering different kinds of negative triples to tailor specific aspects of model performance to the requirements of the application. For instance, based on the results of our experiments, we could envision differently considering negatives for an e-commerce application to obtain a high Sem@K while an aerospace application would require the contrary.

7   Conclusion

In this work, we focus on the main loss functions used for link prediction in knowledge graphs. Building on the assumption that negative triples are not all equally good for learning better embeddings, we propose to differentiate them based on their semantic validity w.r.t. the domain and range of relations by including relation signature information into loss functions. A wide range of KGEMs are subsequently trained under both the vanilla and signature-driven loss functions. In our experiments on three public KGs with different characteristics, the proposed signature-driven loss functions lead to promising results: in most cases, they do not only lead to better MRR and Hits@10 values, but also drive KGEMs towards better semantic correctness as measured with Sem@10. This advocates for the further injection of semantic information into loss functions whenever such information is available. In future work, we will study how the proposed loss functions can accommodate other types of ontological constraints as well as literal nodes.

A   Modified Versions of $\mathcal{L}^S_{BCEL}$ and $\mathcal{L}^S_{PLL}$

In Table 4, we report results using KGEMs trained with $\mathcal{L}^{S'}_{BCEL}$ and $\mathcal{L}^{S'}_{PLL}$, where the superscript S' denotes a different way to include semantic information. $\mathcal{L}^{S'}_{PLL}$ uses the modified labeling function as used in $\mathcal{L}^S_{BCEL}$ and defined in Eq. (5). Conversely, $\mathcal{L}^{S'}_{BCEL}$ uses the binary (unmodified) labelling function ℓ but adopts the same procedure as $\mathcal{L}^S_{PLL}$: semantically valid negative triples are considered as positive with probability ϵ %. Hyperparameters for $\mathcal{L}^S_{BCEL}$, $\mathcal{L}^S_{PLL}$, $\mathcal{L}^{S'}_{BCEL}$, and $\mathcal{L}^{S'}_{PLL}$ are reported in https://github.com/nicolas-hbt/semantic-lossfunc. Results achieved with all the aforementioned loss functions are provided in Table 4. It shows that the signature-driven loss functions presented in the paper are the best performing ones.

Table 4. Rank-based and semantic-based results on FB15k187, DBpedia77k, and Yago14k. Bold fonts indicate which model performs best w.r.t. a given metric. Suffixes S and S' indicate whether the model is trained under the best (as presented in the paper) or the worst (as presented here) signature-driven version of the loss function, respectively.

	FB15k187			DBpedia77k			Yago14k		
	MRR	H@10	S@10	MRR	H@10	S@10	MRR	H@10	S@10
ComplEx-S	**.316**	**.476**	.796	**.297**	**.409**	.897	**.923**	**.931**	.667
ComplEx-S'	.227	.384	.777	.252	.350	.918	.907	.930	.603
SimplE-S	**.268**	**.409**	.759	**.230**	**.302**	.850	**.924**	**.927**	.769
SimplE-S'	.169	.288	**.827**	.230	.297	.583	.885	.915	.290
ConvE-S	**.283**	**.476**	**.996**	**.283**	**.405**	.985	**.933**	.940	**.997**
ConvE-S'	.271	.472	.975	.273	.383	.935	.933	**.941**	.894
TuckER-S	**.320**	**.522**	**.996**	**.312**	**.421**	**.969**	**.931**	**.943**	**.929**
TuckER-S'	.316	.517	.983	.311	.412	.912	.918	.938	.867
RGCN-S	**.260**	**.415**	**.860**	**.197**	**.320**	**.957**	**.927**	**.934**	**.828**
RGCN-S'	.243	.391	.780	.146	.246	.862	.912	.922	.385

B   Bucket Analysis

Relations are separated into three non-intersecting buckets : relations that feature narrow (B1), intermediate (B2), and large (B3) sets of semantically valid heads or tails, respectively. Cut-offs are manually defined for placing a given relation in its corresponding bucket. Such buckets are reported in https://github. com/nicolas-hbt/semantic-lossfunc. Results achieved on B1 are reported in the paper, while results for buckets B2 and B3 for DBpedia77k, FB15k187, and Yago14k are reported in https://github.com/nicolas-hbt/semantic-lossfunc.

References

1. Ali, M., et al.: Bringing light into the dark: a large-scale evaluation of knowledge graph embedding models under a unified framework. IEEE Trans. Pattern Anal. Mach. Intell. **44**(12), 8825–8845 (2022). https://doi.org/10.1109/TPAMI. 2021.3124805
2. Balazevic, I., Allen, C., Hospedales, T.M.: TuckER: tensor factorization for knowledge graph completion. In: Proceedings of the 2019 Conference on Empirical Methods in Natural Language Processing and the 9th International Joint Conference on Natural Language Processing, EMNLP-IJCNLP 2019, Hong Kong, China, November 3-7, 2019, pp. 5184–5193. Association for Computational Linguistics (2019). https://doi.org/10.18653/v1/D19-1522
3. Bordes, A., Usunier, N., García-Durán, A., Weston, J., Yakhnenko, O.: Translating embeddings for modeling multi-relational data. In: Conference on Neural Information Processing Systems (NeurIPS), pp. 2787–2795 (2013)

4. Cai, L., Wang, W.Y.: KBGAN: adversarial learning for knowledge graph embeddings. In: Proceedings of the 2018 Conference of the North American Chapter of the Association for Computational Linguistics: Human Language Technologies, NAACL-HLT 2018, New Orleans, Louisiana, USA, June 1-6, 2018, Volume 1 (Long Papers), pp. 1470–1480. Association for Computational Linguistics (2018). https:// doi.org/10.18653/v1/n18-1133

5. Cao, Z., Xu, Q., Yang, Z., Huang, Q.: ER: equivariance regularizer for knowledge graph completion. In: Thirty-Sixth AAAI Conference on Artificial Intelligence, AAAI 2022, Thirty-Fourth Conference on Innovative Applications of Artificial Intelligence, IAAI 2022, The Twelveth Symposium on Educational Advances in Artificial Intelligence, EAAI 2022 Virtual Event, February 22 - March 1, 2022, pp. 5512–5520. AAAI Press (2022)

6. Cui, Z., Kapanipathi, P., Talamadupula, K., Gao, T., Ji, Q.: Type-augmented relation prediction in knowledge graphs. In: Thirty-Fifth AAAI Conference on Artificial Intelligence, AAAI 2021, Thirty-Third Conference on Innovative Applications of Artificial Intelligence, IAAI 2021, The Eleventh Symposium on Educational Advances in Artificial Intelligence, EAAI 2021, Virtual Event, February 2-9, 2021, pp. 7151–7159. AAAI Press (2021)

7. d'Amato, C., Quatraro, N.F., Fanizzi, N.: Injecting background knowledge into embedding models for predictive tasks on knowledge graphs. In: Verborgh, R., et al. (eds.) The Semantic Web. Lecture Notes in Computer Science(), vol. 12731, pp. 441–457. Springer, Cham (2021). https://doi.org/10.1007/978-3-030-77385-4_26

8. Dettmers, T., Minervini, P., Stenetorp, P., Riedel, S.: Convolutional 2D knowledge graph embeddings. In: Proceedings of the Thirty-Second AAAI Conference on Artificial Intelligence, (AAAI-18), the 30th innovative Applications of Artificial Intelligence (IAAI-18), and the 8th AAAI Symposium on Educational Advances in Artificial Intelligence (EAAI-18), New Orleans, Louisiana, USA, February 2-7, 2018, pp. 1811–1818. AAAI Press (2018)

9. Ding, B., Wang, Q., Wang, B., Guo, L.: Improving knowledge graph embedding using simple constraints. In: Proceedings of the 56th Annual Meeting of the Association for Computational Linguistics, ACL 2018, Melbourne, Australia, July 15-20, 2018, Volume 1: Long Papers, pp. 110–121. Association for Computational Linguistics (2018). https://doi.org/10.18653/v1/P18-1011

10. Guo, S., Wang, Q., Wang, B., Wang, L., Guo, L.: Semantically smooth knowledge graph embedding. In: Proceedings of the 53rd Annual Meeting of the Association for Computational Linguistics and the 7th International Joint Conference on Natural Language Processing of the Asian Federation of Natural Language Processing, ACL 2015, July 26-31, 2015, Beijing, China, Volume 1: Long Papers, pp. 84–94. The Association for Computer Linguistics (2015). https://doi.org/10.3115/ v1/p15-1009

11. Hubert, N., Monnin, P., Brun, A., Monticolo, D.: Knowledge graph embeddings for link prediction: beware of semantics! In: Proceedings of the Workshop on Deep Learning for Knowledge Graphs (DL4KG 2022) Co-Located with the 21th International Semantic Web Conference (ISWC 2022). Virtual Conference, online (2022)

12. Hubert, N., Monnin, P., Brun, A., Monticolo, D.: New strategies for learning knowledge graph embeddings: the recommendation case. In: Corcho, O., Hollink, L., Kutz, O., Troquard, N., Ekaputra, F.J. (eds.) Knowledge Engineering and Knowledge Management. Lecture Notes in Computer Science(), vol. 13514, pp. 66–80. Springer, Cham (2022). https://doi.org/10.1007/978-3-031-17105-5_5

13. Hubert, N., Monnin, P., Brun, A., Monticolo, D.: Sem@k: is my knowledge graph embedding model semantic-aware? (2023)
14. Jain, N., Tran, T., Gad-Elrab, M.H., Stepanova, D.: Improving knowledge graph embeddings with ontological reasoning. In: Hotho, A., et al. (eds.) The Semantic Web - ISWC 2021. Lecture Notes in Computer Science(), vol. 12922, pp. 410–426. Springer, Cham (2021). https://doi.org/10.1007/978-3-030-88361-4_24
15. Ji, S., Pan, S., Cambria, E., Marttinen, P., Yu, P.S.: A survey on knowledge graphs: representation, acquisition, and applications. IEEE Trans. Neural Networks Learn. Syst. **33**(2), 494–514 (2022). https://doi.org/10.1109/TNNLS.2021.3070843
16. Kazemi, S.M., Poole, D.: Simple embedding for link prediction in knowledge graphs. In: Advances in Neural Information Processing Systems 31: Annual Conference on Neural Information Processing Systems 2018, NeurIPS 2018, December 3-8, 2018, Montréal, Canada, pp. 4289–4300 (2018)
17. Kotnis, B., Nastase, V.: Analysis of the impact of negative sampling on link prediction in knowledge graphs. arXiv preprint: arXiv:1708.06816 (2017)
18. Krompaß, D., Baier, S., Tresp, V.: Type-constrained representation learning in knowledge graphs. In: Arenas, M., et al. (eds.) The Semantic Web - ISWC 2015. Lecture Notes in Computer Science(), vol. 9366, pp. 640–655. Springer, Cham (2015). https://doi.org/10.1007/978-3-319-25007-6_37
19. Lv, X., Hou, L., Li, J., Liu, Z.: Differentiating concepts and instances for knowledge graph embedding. In: Proceedings of the 2018 Conference on Empirical Methods in Natural Language Processing, Brussels, Belgium, October 31 - November 4, 2018, pp. 1971–1979. Association for Computational Linguistics (2018). https://doi.org/10.18653/v1/d18-1222
20. Minervini, P., Costabello, L., Muñoz, E., Novácek, V., Vandenbussche, P.: Regularizing knowledge graph embeddings via equivalence and inversion axioms. In: Ceci, M., Hollmen, J., Todorovski, L., Vens, C., Dzeroski, S. (eds.) Machine Learning and Knowledge Discovery in Databases. Lecture Notes in Computer Science(), vol. 10534, pp. 668–683. Springer, Cham (2017). https://doi.org/10.1007/978-3-319-71249-9_40
21. Mohamed, S.K., Muñoz, E., Novacek, V.: On training knowledge graph embedding models. Information **12**(4) (2021). https://doi.org/10.3390/info12040147
22. Mohamed, S.K., Novácek, V., Vandenbussche, P., Muñoz, E.: Loss functions in knowledge graph embedding models. In: Proceedings of the Workshop on Deep Learning for Knowledge Graphs (DL4KG2019) Co-located with the 16th Extended Semantic Web Conference 2019 (ESWC 2019), Portoroz, Slovenia, June 2, 2019. CEUR Workshop Proceedings, vol. 2377, pp. 1–10. CEUR-WS.org (2019)
23. Niu, G., Li, B., Zhang, Y., Pu, S., Li, J.: AutoETER: automated entity type representation with relation-aware attention for knowledge graph embedding. In: Findings of the Association for Computational Linguistics: EMNLP 2020, Online Event, 16-20 November 2020. Findings of ACL, vol. EMNLP 2020, pp. 1172–1181. Association for Computational Linguistics (2020). https://doi.org/10.18653/v1/2020.findings-emnlp.105
24. Rossi, A., Barbosa, D., Firmani, D., Matinata, A., Merialdo, P.: Knowledge graph embedding for link prediction: a comparative analysis. ACM Trans. Knowl. Discovery Data **15**(2), 1–49 (2021)
25. Schlichtkrull, M.S., Kipf, T.N., Bloem, P., van den Berg, R., Titov, I., Welling, M.: Modeling relational data with graph convolutional networks. In: Gangemi, A., et al. (eds.) The Semantic Web. Lecture Notes in Computer Science(), vol. 10843, pp. 593–607. Springer, Cham (2018). https://doi.org/10.1007/978-3-319-93417-4_38

26. Sun, Z., Deng, Z., Nie, J., Tang, J.: Rotate: knowledge graph embedding by relational rotation in complex space. In: 7th International Conference on Learning Representations, ICLR (2019)
27. Trouillon, T., Welbl, J., Riedel, S., Gaussier, É., Bouchard, G.: Complex embeddings for simple link prediction. In: Proceedings of the 33rd International Conference on Machine Learning, ICML, vol. 48, pp. 2071–2080 (2016)
28. Wang, P., Zhou, J., Liu, Y., Zhou, X.: TransET: knowledge graph embedding with entity types. Electronics 10(12), 1407 (2021)
29. Wang, Q., Mao, Z., Wang, B., Guo, L.: Knowledge graph embedding: a survey of approaches and applications. IEEE Trans. Knowl. Data Eng. 29(12), 2724–2743 (2017)
30. Wang, Z., Zhang, J., Feng, J., Chen, Z.: Knowledge graph embedding by translating on hyperplanes. In: Proceedings of the Twenty-Eighth AAAI Conference on Artificial Intelligence, pp. 1112–1119 (2014)
31. Weyns, M., Bonte, P., Steenwinckel, B., Turck, F.D., Ongenae, F.: Conditional constraints for knowledge graph embeddings. In: Proceedings of the Workshop on Deep Learning for Knowledge Graphs (DL4KG@ISWC), vol. 2635 (2020)
32. Xie, R., Liu, Z., Sun, M.: Representation learning of knowledge graphs with hierarchical types. In: Proceedings of the Twenty-Fifth International Joint Conference on Artificial Intelligence, IJCAI 2016, New York, NY, USA, 9-15 July 2016, pp. 2965–2971. IJCAI/AAAI Press (2016)
33. Yang, B., Yih, W., He, X., Gao, J., Deng, L.: Embedding entities and relations for learning and inference in knowledge bases. In: 3rd International Conference on Learning Representations, ICLR (2015)
34. Zhang, Y., Yao, Q., Shao, Y., Chen, L.: NSCaching: simple and efficient negative sampling for knowledge graph embedding. In: 35th IEEE International Conference on Data Engineering, ICDE 2019, Macao, China, April 8-11, 2019, pp. 614–625. IEEE (2019). https://doi.org/10.1109/ICDE.2019.00061

QAGCN: Answering Multi-relation Questions via Single-Step Implicit Reasoning over Knowledge Graphs

Ruijie Wang[1,2]($\boxtimes$) [iD], Luca Rossetto[1] [iD], Michael Cochez[3,4] [iD], and Abraham Bernstein[1] [iD]

[1] Department of Informatics, University of Zurich, Zurich, Switzerland
{ruijie,rossetto,bernstein}@ifi.uzh.ch
[2] University Research Priority Program "Dynamics of Healthy Aging", University of Zurich, Zurich, Switzerland
[3] Vrije Universiteit Amsterdam, Amsterdam, The Netherlands
m.cochez@vu.nl
[4] Elsevier Discovery Lab, Amsterdam, The Netherlands

Abstract. Multi-relation question answering (QA) is a challenging task, where given questions usually require long reasoning chains in KGs that consist of multiple relations. Recently, methods with explicit multi-step reasoning over KGs have been prominently used in this task and have demonstrated promising performance. Examples include methods that perform stepwise label propagation through KG triples and methods that navigate over KG triples based on reinforcement learning. A main weakness of these methods is that their reasoning mechanisms are usually complex and difficult to implement or train. In this paper, we argue that multi-relation QA can be achieved via end-to-end single-step implicit reasoning, which is simpler, more efficient, and easier to adopt. We propose QAGCN — a Question-Aware Graph Convolutional Network (GCN)-based method that includes a novel GCN architecture with controlled question-dependent message propagation for the implicit reasoning. Extensive experiments have been conducted, where QAGCN achieved competitive and even superior performance compared to state-of-the-art explicit-reasoning methods. Our code and pre-trained models are available in the repository: https://github.com/ruijie-wang-uzh/QAGCN.

Keywords: Multi-relation Question Answering · Knowledge Graph · Graph Convolutional Network

1 Introduction

Question answering (QA) over knowledge graphs (KGs)—a task that has been pursued almost since the inception of the Semantic Web (cf. [13])—aims to automatically retrieve answers from KGs for given natural language questions.

© The Author(s), under exclusive license to Springer Nature Switzerland AG 2024
A. Meroño Peñuela et al. (Eds.): ESWC 2024, LNCS 14664, pp. 41–58, 2024.
https://doi.org/10.1007/978-3-031-60626-7_3

In recent years, multi-relation questions that require reasoning chains over multiple KG triples have been a focus of this task. An example is *"who is the mayor of the city where the director of Sleepy Hollow was born,"* which mentions one topic entity (*"Sleepy Hollow"*) and three relations (*"mayor of"*, *"director of"*, and *"was born"*). Correspondingly, starting from the entity Sleepy Hollow, the expected answer Jess Talamantes can be inferred from a reasoning chain that consists of three triples: [(Sleepy Hollow, director, Tim Burton), (Tim Burton, birthplace, Burbank), (Burbank, mayor, Jess Talamantes)]. To answer this kind of question, methods with reasoning mechanisms [5,21,24,31,32] have been proposed to infer the reasoning chains over KGs step by step. They typically commence with topic entities in given questions as anchors and try to infer reasoning chains by extending triples according to the semantics of the questions. This extension is usually performed in two ways: label propagation from entities in the reasoning chain of the current step to other entities that can be added in the next step, and reinforcement learning-based decision-making on choosing entities to add. Given that the reasoning involves multiple steps and produces explicit states of reasoning chains at each step, we refer to these methods as explicit multi-step reasoning-based. Recently, they have achieved state-of-the-art (SOTA) performance in the task. However, their reasoning mechanisms are often complex and difficult to implement or train. Furthermore, for better supervision during training, some of these methods [32] require annotations of reasoning paths, which are usually unavailable in real-world scenarios.

Inspired by the promising performance that graph convolutional networks (GCNs) [22] achieved in learning semantic representations of KG entities, we propose to adapt GCNs for answering multi-relation questions with single-step implicit reasoning that is simpler, more efficient, and easier to adopt than existing reasoning mechanisms. Specifically, in this paper, we propose a novel Question-Aware GCN-based QA method, called QAGCN, which encodes questions and KG entities in a joint embedding space where questions are close to correct answers (entities). The intuition of our method is as follows: Given a question, if an entity is the correct answer, the reasoning chain of this question would be part of the KG context (i.e., the neighboring triples) of that entity. For example, considering the above question, the reasoning chain is part of the context of Jess Talamantes within three hops in the KG. During encoding, we expect the GCN to focus only on messages that pass through the reasoning path and accumulate the messages in the representation of this entity. We refer to this process as question-aware message propagation. In this case, the representation of the entity would contain information that is semantically consistent with the given question and can be aligned with the question in the embedding space. The GCN directly generates semantic representations of the question and KG entities in an end-to-end fashion. Therefore, we classify it as single-step reasoning. Also, the reasoning is implicit, given that it is hidden in the message propagation of the GCN. It is very challenging to achieve this GCN-based reasoning, as the potential KG context of entities could be very large, especially when several hops need to be considered for multi-relation questions.

In summary, we make the following contributions in this paper:

- We propose a novel QA method called QAGCN that can answer multi-relation questions via single-step implicit reasoning. The method is simpler, more efficient, and easier to adopt than existing reasoning-based methods.
- We propose a novel question-aware GCN architecture with contextually controlled message propagation for implicit reasoning.
- We conducted extensive experiments to evaluate the effectiveness and efficiency of QAGCN and demonstrate its competitive and even superior performance compared with recent SOTA reasoning-based methods.

2   Related Work

In this section, we first give a detailed introduction to existing reasoning-based methods. Then, we provide a general overview of other QA methods.

As aforementioned, there are primarily two types of reasoning-based methods: label propagation-based and reinforcement learning-based. Examples of label propagation-based methods include NSM [9], SR+NSM [30], and TransferNet [24]. NSM and SR+NSM label entities in the underlying KG with their probabilities of being part of the reasoning chain of a given question. In the first step, only topic entities are labeled with positive probabilities. During reasoning, the probabilities are expected to be passed through the reasoning chain to final answers. TransferNet also starts with labeling topic entities. It leverages the semantics of relations in the KG to perform the transfer of labels between entities. Examples of reinforcement learning-based methods include MINERVA [5], IRN [32], and SRN [21]. MINERVA trains a reinforcement learning agent that walks on KGs from entity to entity through the edges linking them. Starting from topic entities, the agent is trained to arrive at correct answers after several walking steps. IRN tries to infer chains of relations that could link topic entities to correct answers based on jointly learned KG embeddings and question encodings. SRN employs attentional question encoding and potential-based reward shaping in the training of the reinforcement learning agent. These methods are the most relevant to our work and will be used as baselines in experiments.

Other important types of QA methods include parsing-based [1,11,13,27,29] and embedding-based [3,4,12,17,23,26]. Parsing-based methods construct logical queries, such as SPARQL [8] and λ-DCS [16], that represent the semantics of given questions and can be evaluated to retrieve answers from KGs. They can provide parsing results of questions (e.g., dependency tree and intermediate query structures) that can be used to comprehend and assess the QA process of the model. However, their performance is usually limited by available resources for question parsing, such as dictionaries, corpora of queries, templates, and heuristics. Embedding-based methods learn numerical representations of questions and KGs that can be used for direct answer retrieval without logical query construction. They can learn the knowledge that parsing-based methods require people to manually define. However, most existing methods are limited to answering single-hop questions due to the lack of multi-step reasoning ability.

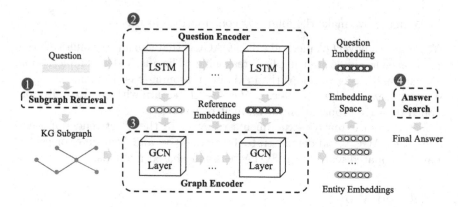

Fig. 1. An overview of the QAGCN model.

3   QAGCN – A Question-Aware GCN-Based Question Answering Model

In this section, we define the multi-relation QA task and elaborate on the proposed QAGCN model.

3.1   Task Definition

A **knowledge graph (KG)** is denoted as $\mathcal{G} = (E, R, T)$, where E and R denote the entity set and relation set of $\mathcal{G}$, and $T \subseteq E \times R \times E$ denotes the triple set of $\mathcal{G}$. A triple $(e_h, r, e_t) \in T$ denotes that there is a relation $r \in R$ linking the head entity $e_h \in E$ to the tail entity $e_t \in E$. Entities and relations are labeled by their names in natural language. We denote the set of all labels as $\mathcal{L}$ and define a function $f_l : E \cup R \rightarrow \mathcal{L}$ that returns the label of a given entity or relation.

The multi-relation QA task is to answer unseen questions from a test set Q over a given KG $\mathcal{G} = (E, R, T)$, with a disjoint training set $\hat{Q}$ provided. Each question $q \in Q \cup \hat{Q}$ is expressed in natural language and can be answered by entities in E. Also, in line with existing work [9,21,25,32], a topic entity is assumed to be annotated.[1] Finally, to simplify the following discussion, we define the reasoning chain of q as a path $P \subseteq T$ that links the topic entity to the final answers using relations specified in q. Note that our method does not require annotations of reasoning chains.

3.2   The Question-Aware GCN

We present an overview of the proposed question-aware GCN in Fig. 1. Given a multi-relation question, we feed it together with a subgraph extracted for the question from the underlying KG as input. The model consists of a question

---

[1] If topic entities are not annotated, they can still be easily obtained via named entity recognition, which has been widely studied for decades.

encoder and a graph encoder that compute semantic representations (embeddings) of the question and subgraph entities, respectively, through several layers of encoding. We select answers from subgraph entities according to their distances to the question in the output embedding space and further improve the precision of final answers based on relation labels. The entire process can be segmented into four components: subgraph extraction, question encoding, subgraph encoding, and answer search. We present detailed elaboration on each of these components in the following.

Subgraph Extraction. In this component, we aim to extract a more tractable subgraph, which encompasses the reasoning chain of the given question, from the potentially very large underlying KG. Specifically, given the question $q \in Q \cup \hat{Q}$ posed over the KG $\mathcal{G} = (E, R, T)$, we aim to extract a subgraph $\mathcal{G}_q = (E_q, R_q, T_q)$ that covers the reasoning chain P, i.e., $P \subseteq T_q \subseteq T$. If q is a x-relation question (i.e., x denotes the number of hops of the question), P would consist of x triples linking the topic entity e_q to the answers. Therefore, we first train a question classifier to predict x for q. The classifier consists of three linear layers with two rectified linear unit (ReLU) layers in between and a softmax output layer. For the input of the classifier, we filter out stop words and annotated topic entities from questions. Then, the rest words are represented as bag-of-words vectors with two additional entries denoting the number of words and the number of out-of-vocabulary (OOV) words. Questions in the training set $\hat{Q}$ are used for vocabulary construction and model training. In addition, we remove 15 least used words from the vocabulary to provide training questions containing OOV words. Based on the estimated x for q, in this step, we add all paths of length x or shorter in $\mathcal{G}$ that start from the topic entity e_q to $\mathcal{G}_q$. Please note that we ignore triple directions to ensure that our method is applicable to different KG schemas. For example, the information that Tim Burton directed Sleepy Hollow can be represented as (`Sleepy Hollow, director, Tim Burton`) or (`Tim Burton, directed, Sleepy Hollow`), depending on the schema.

Question Encoding. In this component, we filter out stop words from the question q and segment q into a sequence of n chunks $[c_1, c_2, ..., c_n]$. The segmentation is performed by splitting on whitespace, with the exception that the mention of the topic entity e_q remains complete in one chunk. Then, a pre-trained BERT model [6] (bert-base-uncased[2]) is used to encode each chunk into an embedding vector. As suggested by bert-as-service[3], the average pooling of hidden embeddings of the second-to-last BERT layer is computed as the embedding of each chunk. We denote the embeddings of $[c_1, c_2, ..., c_n]$ as $[\mathbf{c_1}, \mathbf{c_2}, ..., \mathbf{c_n}]$, where $\mathbf{c_i} \in \mathbb{R}^{d_0}, i = 1, ..., n$, and d_0 is the initial embedding size. The BERT model is only used for the initial encoding of texts. It is not fine-tuned in training and could be replaced by other methods that provide contextual text embeddings.

---

[2] https://huggingface.co/bert-base-uncased.
[3] https://bert-as-service.readthedocs.io/.

These embeddings are subsequently fed to L stacked LSTM [10] layers, as depicted in Fig. 1. In each layer, there are n LSTM units that sequentially update a cell state passing through them given a sequence of embeddings from the previous layer (or the initial embeddings $[\mathbf{c_1}, \mathbf{c_2}, ..., \mathbf{c_n}]$ in the first layer). The computed cell state of the last LSTM layer (denoted as $\mathbf{q_L} \in \mathbb{R}^{d_L}$) is used as the final embedding of q, while the cell state of the l-th layer, i.e., $\mathbf{q_l} \in \mathbb{R}^{d_l}$, $l = 1, ..., L$, is used as a reference embedding for the computation of attention weights in the l-th GCN layer of the graph encoder, which we introduce below.

Subgraph Encoding. We analogously use a pre-trained BERT model to initialize embeddings of entities and relations in $\mathcal{G}_q$ based on their labels. Specifically, each entity $e \in E_q$ and each relation $r \in R_q$ are encoded into $\mathbf{e_0}, \mathbf{r_0} \in \mathbb{R}^{d_o}$ based on their labels $f_l(e)$ and $f_l(r)$, respectively. Following other GCN work [15,28], we add inverse edges and self-loops to $\mathcal{G}_q$, i.e., $T_q \leftarrow T_q \cup \{(e_t, r^{-1}, e_h) | (e_h, r, e_t) \in T_q\} \cup \{(e, r^s, e) | e \in E_q\}$ and $R_q \leftarrow R_q \cup \{r^{-1} | r \in R_q\} \cup \{r^s\}$, where r^{-1} denotes the inverse of r, and r^s denotes the self-loop relation. The inverse relation r^{-1} allows information to be passed from the tail entity e_t to the head entity e_h during graph encoding, and its initial embedding $\mathbf{r_0^{-1}} = -1 \cdot \mathbf{r_0}$. The self-loop relation r^s allows entities to receive information from themselves. Also, its initial embedding is set to a zero vector, i.e., $\mathbf{r_0^s} = \mathbf{0}$.

The subgraph $\mathcal{G}_q$ with initialized entity and relation embeddings are fed to L stacked GCN layers. And the graph encoding in each layer includes two steps: *message passing* and *message aggregation*. In the *message passing step*, we employ a linear transformation to compute messages that each entity receives from its context. Specifically, for an entity e in the subgraph $\mathcal{G}_q$, its context is defined as a set of all incoming triples $C(e) = \{(\hat{e}, r, e) | (\hat{e}, r, e) \in T_q\}$. In the l-th GCN layer, the message that e receives from an incoming triple $(\hat{e}, r, e) \in C(e)$ is computed as follows:

$$m_l(\hat{e}, r, e) = \mathbf{W_l} \cdot [\hat{\mathbf{e}}_{l-1} \| \mathbf{r}_{l-1}] + \mathbf{b_l}, \tag{1}$$

where $\mathbf{W_l} \in \mathbb{R}^{d_l \times 2d_{l-1}}$ and $\mathbf{b_l} \in \mathbb{R}^{d_l}$ denote the weight and bias for the message computation, $\hat{\mathbf{e}}_{l-1}, \mathbf{r}_{l-1} \in \mathbb{R}^{d_{l-1}}$ are embeddings of $\hat{e}$ and r computed by the previous GCN layer (or initial embeddings if $l = 1$), and $[\hat{\mathbf{e}}_{l-1} \| \mathbf{r}_{l-1}]$ denotes the concatenation of $\hat{\mathbf{e}}_{l-1}$ and $\mathbf{r}_{l-1}$.

Depending on the size of $C(e)$, the entity e may receive a large number of messages in each GCN layer. However, we expect the model to only consider messages related to the given question, i.e., question-aware message propagation. For example, given "*who is the director of Sleepy Hollow,*" only the message from Sleepy Hollow should be considered when encoding Tim Burton. To this end, in the *message aggregation step*, we propose an attention mechanism to compute weights for passed messages with reference to the given question. Specifically, in the l-th GCN layer, the weight of the message that e receives from $(\hat{e}, r, e) \in C(e)$ is computed as follows:

$$w_l(\hat{e}, r, e) = \frac{\exp\big(\tanh(\mathbf{A_l} \cdot [m_l(\hat{e}, r, e) \| \mathbf{q_l}])\big)}{\sum_{(\hat{e}', r', e) \in C(e)} \exp\big(\tanh(\mathbf{A_l} \cdot [m_l(\hat{e}', r', e) \| \mathbf{q_l}])\big)}, \tag{2}$$

where $\mathbf{q_l} \in \mathbb{R}^{d_l}$ is the reference embedding computed by the l-th LSTM layer of the question encoder, $m_l(\cdot)$ computes the message passed to e from an incoming triple according to Eq. (1), and $\mathbf{A_l} \in \mathbb{R}^{1 \times 2d_l}$ is the weight to learn for the attention mechanism in the l-th GCN layer. The goal of Eq. (2) is to compute higher weights for messages that are related to the given question while lower weights for irrelevant messages.

Based on computed weights, the embedding of e in the l-th GCN layer is updated as follows:

$$\mathbf{e_l} = \sum_{(\hat{e},r,e) \in C(e)} w_l(\hat{e}, r, e) \cdot m_l(\hat{e}, r, e), \tag{3}$$

where $w_l(\hat{e}, r, e) \in \mathbb{R}$ and $m_l(\hat{e}, r, e) \in \mathbb{R}^{d_l}$.

Answer Search. After L GCN layers, each entity $e \in E_q$ is encoded as an embedding $\mathbf{e_L} \in \mathbb{R}^{d_L}$. The likelihood of e being the answer of the question q is measured by the Euclidean distance between $\mathbf{e_L}$ and the question embedding $\mathbf{q_L}$. We rank all entities in the subgraph $\mathcal{G}_q$ according to computed Euclidean distances to retrieve top-ranked candidate answers, which are denoted as a set E_c. Then, we leverage label information of relations in the KG to filter out outliers in E_c and further improve the precision of our results via the reranking of top-ranked candidate answers. Specifically, for each entity $e_c \in E_c$, we employ NetworkX and Graph Tool[4] to extract relation paths of the predicted length x that link e_c to the topic entity e_q in the KG subgraph. The semantics of the relation paths can be assessed based on the labels of constituent relations. For example, regarding a question estimated to be 2-hop *"where was the director of Sleepy Hollow born,"* we can extract a 2-hop relation path [birthplace, director] for the candidate answer Burbank given triples {(Burbank, birthplace, Tim Burton), (Tim Burton, director, Sleepy Hollow)}[5] in the subgraph. Then, the semantics of this path can be examined based on *"birthplace"* and *"director."*

Specifically, we denote the set of extracted relation paths for each candidate answer $e_c \in E_c$ as $\mathcal{P}_r(e_c)$. The plausibility of e_c is evaluated by measuring the semantic similarities between its relation paths and the given question q. We denote a relation path $P_r \in \mathcal{P}_r(e_c)$ as a sequence of relations, i.e., $P_r = [r_1, ..., r_x]$, where $r_i \in R_q, i = 1, ..., x$. Analogous to the initial encoding of question chunks, relations in P_r are also encoded by the pre-trained BERT model, i.e., P_r is encoded as $[\mathbf{r_1}, ..., \mathbf{r_x}]$, where $\mathbf{r_i} \in \mathbb{R}^{d_o}, i = 1, ..., x$. Then, $[\mathbf{r_1}, ..., \mathbf{r_x}]$ is fed to a single-layer LSTM, and the cell state of the LSTM is used as the embedding of P_r, denoted as $\mathbf{P_r} \in \mathbb{R}^{d_p}$, where d_p is the embedding size. For the given question q, we reuse the segmented chunks $[c_1, ..., c_n]$ and their initial embeddings computed in the question encoder, i.e., $[\mathbf{c_1}, ..., \mathbf{c_n}]$, where $\mathbf{c_i} \in \mathbb{R}^{d_o}, i = 1, ..., n$. The chunk embeddings are fed to another single-layer LSTM, and its cell state is used as the embedding of q, i.e., $\mathbf{q_p} \in \mathbb{R}^{d_p}$. The Euclidean distance between $\mathbf{P_r}$ and

---

[4] https://networkx.org/ and https://graph-tool.skewed.de.
[5] We ignore relation directions in this process.

Table 1. Statistics of adopted benchmark datasets.

Datasets	Underlying KG			Question Sets		
	#Entity	#Relation	#Triple	#Train	#Valid	#Test
PQ-2hop	1,056	13	1,211	1,526	190	192
PQ-3hop	1,836	13	2,839	4,158	519	521
PQL-2hop	5,034	363	4,247	1,276	159	159
PQL-3hop	6,505	411	5,597	825	103	103
MetaQA 1-hop	43,234	9	134,741	96,106	9,992	9,947
MetaQA 2-hop	43,234	9	134,741	118,980	14,872	14,872
MetaQA 3-hop	43,234	9	134,741	114,196	14,274	14,274

$\mathbf{q_p}$ is computed to measure the semantic similarity between P_r and q. We train the above two LSTMs based on training questions in $\hat{Q}$. Specifically, for each question $\hat{q} \in \hat{Q}$, relation paths between its final answer and topic entity are used as positive samples. Using other existing relations in the KG, we randomly generate paths different from the positive ones as negative samples. The distances between positive samples and $\hat{q}$ are trained to be close to zero, while those for negative samples are trained to be close to one. For each candidate answer e_c, we take the minimum computed distance of all its relation paths in $P_r(e_c)$ as the final distance of e_c. Finally, the candidate answer in E_c with the minimum final distance is selected as the final answer.

4   Experiments

In this section, we evaluate the effectiveness and efficiency of our model on widely used benchmark datasets, scrutinize the contribution of each component of the model in an ablation study, and present a case study for a better understanding of our model.

4.1   Effectiveness Evaluation

Baselines. We first evaluate the overall effectiveness of QAGCN in the multi-relation QA task. Given that the main goal of this paper is to propose a simple method that is competitive with existing reasoning-based methods that rely on complex reasoning mechanism, we mainly choose reasoning-based QA methods as baselines: MINERVA [5], IRN [32], SRN [21], TransferNet [24], and NSM [9]. Among these, MINERVA, IRN, and SRN are reinforcement learning-based. TransferNet and NSM are label propagation-based. Also, TransferNet and NSM are state-of-the-art (SOTA) reasoning-based methods in multi-relation QA. Furthermore, considering that our answer search in the learned embedding space is similar to embedding-based methods, we also select a prominently adopted embedding-based method: EmbedKGQA [23].

Datasets. In line with the baselines, we use three collections of datasets that are particularly constructed for the multi-relation QA task: PathQuestion (PQ), PathQuestion-Large (PQL) [32] and MetaQA [31]. They consist of QA sets that are named according to the complexity (number of required reasoning hops) of respectively included questions: PQ-2hop, PQ-3hop, PQL-2hop, PQL-3hop, MetaQA 1-hop, MetaQA 2-hop, and MetaQA 3-hop. PQ and PQL are both open-domain QA datasets based on Freebase [2], while PQL is a more challenging version of PQ with less training data and larger KGs. One example question in PQ is: *what is the place of birth of Marguerite Louise Dorleans's other half's kid?*[6] MetaQA is constructed based on WikiMovies [19], mainly including questions in the movie domain. One example is: *what genres do the films that share directors with Scarlet Street fall under?* Table 1 reports statistics of the QA sets, including the number of included training/validation/test questions and the number of entities/relations/triples in the underlying KG for each question set.

Experimental Details. The experiments on MetaQA and PQ were conducted on a Linux server with two Intel Xeon Gold 6230 CPUs and one NVIDIA GeForce RTX 2080 Ti GPU being used. The experiments on PQL were conducted on another Linux server with two AMD EPYC 9124 16-Core CPUs and one NVIDIA GeForce RTX 4090 being used. We set hyper-parameters based on grid-search: Learning rates are set to 0.0002, 0.0005, 0.001, 0.0005, 0.0005, 0.0005, and 0.001 for MetaQA 1,2,3-hop, PQ 2,3-hop, and PQL 2,3-hop, respectively. Dropouts of 0.1 are applied in each of the GCN layers, excluding the output layer.

Question Classification. The classifier for predicting the complexity of a given question (cf. Subgraph Extraction in Sect. 3.2) was trained and evaluated on MetaQA, PQ, and PQL respectively using the mix of their QA sets (e.g., the mix of 2-hop and 3-hop sets of PQ). The ground-truth complexity of a question (i.e., 1, 2, or 3 hops) is determined by the QA set from which the question originates. We used the negative log-likelihood loss and the Adam optimizer [14] to train the classifier with a learning rate of 5×10^{-4} on each dataset. This classifier achieved 100% accuracy on the test sets of all three datasets.

Overall Results. Table 2 reports the overall QA performance of our model and other baselines on PQ and PQL, while Table 3 reports the results on MetaQA. We follow the baselines to measure the performance by Hits@1, which is the percentage of test questions that have been correctly answered by the returned *top-1* answers. In the above tables, the results of all baselines on MetaQA and the results of MINERVA, IRN, and SRN on PQ and PQL are from existing publications [9,21]. The results of EmbedKGQA, TransferNet, and NSM on PQ and PQL are computed in our experiments with the source code and training configurations released by the original authors. The following can be observed:

---

[6] In the original form of this question, all letters are lowercase, the entity phrase is connected by underlines (e.g., marguerite_louise_dorleans). We slightly change the format for better readability.

Table 2. Effectiveness performance on PathQuestion (PQ) and PathQuestion-Large (PQL). (% Hits@1, best performance in **bold**, second best underlined)

	PQ-2hop	PQ-3hop	PQL-2hop	PQL-3hop
MINERVA [5]	75.9	71.2	71.8	65.7
IRN [32]	91.9	83.3	63.0	61.8
EmbedKGQA [23]	90.1	86.2	79.2	61.2
SRN [21]	96.3	89.2	78.6	**77.5**
TransferNet [24]	91.1	96.5	54.7	62.1
NSM [9]	94.2	**97.1**	74.2	67.0
QAGCN	**98.5***	90.6*	**87.5**	70.9

Results marked with * are the average results of five runs using different random training/validation/test splits.

Table 3. Effectiveness performance on MetaQA. (% Hits@1)

	MetaQA 1-hop	MetaQA 2-hop	MetaQA 3-hop
MINERVA [5]	96.3	92.9	55.2
IRN [32]	85.9	71.3	35.6
EmbedKGQA [23]	**97.5**	98.8	94.8
SRN [21]	97.0	95.1	75.2
TransferNet [24]	**97.5**	**100**	**100**
NSM [9]	97.3	99.9	98.9
QAGCN	97.3	99.9	97.6

- QAGCN achieved superior performance than existing SOTA methods on PQ-2hop and PQL-2hop with relative improvements of 2.3% and 10.5%, respectively. Also, on MetaQA 1-hop and MetaQA 2-hop, QAGCN achieved the second-best performance, which is very close to the SOTA methods with only relative drops of −0.2% and −0.1%. This demonstrates the competitiveness of QAGCN, considering its simplicity in comparison with the SOTA methods.
- On PQL-3hop, QAGCN is ranked second with a drop of −8.5% in comparison with SRN. This demonstrates the strength of SRN's reinforcement learning(RL)-based policy when answering complex questions. However, please note that QAGCN also has a large margin of 5.8% in comparison with the third-best method NSM. This demonstrates that, on complex questions, the simple single-step reasoning of QAGCN could perform better than the SOTA methods with complex multi-step label propagation.
- On PQ-3hop and MetaQA-3hop, QAGCN are both ranked third. This reflects that it is indeed challenging for the simple reasoning of QAGCN to answer complex 3-hop questions. However, the performance of QAGCN is better than most reasoning-based methods, e.g., 29.8% and 1.6% higher than the best-performing RL-based method SRN on MetaQA 3-hop and PQ-3hop, respectively.

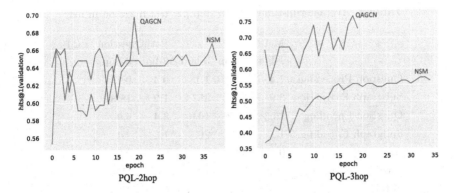

Fig. 2. Training curves of QAGCN and NSM on PQL-2hop and PQL-3hop.

- In general, QAGCN exhibits less promising performance on 3-hop questions compared to 2-hop questions. We could see three possible explanations: First, for 3-hop questions, one more layer of GCN/LSTM would be needed in the graph/question encoders. Hence, it is more challenging to train the model with limited data. Second, the message passing in the GCNs and cell state passing in the LSTMs become longer for 3-hop questions, which are more prone to errors. Third, the search space substantially grows with 3-hop questions. As shown in Table 1, the KGs of 3-hop PQ/PQL questions are substantially larger than those of 2-hop PQ/PQL questions.

4.2   Comparison with SOTA Reasoning-Based Method — NSM

In this section, we compare with the SOTA reasoning-based method NSM to demonstrate that QAGCN is simpler and easier to train.

The training curves of QAGCN and NSM on PQL-2hop and PQL-3hop are shown in Fig. 2. For clarity, we truncate the curves from the second epoch after the optimal epoch of each training. It can be observed that QAGCN only requires about half of the epochs needed by NSM, specifically 51.4% and 54.5% on PQL-2hop and PQL-3hop, respectively. This demonstrates that, while maintaining the above-demonstrated effectiveness, QAGCN is easier to train than NSM. The reason is two-fold: First, NSM relies on a teacher network to provide intermediate supervision for the multi-step reasoning of its student network. Within the teacher network, two additional multi-step reasoning processes over KGs are performed. Therefore, for each training question, NSM needs to reason three times over KGs (once for the student network and twice for the teacher network), which is inefficient given that our model only needs to reason once. Second, the intermediate supervision of the teacher network is represented as probability distributions of entities. The Kullback-Leibler divergence and Jensen-Shannon divergence [7] need to be additionally computed for several pairs of distributions in the loss function of NSM, which makes the model difficult to implement and optimize.

Table 4. Average run-time of each step per test question in ms.

	MetaQA			PathQuestion (PQ)	
	1-hop	2-hop	3-hop	2-hop	3-hop
Question Processing	1.4	1.6	1.1	0.4	0.7
Subgraph Extraction	3.9	6.6	357.7	1.2	18.3
Question Encoding	16.3	16.3	10.0	8.4	9.9
Subgraph Encoding	1.1	2.9	26.5	1.3	2.4
Answer Search	0.6	5.2	287.1	0.4	2.3
Answer Re-ranking	79.5	123.2	1012.1	8.0	15.4
Total	102.8	155.8	1694.5	19.7	49.0

4.3   Efficiency Evaluation

Question answering is supposed to be an online service that answers questions in real-time. We evaluated the average time cost of each step of our model on MetaQA and PathQuestion. The results are reported in Table 4. In this table, Question Processing includes the segmentation and classification of questions. The others are in accordance with the steps introduced in Sect. 3. Please note that we list the path extraction-based answer refinement separately as Answer Re-ranking, given that the explicit path extraction in this step could be particularly time-consuming and is worth specific attention. Most time costs are below or near 100ms, which can be seen as instantaneous in user interfaces [20]. Only 3-hop questions in MetaQA require an average of 1.7 s, which we consider acceptable when contrasted with the complexity of 3-hop questions. It is worth mentioning that our implementation can be further optimized for a real-world deployment. For example, the subgraph extraction can be performed offline given a KG, and the extraction and encoding of relation paths for candidate answers can be performed in a completely parallel way.

4.4   Ablation Study

To examine the contribution of each component of our model, we evaluate the model in a series of degraded scenarios and report the performance changes in Table 5. **Complete Model** reports the results of the complete QAGCN model. **Re-ranking Removed** shows the performance when candidate answers are not re-ranked in Answer Search. The importance of the re-ranking is demonstrated by the performance degradation (e.g., −43.3% on MetaQA 3-hop and −44.2% on PathQuestion 3-hop). In **Q&G-Encoders Removed**, the question and graph encoders are removed, and we use the pre-trained BERT model to directly encode questions and entities. Top-k candidate answers are retrieved and then re-ranked by the original re-ranking module.[7] In this scenario, there is

---

[7] The values of k were set to be consistent with those used in the complete model.

Table 5. Model performance changes in the ablation study. (% Hits@1).

	MetaQA			PathQuestion (PQ)	
	1-hop	2-hop	3-hop	2-hop	3-hop
Complete Model	97.3	99.9	97.6	98.4[*]	92.1[*]
Re-ranking Removed	95.8	87.4	55.3	72.9	51.4
Q&G-Encoders Removed	97.0	91.3	86.9	74.5	43.2
G-Encoder (Conv)	64.6	63.9	40.8	67.7	28.0
G-Encoder (Linear)	72.0	55.6	15.2	71.4	43.0
Q&G-Encoders (Linear)	70.7	52.4	0.1	63.5	16.9

* We report the results of QAGCN on the first run of PQ-2/3hop.
Hence, these results differ from the averages reported in Table 2.

also a significant performance drop (e.g., -11.0% on MetaQA 3-hop and -53.1% on PathQuestion 3-hop). From the above three scenarios, we observe that both the re-ranking module and question/graph encoders are necessary, and that they are complementary to each other.

In **G-Encoder (Conv)**, we replace the proposed attentional GCNs with conventional GCNs [15]. The question encoder remains unchanged, and we report direct results without re-ranking. Compared to Re-ranking Removed, i.e., the original encoders, performance degradations can be observed on all datasets (e.g., -26.2% on MetaQA 3-hop and -45.5% on PathQuestion 3-hop). In **G-Encoder (Linear)**, we further replace the conventional GCNs with feedforward neural networks. In this scenario, entities are encoded without considering their contexts in KGs. Compared with G-Encoder (Conv), performance drops can be observed on questions with large subgraphs (e.g., -13.0% on MetaQA 2-hop and -62.7% on MetaQA 3-hop). While for questions with small subgraphs, in which entities do not have sufficient contexts, improvements are observed (e.g., $+11.5\%$ on MetaQA 1-hop and $+5.5\%$ on PathQuestion 2-hop). In **Q&G-Encoders (Linear)**, based on G-Encoder (Linear), we further replace the LSTMs in the question encoder with feedforward neural networks. In this case, the model mainly relies on the initial encoding of questions and entities computed by the BERT model. Since the BERT model was pre-trained on other tasks, the performance on these QA tasks is expected to be low. In Table 5, we can observe substantial performance drops on all datasets compared to the last scenario (e.g., -99.3% on MetaQA 3-hop and -60.7% on PathQuestion 3-hop).

4.5 Case Study with Embedding Visualization

For an intuitive understanding of the proposed QAGCN model, we select an example question to conduct a case study with the visualization of question and entity embeddings. Specifically, we picked the following 3-hop test question from PathQuestion: *Isabella of Portugal's child's spouse's place of death?* The topic entity is `Isabella of Portugal` with a support path of three triples:

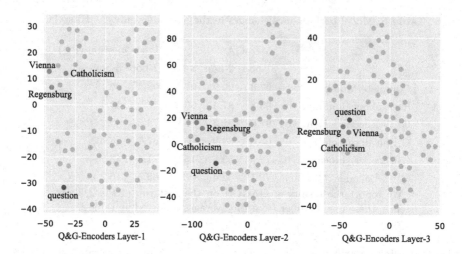

Fig. 3. Visualization of the embedding space after each layer of the Q&G-Encoders. Please note that the axes do not have a same scale. The ■ purple point indicates the question embedding after each layer. The ■ blue points indicate the top-3 candidates finally retrieved. The ■ orange points indicate all other entities in the subgraph.

$$\{(\texttt{Isabella of Portugal}, \texttt{children}, \texttt{Maria of Spain}),$$
$$(\texttt{Maria of Spain}, \texttt{spouse}, \texttt{Maximilian II Holy Roman Emperor}),$$
$$(\texttt{Maximilian II Holy Roman Emperor}, \texttt{place of death}, \texttt{Regensburg})\}.$$

For this question, three GCN layers are used in the graph encoder to encode entities in the subgraph that covers all paths of length one, two, and three starting from the topic entity Isabella of Portugal. The question is accordingly encoded by the question encoder with three layers of LSTMs. Then, according to Euclidean distances between the question embedding and entity embeddings, we obtain the top-3 candidate answers: Vienna (distance: 1.53), Regensburg (distance: 1.89), and Catholicism (distance: 3.17). The correct answer Regensburg is ranked after Vienna. Since Vienna is the place of birth of Maximilian II Holy Roman Emperor, it is likely that this trained model struggles to differentiate between "birth" and "death." A second possible reason is that this trained model is biased towards Vienna given that Vienna is the answer of 12 training questions that ask for the place of death of people, while Regensburg is the answer of only one training question.

To visualize the joint embedding space of the LSTM-based question encoder and GCN-based graph encoder for each of the three layers, we use t-SNE [18] to compute dimension-reduced entity and question embeddings and plot them in Fig. 3. We highlight the final three top candidate answers in blue and plot all other entities in the subgraph as orange points. In addition, we plot the location of the question embedding after each of the LSTM layers with a purple point. It can be observed that initially (i.e., in Layer-1) the correct answer is not at all close to the question in the embedding space. However, after more layers of encoding, more information consistent with the given question is passed to

Regensburg by the graph encoder, and the question embedding is accordingly transformed by the question encoder. Therefore, we can observe that Regensburg and the question approach each other gradually in the embedding space.

Then, we extract relation paths between the top-3 candidate answers and the topic entity, which are respectively {[place of birth, spouse, children]}, {[place of death, spouse, children]}, and {[religion, spouse, children], [religion, parents, children]} for Vienna, Regensburg, and Catholicism. The minimum path-question distances of Vienna, Regensburg, and Catholicism computed by the re-ranking module are respectively 0.99, 0.69, and 1.00. Finally, with the minimum distance 0.69, Regensburg is correctly returned as the final answer.

5   Conclusion and Outlook

In this paper, we propose a novel QA model called QAGCN, which answers multi-relation questions via single-step implicit reasoning over KGs. A novel question-aware GCN is proposed to perform question-dependent message propagation while encoding KGs. It is able to represent questions and KG entities in a joint embedding space, where questions are close to their answers. Our model has been demonstrated to be competitive and even superior against SOTA reasoning-based methods in experiments, while our model is simpler and easier to train and implement. The efficiency and contribution of each component of our model have also been examined and analyzed in detail.

The task we solve in this paper is in line with other existing QA work that assumes all given questions are answerable. Therefore, if the answer to a question does not exist in the accompanied KG, our model will still confidently report an answer, which is actually incorrect. In future work, we plan to investigate a method that can detect if a given question is answerable regarding a given KG.

Acknowledgement. This work has been partially supported by the University Research Priority Program "Dynamics of Healthy Aging" at the University of Zurich and the Swiss National Science Foundation through project MediaGraph (contract no. 202125). Michael Cochez is partially funded by the Graph-Massivizer project, funded by the Horizon Europe programme of the European Union (grant 101093202).

References

1. Berant, J., Liang, P.: Semantic parsing via paraphrasing. In: Proceedings of the 52nd Annual Meeting of the Association for Computational Linguistics, ACL 2014, June 22-27, 2014, Baltimore, MD, USA, Volume 1: Long Papers, pp. 1415–1425. The Association for Computer Linguistics (2014). https://doi.org/10.3115/v1/p14-1133

2. Bollacker, K.D., Evans, C., Paritosh, P.K., Sturge, T., Taylor, J.: Freebase: a collaboratively created graph database for structuring human knowledge. In: Wang, J.T. (ed.) Proceedings of the ACM SIGMOD International Conference on Management of Data, SIGMOD 2008, Vancouver, BC, Canada, June 10-12, 2008, pp. 1247–1250. ACM (2008). https://doi.org/10.1145/1376616.1376746

3. Bordes, A., Chopra, S., Weston, J.: Question answering with subgraph embeddings. In: Moschitti, A., Pang, B., Daelemans, W. (eds.) Proceedings of the 2014 Conference on Empirical Methods in Natural Language Processing, EMNLP 2014, October 25-29, 2014, Doha, Qatar, A meeting of SIGDAT, a Special Interest Group of the ACL, pp. 615–620. ACL (2014). https://doi.org/10.3115/v1/d14-1067

4. Bordes, A., Usunier, N., Chopra, S., Weston, J.: Large-scale simple question answering with memory networks. CoRR **abs/1506.02075** (2015). http://arxiv.org/abs/1506.02075

5. Das, R., et al.: Go for a walk and arrive at the answer: reasoning over paths in knowledge bases using reinforcement learning. In: 6th International Conference on Learning Representations, ICLR 2018, Vancouver, BC, Canada, April 30 - May 3, 2018, Conference Track Proceedings. OpenReview.net (2018). https://openreview.net/forum?id=Syg-YfWCW

6. Devlin, J., Chang, M., Lee, K., Toutanova, K.: BERT: pre-training of deep bidirectional transformers for language understanding. In: Burstein, J., Doran, C., Solorio, T. (eds.) Proceedings of the 2019 Conference of the North American Chapter of the Association for Computational Linguistics: Human Language Technologies, NAACL-HLT 2019, Minneapolis, MN, USA, June 2-7, 2019, Volume 1 (Long and Short Papers), pp. 4171–4186. Association for Computational Linguistics (2019). https://doi.org/10.18653/v1/n19-1423

7. Fuglede, B., Topsøe, F.: Jensen-shannon divergence and Hilbert space embedding. In: Proceedings of the 2004 IEEE International Symposium on Information Theory, ISIT 2004, Chicago Downtown Marriott, Chicago, Illinois, USA, June 27 - July 2, 2004, p. 31. IEEE (2004). https://doi.org/10.1109/ISIT.2004.1365067

8. Harris, S., Seaborne, A., Prud'hommeaux, E.: SPARQL 1.1 query language. w3c recommendation (2013). https://www.w3.org/TR/sparql11-query/

9. He, G., Lan, Y., Jiang, J., Zhao, W.X., Wen, J.: Improving multi-hop knowledge base question answering by learning intermediate supervision signals. In: Lewin-Eytan, L., Carmel, D., Yom-Tov, E., Agichtein, E., Gabrilovich, E. (eds.) WSDM '21, The Fourteenth ACM International Conference on Web Search and Data Mining, Virtual Event, Israel, March 8-12, 2021, pp. 553–561. ACM (2021). https://doi.org/10.1145/3437963.3441753

10. Hochreiter, S., Schmidhuber, J.: Long short-term memory. Neural Comput. **9**(8), 1735–1780 (1997)

11. Hu, S., Zou, L., Yu, J.X., Wang, H., Zhao, D.: Answering natural language questions by subgraph matching over knowledge graphs. IEEE Trans. Knowl. Data Eng. **30**(5), 824–837 (2018). https://doi.org/10.1109/TKDE.2017.2766634

12. Huang, X., Zhang, J., Li, D., Li, P.: Knowledge graph embedding based question answering. In: Culpepper, J.S., Moffat, A., Bennett, P.N., Lerman, K. (eds.) Proceedings of the Twelfth ACM International Conference on Web Search and Data Mining, WSDM 2019, Melbourne, VIC, Australia, February 11-15, 2019, pp. 105–113. ACM (2019). https://doi.org/10.1145/3289600.3290956

13. Kaufmann, E., Bernstein, A.: How useful are natural language interfaces to the semantic web for casual end-users? In: Aberer, K., et al. (eds.) The Semantic Web, pp. 281–294. Springer, Berlin Heidelberg, Berlin, Heidelberg (2007). https://doi.org/10.1007/978-3-540-76298-0_21

14. Kingma, D.P., Ba, J.: Adam: a method for stochastic optimization. In: Bengio, Y., LeCun, Y. (eds.) 3rd International Conference on Learning Representations, ICLR 2015, San Diego, CA, USA, May 7-9, 2015, Conference Track Proceedings (2015). http://arxiv.org/abs/1412.6980

15. Kipf, T.N., Welling, M.: Semi-supervised classification with graph convolutional networks. In: 5th International Conference on Learning Representations, ICLR 2017, Toulon, France, April 24-26, 2017, Conference Track Proceedings. OpenReview.net (2017). https://openreview.net/forum?id=SJU4ayYgl

16. Liang, P.: Lambda dependency-based compositional semantics. CoRR **abs/1309.4408** (2013). http://arxiv.org/abs/1309.4408

17. Lukovnikov, D., Fischer, A., Lehmann, J., Auer, S.: Neural network-based question answering over knowledge graphs on word and character level. In: Barrett, R., Cummings, R., Agichtein, E., Gabrilovich, E. (eds.) Proceedings of the 26th International Conference on World Wide Web, WWW 2017, Perth, Australia, April 3-7, 2017, pp. 1211–1220. ACM (2017). https://doi.org/10.1145/3038912.3052675

18. van der Maaten, L., Hinton, G.: Visualizing data using t-SNE. J. Mach. Learn. Res. **9**(86), 2579–2605 (2008), http://jmlr.org/papers/v9/vandermaaten08a.html

19. Miller, A.H., Fisch, A., Dodge, J., Karimi, A., Bordes, A., Weston, J.: Key-value memory networks for directly reading documents. In: Su, J., Carreras, X., Duh, K. (eds.) Proceedings of the 2016 Conference on Empirical Methods in Natural Language Processing, EMNLP 2016, Austin, Texas, USA, November 1-4, 2016, pp. 1400–1409. The Association for Computational Linguistics (2016). https://doi.org/10.18653/v1/d16-1147

20. Nielsen, J.: Response times: the 3 important limits (1991). https://www.nngroup.com/articles/response-times-3-important-limits/

21. Qiu, Y., Wang, Y., Jin, X., Zhang, K.: Stepwise reasoning for multi-relation question answering over knowledge graph with weak supervision. In: Caverlee, J., Hu, X.B., Lalmas, M., Wang, W. (eds.) WSDM '20: The Thirteenth ACM International Conference on Web Search and Data Mining, Houston, TX, USA, February 3-7, 2020, pp. 474–482. ACM (2020). https://doi.org/10.1145/3336191.3371812

22. Ren, H., et al.: Graph convolutional networks in language and vision: a survey. Knowl. Based Syst. **251**, 109250 (2022). https://doi.org/10.1016/J.KNOSYS.2022.109250

23. Saxena, A., Tripathi, A., Talukdar, P.P.: Improving multi-hop question answering over knowledge graphs using knowledge base embeddings. In: Proceedings of the 58th Annual Meeting of the Association for Computational Linguistics, ACL 2020, Online, July 5-10, 2020, pp. 4498–4507. Association for Computational Linguistics (2020). https://doi.org/10.18653/v1/2020.acl-main.412, https://doi.org/10.18653/v1/2020.acl-main.412

24. Shi, J., Cao, S., Hou, L., Li, J., Zhang, H.: TransferNet: an effective and transparent framework for multi-hop question answering over relation graph. In: Proceedings of the 2021 Conference on Empirical Methods in Natural Language Processing, EMNLP 2021, Virtual Event / Punta Cana, Dominican Republic, 7-11 November, 2021, pp. 4149–4158. Association for Computational Linguistics (2021). https://doi.org/10.18653/v1/2021.emnlp-main.341

25. Sun, H., Bedrax-Weiss, T., Cohen, W.W.: PullNet: open domain question answering with iterative retrieval on knowledge bases and text. In: Inui, K., Jiang, J., Ng, V., Wan, X. (eds.) Proceedings of the 2019 Conference on Empirical Methods in Natural Language Processing and the 9th International Joint Conference on Natural Language Processing, EMNLP-IJCNLP 2019, Hong Kong, China, November 3-7, 2019, pp. 2380–2390. Association for Computational Linguistics (2019). https://doi.org/10.18653/v1/D19-1242

26. Sun, H., Dhingra, B., Zaheer, M., Mazaitis, K., Salakhutdinov, R., Cohen, W.W.: Open domain question answering using early fusion of knowledge bases and text.

In: Riloff, E., Chiang, D., Hockenmaier, J., Tsujii, J. (eds.) Proceedings of the 2018 Conference on Empirical Methods in Natural Language Processing, Brussels, Belgium, October 31 - November 4, 2018, pp. 4231–4242. Association for Computational Linguistics (2018). https://doi.org/10.18653/v1/d18-1455

27. Thai, D., et al.: CBR-iKB: a case-based reasoning approach for question answering over incomplete knowledge bases. CoRR **abs/2204.08554** (2022). https://doi.org/10.48550/ARXIV.2204.08554

28. Vashishth, S., Sanyal, S., Nitin, V., Talukdar, P.P.: Composition-based multi-relational graph convolutional networks. In: 8th International Conference on Learning Representations, ICLR 2020, Addis Ababa, Ethiopia, April 26-30, 2020. OpenReview.net (2020). https://openreview.net/forum?id=BylA_C4tPr

29. Wang, R., Wang, M., Liu, J., Chen, W., Cochez, M., Decker, S.: Leveraging knowledge graph embeddings for natural language question answering. In: Li, G., Yang, J., Gama, J., Natwichai, J., Tong, Y. (eds.) Database Systems for Advanced Applications. Lecture Notes in Computer Science(), vol. 11446, pp. 659–675. Springer, Cham (2019). https://doi.org/10.1007/978-3-030-18576-3_39

30. Zhang, J., et al.: Subgraph retrieval enhanced model for multi-hop knowledge base question answering. In: Muresan, S., Nakov, P., Villavicencio, A. (eds.) Proceedings of the 60th Annual Meeting of the Association for Computational Linguistics (Volume 1: Long Papers), ACL 2022, Dublin, Ireland, May 22-27, 2022, pp. 5773–5784. Association for Computational Linguistics (2022). https://doi.org/10.18653/V1/2022.ACL-LONG.396

31. Zhang, Y., Dai, H., Kozareva, Z., Smola, A.J., Song, L.: Variational reasoning for question answering with knowledge graph. In: McIlraith, S.A., Weinberger, K.Q. (eds.) Proceedings of the Thirty-Second AAAI Conference on Artificial Intelligence, (AAAI-18), the 30th innovative Applications of Artificial Intelligence (IAAI-18), and the 8th AAAI Symposium on Educational Advances in Artificial Intelligence (EAAI-18), New Orleans, Louisiana, USA, February 2-7, 2018, pp. 6069–6076. AAAI Press (2018). https://www.aaai.org/ocs/index.php/AAAI/AAAI18/paper/view/16983

32. Zhou, M., Huang, M., Zhu, X.: An interpretable reasoning network for multi-relation question answering. In: Bender, E.M., Derczynski, L., Isabelle, P. (eds.) Proceedings of the 27th International Conference on Computational Linguistics, COLING 2018, Santa Fe, New Mexico, USA, August 20-26, 2018, pp. 2010–2022. Association for Computational Linguistics (2018), https://aclanthology.org/C18-1171/

Leveraging Pre-trained Language Models for Time Interval Prediction in Text-Enhanced Temporal Knowledge Graphs

Duygu Sezen Islakoglu[1]([✉]) [ID], Melisachew Wudage Chekol[1] [ID], and Yannis Velegrakis[1,2] [ID]

[1] Utrecht University, Utrecht, The Netherlands
{d.s.islakoglu,m.w.chekol,i.velegrakis}@uu.nl
[2] University of Trento, Trento, Italy

Abstract. Most knowledge graph completion (KGC) methods rely solely on structural information, even though a large number of publicly available KGs contain additional temporal (validity time intervals) and textual data (entity descriptions). While recent temporal KGC methods utilize time information to enhance link prediction, they do not leverage textual descriptions or support inductive inference (prediction for entities that have not been seen during training).

In this work, we propose a novel framework called TEMT that exploits the power of pre-trained language models (PLMs) for temporal KGC. TEMT predicts time intervals of facts by fusing their textual and temporal information. It also supports inductive inference by utilizing PLMs. In order to showcase the power of TEMT, we carry out several experiments including time interval prediction, both in transductive and inductive settings, and triple classification. The experimental results demonstrate that TEMT is competitive with the state-of-the-art, while also supporting inductiveness.

Keywords: Knowledge Graph Completion · Temporal Knowledge Graphs · Pre-trained Language Models

1 Introduction

Knowledge graphs are often incomplete, meaning that some elements of the facts are not available. To this end, knowledge graph completion (KGC) methods aim at finding missing links between the entities. Most of these studies focus on static knowledge graphs where the graph remains unchanged over time [15]. However, in real life, facts are not always valid throughout time, but only in specific time periods. For instance, presidents of a country are valid only throughout their term.

To model this, triples in a knowledge graph may have validity time intervals associated with them. This additional information converts a triple into

© The Author(s), under exclusive license to Springer Nature Switzerland AG 2024
A. Meroño Peñuela et al. (Eds.): ESWC 2024, LNCS 14664, pp. 59–78, 2024.
https://doi.org/10.1007/978-3-031-60626-7_4

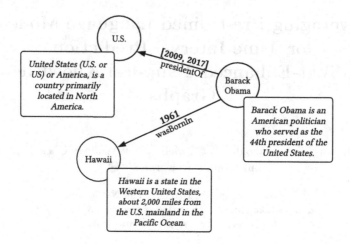

Fig. 1. An example of a text-enhanced temporal knowledge graph.

a quadruple form ⟨subject, relation, object, time interval⟩, e.g. ⟨Obama, presidentOf, U.S., [2009, 2017]⟩. A graph consisting of a set of such temporal facts -quadruples- is referred to as a *Temporal Knowledge Graph* (TKG).

Alike traditional KGs, temporal knowledge graphs suffer inherently from incompleteness due to their dynamic nature. In particular, the temporal information (validity time interval) is often missing after an automatic knowledge graph construction. To this end, the temporal knowledge graph completion (TKGC) task aims to predict a missing quadruple element, such as time interval prediction ⟨s, r, o, ?⟩. The time interval prediction task is useful for temporal question answering, automatic TKG construction, and verification of TKGs that have temporal constraints [35].

(Temporal) knowledge graphs completion methods [36] can benefit from textual descriptions of entities and relations since they contain valuable information regarding semantic relationships across entities. For instance, in a description of an entity, there may be a reference to some other entity, despite the absence of any type of relationship (edge) in the knowledge graph among them. Furthermore, by considering the semantics of the descriptions, one may gain insight into the validity time of the facts. As an example, if an entity description contains elements from a certain century or a period like Renaissance, facts involving that entity may be valid for that period. Figure 1 illustrates a temporal knowledge graph with entities that have textual descriptions associated with them.

The availability of textual descriptions in knowledge graphs provides an excellent opportunity for exploiting the benefits that both knowledge graphs and language models can offer. Recent works has shown that language models store real-world knowledge in their parameters and can potentially be used as knowledge graphs [3,27], and that textual information can improve link prediction for static knowledge graphs [20,38]. Moreover, it has been shown that entity descriptions and pre-trained language models can model facts that involve unseen entities (inductiveness) [9,34]. Supporting inductive reasoning is crucial since most real-world knowledge graphs are often continuously extended with new entities. Unfortunately, most temporal knowledge graph completion mechanisms are transductive, which means that they can only perform predictions on the entities they have already seen during training, i.e., are part of the training set.

We provide a novel temporal knowledge graph completion framework called TEMT (**T**ext **E**ncoder **M**eets **T**ime)[1] that combines the available textual and temporal information for **time interval prediction**. TEMT is able to predict time intervals for unseen entities by leveraging a pre-trained language model. We extend the two commonly used datasets, i.e., YAGO11k and Wikidata12k [8], with textual descriptions. In addition, we create new train/valid/test splits for experiments on time interval prediction in inductive setting. Our experiments show that TEMT is competitive with state-of-the-art and able to reason on unseen entities (even when both the subject and object of a quadruple are unseen in training).

2 Preliminaries

A temporal knowledge graph (TKG) is a directed graph $\mathcal{G} = (\mathcal{E}, \mathcal{R}, \mathcal{T}, \mathcal{Q})$ where $\mathcal{E}, \mathcal{R}, \mathcal{T}$ are sets of entities, relations, time points and $\mathcal{Q}$ represents the set of quadruples (or temporal facts) in the format ⟨subject s, relation r, object o, time interval t_I⟩ where $s, o \in \mathcal{E}$, $r \in \mathcal{R}$, $t_I = [t_{start}, t_{end}]$ and $t_{start}, t_{end} \in \mathcal{T}$. Note that t_I can also be a single time point if $t_{start}{=}t_{end}$.

Temporal knowledge graphs can be grouped into two types: event TKGs and interval-based TKGs. The former refers to TKGs in which $t_{start} = t_{end}$ for every quadruple. On the other hand, interval-based TKGs, which are the focus of this paper, represent TKGs where each quadruple has a validity time interval $t_I = [t_{start}, t_{end}]$. An interval is called left-open if t_{start} is unknown, right-open if t_{end} is unknown, and closed interval if both the start and end points are known. The format of time points depends on the chosen time granularity, such as years, months, or days.

Text-enhanced temporal knowledge graphs are TKGs where each entity and relation is associated with a name and each entity has some natural language text that describes its meaning. This additional information provides a context and attaches a semantic meaning to the facts, which can be informative for predicting the validity time interval of the fact. More formally, a text-enhanced temporal knowledge graph is a directed graph $\mathcal{G} = (\mathcal{E}, \mathcal{R}, \mathcal{T}, \mathcal{Q}, \mathcal{N}, \mathcal{D})$ where $\mathcal{E}, \mathcal{R}, \mathcal{T}$ are sets of entities, relations, time points and $\mathcal{Q}$ represents the set of quadruples, $\mathcal{N}$ denotes the set of entity and relation names, and $\mathcal{D}$ denotes the set of entity descriptions. An example for text-enhanced TKG is given in Fig. 1.

We can split $\mathcal{Q}$ into three disjoint sets as train, validation, and test sets. Formally, $\mathcal{Q} = \mathcal{Q}_{train} \cup \mathcal{Q}_{val} \cup \mathcal{Q}_{test}$ where $\mathcal{Q}_{train}$ represents the set of train

---

[1] The datasets and the source code are available at https://github.com/duyguisla koglu/TEMT.

quadruples, $\mathcal{Q}_{val}$ represents the set of validation quadruples and $\mathcal{Q}_{test}$ represents the set of test quadruples. Similarly, we can specify the set of entities as $\mathcal{E}_{train}$, $\mathcal{E}_{val}$ and $\mathcal{E}_{test}$.

Problem Statement. Time interval prediction is the task of predicting the validity time interval of a triple. More formally, given a quadruple $\langle s, r, o, ?\rangle$ with unknown validity time interval, the objective is to predict a time interval t_I. We can reformulate the question as follows: given training and validation sets, $\mathcal{Q}_{train}$ and $\mathcal{Q}_{val}$, the time interval prediction is to output t_I for each test quadruple $\langle s, r, o, ?\rangle \in \mathcal{Q}_{test}$.

In this paper, we focus on two variants of this problem: transductive time interval prediction and inductive time interval prediction. The transductive case aims to predict t_I for each test quadruple in $\mathcal{Q}_{test}$ that does not contain any new entities, i.e., $\mathcal{Q}_{test} = \{\langle s, r, o, ?\rangle | s, o \in \mathcal{E}_{train}\}$. In other words, all entities in the test set are included in the training set, i.e., $\mathcal{E}_{test} \subseteq \mathcal{E}_{train}$. Furthermore, inductive time interval prediction aims to predict the time interval of facts where the test set contains previously unseen entities [9]. For inductive inference, we create splits as follows: for each quadruple in the test set, the subject or the object (or both) does not appear in $\mathcal{E}_{train}$. Therefore, the test quadruples $\mathcal{Q}_{test} = \{\langle s, r, o, ?\rangle | s \notin \mathcal{E}_{train}\} \cup \{\langle s, r, o, ?\rangle | o \notin \mathcal{E}_{train}\}$.

Pre-trained Language Models. We exploit the representational power of pre-trained language models to capture the semantics of facts and to deal with unseen entities. Language models assign a probability to a word by taking into account the other words in a sentence and can predict the next word given a sequence of words. This can be done by learning a latent representation of words in a vector space. Moreover, models such as bidirectional encoder representations from transformers (BERT) [11] do not only consider the previous words but also take the subsequent words into account. BERT generates a contextual word embedding where the representation of a word depends on the whole context.

A pre-trained language model (PLMs) is a language model that is trained on a large text corpora including books, encyclopedias, and web data. PLMs can be used for many downstream tasks such as question-answering and text summarization. A pre-trained model, such as pre-trained BERT, can be further fine-tuned for a specific task or can be used for feature extraction of a sentence. However, since BERT is designed for word-level tasks and not optimized for sentence-level tasks, it performs poorly in semantic textual similarity tasks [28]. On the other hand, Sentence-BERT [28], a language model built on top of BERT, is explicitly trained to generate sentence embeddings where semantically similar sentences are closer in the embedding space. In the next section, we explain how Sentence-BERT can be used to generate an embedding for a triple.

3   The Framework

We materialize the idea of using pre-trained language models for temporal knowledge graph completion into a framework called TEMT (**T**ext **E**ncoder **M**eets

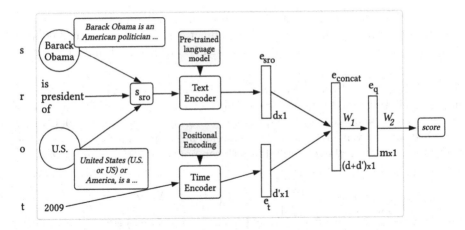

Fig. 2. Overview of TEMT's architecture for quadruple scoring: the TextEncoder produces *triple embedding* e_{sro}, and the TimeEncoder generates *time point embedding* e_t. These representations are fused to compute a plausibility score.

Time). It learns a scoring function that takes a quadruple $\langle s, r, o, t \rangle$ where t is a single time point and outputs a plausibility score. In the inference time, we utilize the learned scoring function to output a time interval for a given triple $\langle s, r, o, ? \rangle$. Figure 2 gives an overview of the TEMT framework.

Embedding Triples with a Text Encoder. The text encoder packs the names and descriptions of triple elements as a single sentence and returns a vector. As our text encoder, we leverage a pre-trained language model Sentence-BERT [28][2] to benefit from its representation power. Inspired by [38], we form a single textual sentence S_{sro} for a triple to feed Sentence-BERT.

$$S_{sro} = \mathcal{N}_s + \mathcal{N}_r + \mathcal{N}_o + (\mathcal{D}_s + \mathcal{D}_o), \tag{1}$$

where S_{sro} is a string concatenation of the names and descriptions of the entities and relations, $\mathcal{N}$ refers to names and $\mathcal{D}$ refers to entity descriptions. The text encoder then outputs $e_{sro} \in \mathbb{R}^d$ which we call *triple embedding*:

$$e_{sro} = \text{TextEncoder}(S_{sro}) \tag{2}$$

Our main motivation to leverage a language model as a text encoder is two-fold. Firstly, the language model captures the interactions between the subject, relation, and the object and outputs a semantically rich contextualized embedding of the fact. Secondly, language models can model unobserved entities and therefore support inductive reasoning. Moreover, not only Sentence-BERT, but also any encoder-only model (e.g. BERT) can be used as our text encoder. This is possible by adding a special [CLS] token to the beginning of the S_{sro}. In this

---

[2] The name of the model used is all-mpnet-base-v2.

way, the token embedding for [CLS] captures the possible interactions between the subject, relation and object [38] and functions as a sentence embedding. Thus, TEMT is not dependent on Sentence-BERT; instead, it is model-agnostic.

Embedding Time Points with a Time Encoder. We use positional encoding method [33] to produce vector representations of time points. This method is widely used in (knowledge graph) question answering systems [16,23,39] in order to embed time. Given a time point t and a reference time point t_{min} (which is the earliest time point in the dataset), the j-th component of a *time point embedding* for t is defined as follows:

$$\text{TimeEncoder}(t, t_{min})[j] = \begin{cases} \sin\left(\frac{t-t_{min}}{10000^{2i/d'}}\right) & \text{if } j = 2i \\ \cos\left(\frac{t-t_{min}}{10000^{2i/d'}}\right) & \text{if } j = 2i+1 \end{cases} \tag{3}$$

where the term $t-t_{min}$ refers to the position of t relative to the earliest time point t_{min} in $\mathcal{T}$ and d' is the dimension of the *time point embedding*. Intuitively, the *time point embedding* can be thought of as a position in time. The time encoder requires a first time point t_{min} as a reference point, then the other time points will be positioned relative to this reference point. For the sake of brevity, we omit t_{min} from the time encoder function and simply write as $\text{TimeEncoder}(t)$. The time encoder outputs $e_t \in \mathbb{R}^{d'}$ which we call *time point embedding*.

$$e_t = \text{TimeEncoder}(t) \tag{4}$$

We emphasize mainly the two properties of positional encodings: each time point corresponds to a unique vector and the vectors of close time points are closer in the vector space [16]. This enables us to model the dependencies across different time points and the notion of ordering. On the contrary, many previous works [14,19] learn the time vectors within the same space as entities and relations, and they cannot model the time point dependencies. For instance, they learn the consecutive years independently which may not capture ordering.

Moreover, TEMT can be adapted to different time granularities. For instance, one can set months as granularity by converting the years into corresponding number of months. Lastly, in contrast to previous work, the time encoder can represent unobserved time points. Although we do not focus on temporal inductiveness in this paper, the time encoder can potentially be used for performing predictions on future or unseen time points.

Fusing Triple and Time Point Embeddings. In the previous sections, we introduced two functions, namely TextEncoder and TimeEncoder, that allow us to produce embeddings of triples and time points respectively. We are now ready to discuss how these embeddings from different spaces can be combined (or fused) together. Similar to [12,25], by treating the textual and temporal features as different modalities, TEMT combines triple and time embeddings using a multi-layer perceptron (MLP). The fusion of these two embeddings produces a representation of a quadruple. Formally, given a quadruple q, the time-aware quadruple representation e_q is obtained as follows:

$$e_q = (W_1 v_{concat} + b_1) \in \mathbb{R}^m \tag{5}$$

$$v_{concat} = [e_{sro}; e_t] \in \mathbb{R}^{d+d'} \tag{6}$$

where $[\,;\,]$ denotes concatenation operation, $W_1 \in \mathbb{R}^{m \times (d+d')}$ and $b_1 \in \mathbb{R}^m$ denote the learnable parameters and m is the dimension of e_q where $m < (d + d')$.

Another approach for embedding a quadruple is to append the time point to the textual sentence S_{sro} in Eq. (1) and feed pre-trained language model with this new sentence. This would represent all quadruple elements in a single space, which is the text space. However, it is shown that PLMs are not good at number representations [31]. Our preliminary analysis also demonstrated that language models are not good at temporal reasoning such as ordering events and interval arithmetic. This motivates the need for an external time encoder.

Parametric Quadruple Scoring Function. Although most methods use a fixed distance function for scoring triples or quadruples, there are some methods such as ConvE [10] and ConvKB [24] that learn the parameters for a scoring function. Similarly, we employ a parametric scoring function to output a plausibility score for a quadruple of a given TKG:

$$f(s, r, o, t) = W_2 e_q + b_2 \tag{7}$$

where $W_2 \in \mathbb{R}^{1 \times m}$ and $b_2 \in \mathbb{R}$ are learnable parameters of the final layer of the neural network. Before feeding the input to this final layer, we use ReLU [1] as an activation function.

Negative Sampling. The model learns by distinguishing valid quadruples from incorrect quadruples. To this end, TEMT employs two different types of negative sampling. The first type is called *entity-corrupted negative sampling*. In this approach, the set of negative quadruples $D^-_{\langle s,r,o,t \rangle}$ is created by corrupting the subject or the object of a given quadruple $\langle s, r, o, t \rangle$ as shown below:

$$D^-_{\langle s,r,o,t \rangle} = \{\langle s', r, o, t \rangle \notin D^+ | s' \in \mathcal{E}\} \cup \{\langle s, r, o', t \rangle \notin D^+ | o' \in \mathcal{E}\}. \tag{8}$$

where $D^+ = \mathcal{Q}_{train}$ denotes the set of positive quadruples.

The second one is called *time-corrupted negative sampling* [6]. In this approach, the set of negative quadruples $D^-_{\langle s,r,o,t \rangle}$ is created by corrupting the time point of a given quadruple $\langle s, r, o, t \rangle$ as the following:

$$D^-_{\langle s,r,o,t \rangle} = \{\langle s, r, o, t' \rangle \notin D^+ | t' \in \mathcal{T}\}. \tag{9}$$

t' is sampled based on the validity time interval t_I of the given quadruple. If t_I is a right-open interval, then $t' < t_{start}$ is sampled (e.g. $t' = 2006$ for $t_I = [2008,$ unknown]); if t_I is left-open interval, then $t' > t_{end}$ is sampled (e.g. $t' = 1950$ for $t_I = [\text{unknown}, 1930]$); if t_I is a closed interval, then $t' \notin [t_{start}, t_{end}]$ is randomly chosen (e.g. $t' = 1750$ for $t_I = [1800, 1810]$).

Training. Similar to [4], we use the following margin-based ranking loss for training:

$$\mathcal{L} = \sum_{q_p \in D^+} \sum_{q_n \in D^-_{q_p}} \max(0, f(q_n) - f(q_p) + \gamma). \tag{10}$$

where q_p is a positive quadruple, q_n is a negative quadruple, γ is the margin value and f is the scoring function from Eq. (7). We train the model to give higher scores to positive quadruples (with a given margin γ) than negative quadruples. In this way, we learn a quadruple scoring function that will be used for predicting time intervals.

Inference. In the previous sections, we discuss how TEMT learns a scoring function that gives plausibility score to an individual quadruple. Now, we will discuss how we go from the individual quadruple scores to a time interval. As discussed in Sect. 2, our main goal is to predict time interval for a given $\langle s, r, o, ?\rangle$. Given the earliest ($t_{min}^{test}$) and the latest time point (t_{max}^{test}) in the test set, we compute the plausibility score of quadruple with each time point in the interval $[t_{min}^{test}, t_{max}^{test}]$. So the list of test scores are defined as the following:

$$\mathbf{S} = \big[f(s, p, o, t) | t \in [t_{min}^{test}, t_{max}^{test}] \big] \tag{11}$$

We turn these scores into probabilities by using the softmax function.

$$\mathbf{P} = \big[P(t | s, r, o) | t \in [t_{min}^{test}, t_{max}^{test}] \big] \tag{12}$$

where

$$P(t_i | s, r, o) = \frac{exp(f(s, p, o, t_i))}{\sum_{s_j \in \mathbf{S}} exp(s_j)}. \tag{13}$$

Lastly, we use greedy-coalescing algorithm from [6] that takes probabilities $\mathbf{P}$ and outputs k time intervals as our predictions.

4   Experiments

4.1   The Datasets

We perform our experiments on two interval-based TKGs: YAGO11k and Wikidata12k [8]. YAGO11k is created from YAGO3 knowledge graph [22] by the meta-facts in the form of (#factID, occurSince, t_{start}) and (#factID, occurUntil, t_{end}) that are available in some of the facts. Wikidata12k is a subgraph of a preprocessed version of Wikidata [19] that contains facts with temporal annotations (e.g. point-in-time, start time, end time) or properties (e.g. "start time", "inception", "demolition time"). In both datasets, each fact has a time interval attached to it and each entity has at least two edges. Furthermore, ind-YAGO11k datasets and ind-Wikidata12k are inductive splits that we generate from YAGO11k and Wikidata12k. The details of the four datasets are given in Table 1.

Table 1. Dataset statistics.

Dataset	Entity	Relation	Train	Valid	Test
YAGO11k	10,623	10	16,408	2,050	2,051
Wikidata12k	12,554	24	32,497	4,062	4,062
ind-YAGO11k	10,623	10	12,330	3,726	4,453
ind-Wikidata12k	12,554	24	27,330	6,354	6,937

We enhance the datasets with the names and descriptions of entities and relations. For Wikidata12k, the entity names and descriptions are extracted from their corresponding Wikipedia pages. For YAGO11k, the entity and relation names are already available in the dataset. We extract the entity descriptions for YAGO11k from Wikipedia pages as well. For both datasets, the entity descriptions are limited to one sentence. Similar to [6,14], we fix the time granularity as "year" and drop the months and days. For each quadruple in the training set that has closed-interval, i.e., $\langle s, r, o, [t_{start}, t_{end}]\rangle$, we get two training data points $\langle s, r, o, t_{start}\rangle$ and $\langle s, r, o, t_{end}\rangle$. An alternative would be to get all intermediate time points between t_{start} and t_{end}. However, this approach would result in oversampling for relations with long duration [14]. Lastly, if either t_{start} and t_{end} is unknown, we only consider the known time point.

To test TEMT's ability to generalize on unseen entities, we design new splits based on YAGO11k and Wikidata12k and refer to them as ind-YAGO11k and ind-Wikidata12k, respectively. For inductive reasoning, the validation and test sets should have some entities that are not in the training set. We employ the algorithm from [9] to create the new splits. The algorithm samples an entity and removes this entity from the graph $\mathcal{G}$ if this removal does not result in any isolated node or any relation type with less than 100 edges in the graph. The removed entity and its edges are then added either to the validation set and or to the test set. Thus, each triple in the test set has either a new subject or a new object. The test set has 1062 and 1255 unseen entities for YAGO11k and Wikidata12k, respectively. The algorithm works in triple level therefore assumes that the underlying graph is static by ignoring the validity time intervals. Therefore, each split has different triples, not quadruples. As a last step, we attach the corresponding time information to each triple to convert it to a quadruple.

4.2 Evaluation Metrics

For time interval prediction, we use three interval metrics that compare the predicted interval $I_p = [t^p_{start}, t^p_{end}]$ and the ground-truth interval $I_g = [t^g_{start}, t^g_{end}]$. Ideally, I_p is completely the same as I_g or they have some overlap. If there is no overlap, at least I_p and I_g should be close to each other. The first metric gIOU (generalized intersection over union) [30] is defined as follows:

$$gIOU\left(I_p, I_g\right) = IOU\left(I_p, I_g\right) - \frac{|gap(I_p, I_g)|}{|hull(I_p, I_g)|} \qquad (14)$$

where

$$IOU\left(I_p, I_g\right) = \frac{|overlap(I_p, I_g)|}{|I_p| + |I_g| - |overlap(I_p, I_g)|} \tag{15}$$

$|gap(I_p, I_g)|$ is the distance between I_p and I_g in non-overlapping case (otherwise 0) and the hull is the shortest interval that covers both I_p and I_g. The length of an interval is defined as $||t_{start}, t_{end}|| = t_{end} - t_{start} + 1$. In Sect. 7.2, we provide a visual illustration demonstrating the terminology used.

The second metric is aeIOU [14], affinity enhanced intersection over union. It is defined as follows:

$$aeIOU\left(I_p, I_g\right) = \begin{cases} \frac{|overlap(I_p, I_g)|}{|hull(I_p, I_g)|} & |overlap(I_p, I_g)| > 0, \\ \frac{1}{|hull(I_p, I_g)|} & \text{otherwise.} \end{cases} \tag{16}$$

The drawback of aeIOU that it outputs the same scores for both I_p and I_{p^*} if hull(I_p, I_g) = hull(I_{p^*}, I_g), ignoring the fact that one of them can be closer to I_g. In order to address this drawback, the study in [6] introduces a new metric called gaeIOU (generalized aeIOU).

$$gaeIOU\left(I_p, I_g\right) = \begin{cases} \frac{|overlap(I_p, I_g)|}{|hull(I_p, I_g)|} & |overlap(I_p, I_g)| > 0, \\ \frac{|gap(I_p, I_g)|^{-1}}{|hull(I_p, I_g)|} & \text{otherwise.} \end{cases} \tag{17}$$

Using the three metrics, given in Eq. 14, 16 and 17, we report time interval prediction results based on gIOU@k, aeIOU@k, and gaeIOU@k. Given the predicted intervals for a test triple $\langle s, r, o, ? \rangle$ and its ground-truth interval I_g, gIOU@k is defined as the following:

$$gIOU@k = \max_{1 \le i \le k} gIOU(I_{p_i}, I_g) \tag{18}$$

where I_{p_i} denotes i^{th} predicted interval for that particular triple. aeIOU@k and gaeIOU@k are defined analogously. We report the results averaged over all the test triples. We present the variance results for each metric in Sect. 7.3. The range for gIOU is $[-1, 1]$. We follow the same procedure as our baselines and report the scaled gIOU values $((gIOU+1)/2)$. The range for aeIOU and gaeIOU is $[0,1]$.

4.3   Experimental Setup

For all experiments, the dimension d for the triple embedding e_{sro} is 768 and the dimension d' for time point embedding is 64. We set m in Eq. (5) to 64. We use 128 time-corrupted negative samples. A further analysis on the effect of the number of negative samples can be found in Sect. 7.1. The results show that time-corrupted negative sampling strategy is more suitable for our problem. We train our model with Adam optimizer [17] for 50 epochs with a learning rate of 0.001 and margin value $\gamma = 2$. We set the threshold for greedy-coalescing to 0.65. We report the effect of hyperparameters in Sect. 4.6.

Model Variants. We use two different variants of TEMT, namely, TEMT_N and TEMT_{ND}. The variant TEMT_N is trained without the entity descriptions but only with the entity and relation names. In order to reflect this change, we modify Eq. (1) as $S_{sro} = N_s + N_r + N_o$. The variant TEMT_{ND} is trained with both entity descriptions and names. Regarding the sequence length for the language model, we keep the default Sentence-BERT setting which is 128 tokens.

Baselines. In order to compare TEMT variants against the state-of-the-art, we identify four different TKGC methods as baselines: HyTE [8], TNT-ComplEx [18], TIMEPLEX-base [14] and TIME2BOX-TNS [6]. To the best of our knowledge, these are the only methods that perform time interval prediction. TIME-PLEX has two variants: TIMEPLEX-base and TIMEPLEX. Unlike the other baselines, TIMEPLEX relies on temporal constraints to improve its performance. However, TIMEPLEX-base does not follow the same. Here, we report the results of the latter. All of the baselines are transductive and do not use pre-trained language models for learning entity and relation embeddings. To the best of our knowledge, TEMT is the only method that supports inductive reasoning for time interval prediction. Following our baselines, we only report on the test instances that contain known time points which is compatible with our metrics so we do not report results on test quadruples that has unknown start or end time points.

4.4   Transductive Time Interval Prediction

In this experiment, the task is to predict the validity time intervals of facts in TKGs, namely predicting $\langle s, r, o, ? \rangle$. We compare the TEMT variants with the baselines and report the transductive time interval prediction results for YAGO11k and Wikidata12k in Table 2. On the YAGO11k dataset, TEMT outperforms the baselines in all the metrics but aeIOU@10. Notably, in the gIOU@1 metric, TEMT achieves 16 points more than the next best competitor TIME2BOX-TNS. For Wikidata12k, we observe that TEMT variants show improvements in the gIOU@1 metric in comparison with the baselines. For aeIOU@1 and gaeIOU@1, which are more stringent metrics as discussed in Sect. 4.2, TEMT variants are outperformed by the baselines. However, we also observe that there is not a significant performance difference across the datasets unlike our baselines. This may indicate that TEMT is not very sensitive to dataset size since YAGO11k is half the size of Wikidata12k. Moreover, TEMT is competitive with the state-of-the-art on the metrics gIOU@10, aeIOU@10 and gaeIOU@10.

Comparing the variants, we observe that TEMT_{ND} performs better than TEMT_N in most cases. This observation supports the claim that entity descriptions improve the context and, therefore, help to create more meaningful semantic triple embeddings. We also observe that TEMT variants are better at capturing the start and the end years compared to intermediate years, which possibly hurts the time interval prediction performance. The possible explanation is that the text corpora that the language model is trained on generally contain either the starting date or end date. In addition, the textual descriptions of entities may

Table 2. Transductive time interval prediction experiment results on YAGO11k and Wikidata12k datasets. The values are expressed in percentages. Results marked (*) are taken from [6], results marked (†) are reproduced by us, and the others are taken from [14]. "–" denotes unavailable results.

Methods	gIOU@1	aeIOU@1	gaeIOU@1	gIOU@10	aeIOU@10	gaeIOU@10
YAGO11k						
HyTE	15.96	5.41	–	–	–	–
TNT-ComplEx	20.78	8.40	–	–	–	–
TIMEPLEX-base†	23.77	12.62	6.92	48.30	**34.63**	26.63
TEMT_N	**39.85**	13.05	**10.05**	58.78	32.89	29.24
TEMT_{ND}	38.60	**13.48**	9.61	**60.65**	34.33	**30.34**
Wikidata12k						
HyTE	14.55	5.41	–	–	–	–
TNT-ComplEx	36.63	23.35	–	–	–	–
TIMEPLEX-base†	39.44	**26.14**	17.23	69.00	46.82	42.98
TIME2BOX-TNS*	42.30	25.78	**17.41**	**70.16**	**50.04**	**47.54**
TEMT_N	39.35	12.90	8.81	61.68	34.97	30.71
TEMT_{ND}	**43.52**	17.13	12.58	65.84	42.00	38.43

Table 3. Inductive time interval prediction experiment results on ind-YAGO11k and ind-Wikidata12k datasets. The values are expressed in percentages.

Methods	gIOU@1	aeIOU@1	gaeIOU@1	gIOU@10	aeIOU@10	gaeIOU@10
ind-YAGO11k						
TEMT_N	**39.07**	14.23	**10.32**	**61.53**	35.90	32.79
TEMT_{ND}	37.20	**14.81**	10.15	60.07	**36.73**	**33.32**
ind-Wikidata12k						
TEMT_N	**39.78**	12.94	9.18	60.88	34.92	31.07
TEMT_{ND}	38.43	**16.43**	**11.01**	**64.50**	**40.06**	**36.63**

contain temporally irrelevant descriptions. For instance, for time point 2000, we may get an entity description from Wikipedia that is updated in 2020. Note that the results of TIME2BOX-TNS are taken from the paper [6]. We could not reproduce the results for Wikidata12k and test the method on YAGO11k as neither the source code nor the details for pre-processing the datasets is available. In addition, TIME2BOX-TNS does not provide results for the YAGO11k dataset.

4.5   Inductive Time Interval Prediction

In this experiment, we perform inductive time interval prediction on newly generated inductive datasets ind-YAGO11k and ind-Wikidata12k. Since all our baselines only support transductive reasoning, they cannot be compared with

TEMT. Hence, we exclude them from this experiment. The inductive time prediction results are reported in Table 3. ind-YAGO11k and ind-Wikidata12k have 1062 and 1255 unseen entities in test set, respectively. The results show that TEMT's performance on inductive datasets is quite close to the transductive setting (Table 2). This demonstrates the generalization power of TEMT on unseen entities by the usage of pre-trained language models.

Another observation is that we do not see any significant drop in performance although the models are trained on $\sim$ 4,000 fewer training points than YAGO11k and $\sim$ 5,000 fewer training points than Wikidata12k. Moreover, similar to Sect. 4.4, TEMT$_{ND}$ variant performs better in most cases. This indicates that the context that entity descriptions provide helps the model to capture the semantics of a triple better. TEMT is also applicable to a fully inductive setting where there is no overlap between train and test set entities, which we leave as future work.

4.6   Fine-Grained Analysis

Triple Classification. We investigate the representation power of triple embeddings by performing triple classification experiment. The motivation is to make sure that our text space is also meaningful like our time space. With this experiment, we predict whether a triple is correct or not. To this end, we train an MLP classifier [26] with *triple embeddings* (e_{sro}). We first convert the train and test quadruples into triples by removing time intervals. For the training set, we corrupt the subject or object randomly and create one negative example to avoid class imbalance. For the test set, we remove the triples that exist in training or validation set to prevent information leakage. For each test triple, we create one negative example that does not appear in training, validation or test set. We create the sentences by applying Eq. (1) and extract the features using our text encoder. The sizes of the training sets are as follows: 32,690 for YAGO11k, 64,980 for Wikidata12k, 24,558 for ind-YAGO11k, and 54,646 for ind-Wikidata12k. Similarly, the sizes of the test sets are: 4,100 for YAGO11k, 5,530 for Wikidata12k, 8,880 for ind-YAGO11k, and 13,872 for ind-Wikidata12k.

We set L2 regularization term alpha to 0.05 and perform maximum 1000 iterations. We keep the default values of the MLP classifier in [26] for the other settings. TEMT achieves an accuracy of 89.12% on YAGO11k, 91.55% on Wikidata12k, 88.64% on ind-YAGO11k and 89.82% on ind-Wikidata12k respectively. It illustrates the effectiveness of the text encoder thus supporting the claim that the triple embeddings are semantically meaningful and potentially capture structural information.

Time Prediction Diagnosis. Table 4 illustrates some examples for time interval prediction experiment on both YAGO11k and Wikidata12k datasets. "Triple" column represents some triples from the test set and the "Gold answer" column represents their correct validity time interval. The table covers the triples that occurred in different centuries and that have varying durations. The next

Table 4. Example predictions from Yago11k and Wikidata12k. The entity descriptions are not displayed although utilized. (T1: "Kaká member of sports team Hertha BSC", T2: "Ippling located in the administrative territorial entity Bezirk Lothringen", T3: "Paulo Lopes (footballer) plays for S.L. Benfica", T4: "Jeff Morrow is married to Anna Karen Morrow", T5: "Henry Clay is affiliated to Whig Party (United States)").

Triple	Gold answer	1st prediction	2nd prediction	3rd prediction
T1	**[2008, 2012]**	[2008, 2014]	[2011, 2012]	[2004, 2011]
T2	**[1871, 1920]**	[1860, 1920]	[1871, 2018]	[1919, 1920]
T3	**[1997, 2002]**	[1997, 2004]	[1999, 2000]	[2000, 2008]
T4	**[1947, 1993]**	[1956, 1992]	[1959, 1960]	[1960, 2013]
T5	**[1833, 1852]**	[1830, 1862]	[1817, 1847]	[1818, 1829]

Table 5. Results of different hyperparameters on the validation set of Wikidata12k.

d'	gaeIOU@1	learning rate	gaeIOU@1	margin	gaeIOU@1	threshold	gaeIOU@1
32	11.76	0.001	11.86	1	12.26	0.4	10.29
64	11.94	0.002	11.52	2	12.33	0.5	11.56
128	11.77	0.003	11.74	5	11.18	0.65	11.87
256	11.78	0.01	11.36	7	10.97	0.7	11.66

columns report the TEMT_{ND} predictions for the corresponding triple. This experiment shows that TEMT_{ND} is able to predict intervals that are close to the ground-truth. In the first row, TEMT successfully predicts the starting point but output a longer interval than the ground-truth. In the second row, the ending time point is predicted correctly with an earlier starting point from the gold answer. The predictions are usually a subset of the gold interval so it shows that the textual information helps to predict the time period of facts.

Ablation Study. We explore the effect of various hyperparameters on the performance of TEMT with a number of experiments. This also allows us to choose the optimal parameters that are discussed in Sect. 4.3. The results are shown in Table 5 and the setting where d' = 64, learning rate = 0.001, margin = 2, and threshold = 0.65 obtains the best results.

5   Related Work

Static KGC methods [15] can be roughly divided into two: knowledge graph embedding (KGE) methods and text-based methods. KGE methods represent entities and relations with low-dimensional vectors. They can be broadly classified into three different types: translational [4], semantic matching [37], and deep learning methods [10]. A common approach for KGE methods is to learn a function to score the plausibility of a triple. These methods perform well on many downstream tasks such as link prediction. However, they only utilize the structure of a graph and cannot easily be adapted to use additional information such as the textual descriptions of entities and relations. Text-based methods, which we will discuss at the end of this section, utilize textual information available in knowledge graphs to infer missing links between entities.

Although the majority of prior research has focused on static KGs, there has been a growing interest in exploring evolving knowledge graphs [5]. In this section, we focus on interval-based TKG completion methods. A common approach is to incorporate time into the scoring functions of static KGE methods. For instance, HyTE [8] learns to assign a hyperplane for each time point. For each hyperplane, it learns the temporal embeddings of entities and relations using the TransE scoring function [4]. Since the hyperplanes are learned independently, it is not able to model the dependencies between the time points. Moreover, both TNT-ComplEx [18] and TIMEPLEX [14] are based on ComplEx [32] and learn complex-valued embeddings for entity, relation and time points. TNT-ComplEx extends ComplEx by adding a new factor and solve a tensor completion problem. TIMEPLEX adds multiple time-dependent components to the scoring function and also takes into account additional learned features such as temporal constraints. TIME2BOX [6] extends the box embedding idea [29] by time-aware boxes and allows atemporal and temporal facts. Unlike TEMT, these models do not benefit from external information such as textual descriptions of entity and relations. Furthermore, these models are transductive so they cannot predict on unseen entities.

Recent works on text-enhanced static KGs employ pre-trained language models for static KGC [2,9,20,34,38]. The textual descriptions are fed into pre-trained language models (PLMs), that store real-world knowledge in their parameters, to obtain rich contextual entity/relation representations. However, they ignore the dynamics in which the relations between entities hold in a time interval. Only a few studies combines LMs and TKGs. ECOLA [13] jointly optimizes the LM and TKG embedding objectives via combining their loss functions. It retrieves textual information from news articles that correspond to specific dates. Moreover, it does not focus on time interval prediction. By contrast, TEMT leverages entity/relation names and descriptions from Wikipedia pages for time interval prediction. Similar to TEMT, SST-BERT [7] combines the textual information of entities/relations with time to get a plausibility score of a temporal fact. However, it utilizes relation paths with a primary focus on relation prediction whereas TEMT focuses on time interval prediction.

6 Conclusion

We propose TEMT, a model for text-enhanced temporal knowledge graph completion. TEMT outperforms state-of-the-art methods on the YAGO11k dataset and achieves competitive results on the Wikidata12k dataset. To the best of our knowledge, TEMT is the first method that is capable of performing time interval prediction on unseen entities. As a future work, we plan to investigate other pre-trained language models e.g. RoBERTa [21] and time encoding methods. We also plan to incorporate structural information into TEMT's fusing function.

7 Appendix

7.1 Effect of Number of Negative Samples

Table 6. Time prediction performance with respect to the number of negative samples.

	gIOU@1	aeIOU@1	gaeIOU@1	gIOU@10	aeIOU@10	gaeIOU@10
YAGO11k						
#Entity-corrupted						
16	14.67	1.37	0.34	42.82	4.73	2
32	21.29	0.65	0.24	44.72	3.14	1.78
64	22.61	5.35	2.49	39.29	11.55	7.34
128	4.77	0.46	0.1	35.09	1.52	0.68
#Time-corrupted						
16	44.24	11.36	9.05	58.71	32.45	28.38
32	41.17	12.78	9.77	59.54	33.38	29.39
64	39.50	13.14	9.59	60	34.02	30.03
128	38.60	13.48	9.61	60.65	34.33	30.34
ind-YAGO11k						
#Entity-corrupted						
16	26.96	4.8	2.22	40.52	10.81	6.69
32	32.75	1.96	1.08	47.55	6.99	4.73
64	11.99	1.29	0.33	33.01	2.11	1.11
128	3.28	0.22	0	20.72	0.39	0.06
#Time-corrupted						
16	47.47	14.39	12.19	61.64	36.23	33
32	41.89	15.31	11.51	62.08	37.50	34.17
64	41.79	14.12	10.71	63.70	37.61	34.6
128	37.20	14.81	10.15	60.07	36.73	33.32

We conduct an empirical study to see how sampling types discussed in Sect. 3 affect the performance of TEMT_{ND}. We analyze different number of entity-corrupted and time-corrupted negative samples on YAGO11k and ind-YAGO11k datasets. The results are reported in Table 6. We perform the same experiments on Wikidata12k and ind-Wikidata12k as well, however, we do not include their results here, for the sake of brevity.

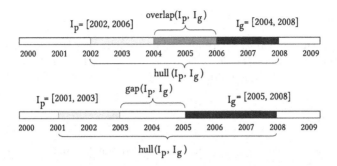

Fig. 3. Time prediction evaluation terms

Table 7. The variance values over the test set

Methods	gIOU@1	aeIOU@1	gaeIOU@1	gIOU@10	aeIOU@10	gaeIOU@10
YAGO11k						
TEMT$_N$	0.0485	0.0286	0.0317	0.0546	0.0894	0.1037
TEMT$_{ND}$	0.0466	0.0246	0.0280	0.0466	0.0832	0.0990
Wikidata12k						
TEMT$_N$	0.0411	0.0203	0.0223	0.0385	0.0670	0.0828
TEMT$_{ND}$	0.0424	0.0281	0.0334	0.0407	0.0716	0.0897
ind-YAGO11k						
TEMT$_N$	0.0478	0.0291	0.0330	0.0522	0.0953	0.1097
TEMT$_{ND}$	0.0499	0.0290	0.0331	0.0636	0.1013	0.1178
ind-Wikidata12k						
TEMT$_N$	0.0411	0.0209	0.0238	0.0453	0.0746	0.0897
TEMT$_{ND}$	0.0482	0.0292	0.0332	0.0450	0.0803	0.0969

The results in Table 6 show that the entity-corrupted negative sampling performs worse than the time-corrupted negative sampling for both datasets. Since the time interval prediction also requires the model to distinguish facts with different time points, this difference is expected. Moreover, in time-corrupted cases, the number of negative samples does not result in marginal changes in gaeIOU@1 metric, which is the most stringent metric.

7.2 Evaluation Terminology

In this section, we illustrate the evaluation terms employed within our interval metrics. In Fig. 3, we demonstrate two scenarios when the predicted interval overlaps with the gold interval (top figure) and when it does not (bottom figure). In the case of the former, given $I_p = [2002, 2006]$ and $I_g = [2004, 2008]$, we get the following: the $hull(I_p, I_g) = [2002, 2008]$, and the $overlap(I_p, I_g) = [2004, 2006]$. In the latter case, given $I_p = [2001, 2003]$ and $I_g = [2005, 2008]$, then the $hull(I_p, I_g) = [2001, 2008]$ and $|gap(I_p, I_g)| = 3$.

7.3 Variance Analysis

Since our experimental results are averaged over the test triples, we report the variance values in Table 7. The results across the datasets and the variants illustrate the effectiveness of our model.

References

1. Agarap, A.F.: Deep learning using rectified linear units (ReLU). ArXiv **abs/1803.08375** (2018)
2. Alam, M.M., et al.: Language model guided knowledge graph embeddings. IEEE Access **10**, 76008–76020 (2022)
3. AlKhamissi, B., Li, M., Celikyilmaz, A., Diab, M.T., Ghazvininejad, M.: A review on language models as knowledge bases. ArXiv **abs/2204.06031** (2022)
4. Bordes, A., Usunier, N., García-Durán, A., Weston, J., Yakhnenko, O.: Translating embeddings for modeling multi-relational data. Neural Inf. Process. Syst. (2013)
5. Cai, B., Xiang, Y., Gao, L., Zhang, H., Li, Y., Li, J.: Temporal knowledge graph completion: a survey. In: International Joint Conference on Artificial Intelligence (2022)
6. Cai, L., Janowicz, K., Yan, B., Zhu, R., Mai, G.: Time in a box: advancing knowledge graph completion with temporal scopes. In: Proceedings of the 11th Knowledge Capture Conference (2021)
7. Chen, Z., Xu, C., Su, F., Huang, Z., Dou, Y.: Incorporating structured sentences with time-enhanced Bert for fully-inductive temporal relation prediction. In: Proceedings of the 46th International ACM SIGIR Conference on Research and Development in Information Retrieval (2023)
8. Dasgupta, S.S., Ray, S.N., Talukdar, P.P.: Hyte: hyperplane-based temporally aware knowledge graph embedding. In: Conference on Empirical Methods in Natural Language Processing (EMNLP) (2018)
9. Daza, D., Cochez, M., Groth, P.T.: Inductive entity representations from text via link prediction. In: Proceedings of the Web Conference 2021 (2020)
10. Dettmers, T., Minervini, P., Stenetorp, P., Riedel, S.: Convolutional 2d knowledge graph embeddings. In: AAAI Conference on Artificial Intelligence (2017)
11. Devlin, J., Chang, M.W., Lee, K., Toutanova, K.: Bert: pre-training of deep bidirectional transformers for language understanding. In: North American Chapter of the Association for Computational Linguistics (2019)
12. Gu, K., Budhkar, A.: A package for learning on tabular and text data with transformers. In: MAIWORKSHOP (2021)
13. Han, Z., et al.: Enhanced temporal knowledge embeddings with contextualized language representations. ArXiv **abs/2203.09590** (2022)
14. Jain, P., Rathi, S., Mausam, Chakrabarti, S.: Temporal knowledge base completion: new algorithms and evaluation protocols. In: Proceedings of the 2020 Conference on Empirical Methods in Natural Language Processing (EMNLP) (2020)
15. Ji, S., Pan, S., Cambria, E., Marttinen, P., Yu, P.S.: A survey on knowledge graphs: representation, acquisition, and applications. IEEE Trans. Neural Netw. Learn. Syst. (2020)
16. Jia, Z., Pramanik, S., Roy, R.S., Weikum, G.: Complex temporal question answering on knowledge graphs. In: Proceedings of the 30th ACM International Conference on Information & Knowledge Management (2021)

17. Kingma, D.P., Ba, J.: Adam: a method for stochastic optimization. In: International Conference on Learning Representations (ICLR) (2015)
18. Lacroix, T., Obozinski, G., Usunier, N.: Tensor decompositions for temporal knowledge base completion. In: International Conference on Learning Representations (ICLR) (2020)
19. Leblay, J., Chekol, M.W.: Deriving validity time in knowledge graph. In: Companion Proceedings of the The Web Conference 2018 (2018)
20. Li, M., Wang, B., Jiang, J.: Siamese pre-trained transformer encoder for knowledge base completion. Neural Process. Lett. (2021)
21. Liu, Y., et al.: Roberta: a robustly optimized bert pretraining approach. ArXiv **abs/1907.11692** (2019)
22. Mahdisoltani, F., Biega, J.A., Suchanek, F.M.: Yago3: a knowledge base from multilingual wikipedias. In: Conference on Innovative Data Systems Research (2015)
23. Mavromatis, C., et al.: Tempoqr: temporal question reasoning over knowledge graphs. In: AAAI Conference on Artificial Intelligence (2021)
24. Nguyen, D.Q., Nguyen, D.Q., Nguyen, T.D., Phung, D.Q.: A convolutional neural network-based model for knowledge base completion and its application to search personalization. Semantic Web (2019)
25. Ostendorff, M., Bourgonje, P., Berger, M., Schneider, J.M., Rehm, G., Gipp, B.: Enriching bert with knowledge graph embeddings for document classification. ArXiv **abs/1909.08402** (2019)
26. Pedregosa, F., et al.: Scikit-learn: machine learning in python. J. Mach. Learn. Res. (2011)
27. Petroni, F., et al.: Language models as knowledge bases? In: Proceedings of the 2019 Conference on Empirical Methods in Natural Language Processing and the 9th International Joint Conference on Natural Language Processing (EMNLP-IJCNLP) (2019)
28. Reimers, N., Gurevych, I.: Sentence-bert: sentence embeddings using Siamese Bert-networks. In: Conference on Empirical Methods in Natural Language Processing (EMNLP) (2019)
29. Ren, H., Hu, W., Leskovec, J.: Query2box: reasoning over knowledge graphs in vector space using box embeddings. In: International Conference on Learning Representations (2020)
30. Rezatofighi, S.H., Tsoi, N., Gwak, J., Sadeghian, A., Reid, I.D., Savarese, S.: Generalized intersection over union: a metric and a loss for bounding box regression. In: 2019 IEEE/CVF Conference on Computer Vision and Pattern Recognition (CVPR) (2019)
31. Rogers, A., Kovaleva, O., Rumshisky, A.: A primer in bertology: what we know about how bert works. Trans. Assoc. Comput. Linguist. (2020)
32. Trouillon, T., Welbl, J., Riedel, S., Gaussier, É., Bouchard, G.: Complex embeddings for simple link prediction. In: International Conference on Machine Learning (2016)
33. Vaswani, A., et al.: Attention is all you need. In: Neural Information Processing Systems (2017)
34. Wang, L., Zhao, W., Wei, Z., Liu, J.: Simkgc: simple contrastive knowledge graph completion with pre-trained language models. In: Proceedings of the 60th Annual Meeting of the Association for Computational Linguistics (2022)
35. Weikum, G., Dong, L., Razniewski, S., Suchanek, F.M.: Machine knowledge: creation and curation of comprehensive knowledge bases. Found. Trends Databases (2020)

36. Xie, R., Liu, Z., Jia, J., Luan, H., Sun, M.: Representation learning of knowledge graphs with entity descriptions. In: AAAI Conference on Artificial Intelligence (2016)
37. Yang, B., Yih, W., He, X., Gao, J., Deng, L.: Embedding entities and relations for learning and inference in knowledge bases. In: International Conference on Learning Representations (ICLR) (2015)
38. Yao, L., Mao, C., Luo, Y.: KG-BERT: bert for knowledge graph completion. ArXiv **abs/1909.03193** (2019)
39. Zhang, X., et al.: Temporal context-aware representation learning for question routing. In: Proceedings of the 13th International Conference on Web Search and Data Mining (2020)

A Language Model Based Framework for New Concept Placement in Ontologies

Hang Dong[1,2(✉)] , Jiaoyan Chen[1,3] , Yuan He[1] , Yongsheng Gao[4] ,
and Ian Horrocks[1]

[1] University of Oxford, Oxford, UK
{yuan.he,ian.horrocks}@cs.ox.ac.uk
[2] University of Exeter, Exeter, UK
h.dong2@exeter.ac.uk
[3] University of Manchester, Manchester, UK
jiaoyan.chen@manchester.ac.uk
[4] SNOMED International, London, UK
yga@snomed.org

Abstract. We investigate the task of inserting new concepts extracted from texts into an ontology using language models. We explore an approach with three steps: *edge search* which is to find a set of candidate locations to insert (i.e., subsumptions between concepts), *edge formation and enrichment* which leverages the ontological structure to produce and enhance the edge candidates, and *edge selection* which eventually locates the edge to be placed into. In all steps, we propose to leverage neural methods, where we apply embedding-based methods and contrastive learning with Pre-trained Language Models (PLMs) such as BERT for edge search, and adapt a BERT fine-tuning-based multi-label Edge-Cross-encoder, and Large Language Models (LLMs) such as GPT series, FLAN-T5, and Llama 2, for edge selection. We evaluate the methods on recent datasets created using the SNOMED CT ontology and the MedMentions entity linking benchmark. The best settings in our framework use fine-tuned PLM for search and a multi-label Cross-encoder for selection. Zero-shot prompting of LLMs is still not adequate for the task, and we propose explainable instruction tuning of LLMs for improved performance. Our study shows the advantages of PLMs and highlights the encouraging performance of LLMs that motivates future studies.

Keywords: Ontology Enrichment · Concept Placement · Pre-trained Language Models · Large Language Models · SNOMED CT

1 Introduction

New concepts appear as they are discovered in the real world, for example, new diseases, species, events, etc. Ontologies are inherently incomplete and require evolution by enriching with new concepts. A main source for concepts is corpora, e.g., new publications that contain mentions of concepts not in an ontology.

A. Meroño Peñuela et al. (Eds.): ESWC 2024, LNCS 14664, pp. 79–99, 2024.
https://doi.org/10.1007/978-3-031-60626-7_5

In this work, we focus on the problem of placing a new concept into an ontology by inserting it into an edge which corresponds to a subsumption relationship between two atomic concepts, or between one atomic concept and one complex concept constructed with logical operators like existential restriction ($\exists r.C$). Distinct from previous work in taxonomy completion (e.g., [30,33,34]), the task allows natural language contexts together with the mention as an input and also considers logically complex concepts by Web Ontology Language (OWL). Distinct from previous work in new entity discovery (e.g., [8]), the task places the new entity into the ontology, a step further to their discovery from the texts. The task is more challenging than entity linking from a mention to a concept, considering that there are many more edges than already the large number of concepts and axioms in an ontology (of a form much more complex than a tree), even by limiting the edges to only those having one-hop or two-hop.

Recently, machine learning, neural network based methods, and especially pre-trained language models (PLM), have been applied to ontology engineering tasks. For new entity discovery tasks, typically, the entity linking or retrieval tasks comprise two steps, the first is to search relevant entities by narrowing down the candidates, and the second is to select the correct one. Previous studies on entity linking and new entity discovery mostly use BERT-based fine-tuning methods [8,31]. We differ Large Language Models (LLMs) from PLMs by their vast difference in scale and language generation capabilities. There is a recent growth of studies using LLMs, e.g., for entity linking [29] and ontology matching [14], but the experimental results are yet to be confirmed and their advantages and drawbacks for concept placement are not clear. A more detailed investigation is needed to compare the methods for the representation, and a framework is needed for their comparison. In the texts below, we use LMs as a general term for both PLMs and LLMs and use more specific terms where necessary.

For concept placement, we propose a framework that extends the two-step process, with another edge enrichment step. After the edge search to narrow the edge candidates to a limited number, we enrich the edges by walking in the ontological graph by extending the parents and children to another layer. Then this enriched set of edges is re-ranked through the edge selection part (which can be modelled as a multi-label classification task). Using this framework, we are able to compare different data representation methods, including traditional inverted index, fixed embedding based similarity, contrastive learning based PLM fine-tuning, and instruction-tuning and prompting of LLMs.

The evaluation is based on the recent datasets in [7], created by using an ontology versioning strategy (i.e., comparing two versions of an evolving ontology) to synthesise new concepts and their gold edges to be placed w.r.t. the older version of the ontology. The ontology is SNOMED CT, under *Disease* and *CPP* (Clinical Finding, Procedure, and Pharmaceutical/biologic products) branches.

Results indicate that edge enrichment by leveraging the structure of ontology greatly improves the performance of new concept placement. Also, among the data representation methods, contrastive learning based PLM fine-tuning generally performed the best in all settings. The inadequate yet encouraging results of LLMs under our experimental setting may be related to the input length restriction and the inherent knowledge deficiencies of LLMs for nuanced concept relations of domain specific ontologies. Instruction-tuning, especially with automated explainable prompts, improves over the zero-shot prompting (i.e., no further instruction-tuning) of LLMs. Our results suggest the potential of LLMs and motivate future studies to leverage them for ontology concept placement.[1]

2   Related Work

2.1   Ontology Concept Placement

Ontology concept placement is a key task in ontology engineering and evolution. It aims to automatically place or insert a new concept, in its natural language form and potentially with contexts in a corpora, to an existing ontology. This automated task helps to reduce the immense initial human effort to discover and insert new concepts, as humans may not be able to review all available new information at the rate when they are available, and the manual process while of high quality, is of high cost and low efficiency [3,11].

The recent study in [7] summarised the related available datasets on ontology concept placement. Datasets for the relevant tasks include taxonomy completion, ontology extension, post-coordination, and new mention and entity discovery. The proposed new datasets in [7] supports a more comprehensive set of characteristics, including NIL entity discovery, contextual mentions, concept placement (under both atomic and complex concepts in ontologies). We extend the datasets in the work [7] and use them for benchmarking in this paper.

Another relevant task is entity linking, which links a textual mention to its concept in a Knowledge Base (KB) or an ontology [24]. Entity linking can be extended to the case for out-of-KB mentions [8]. Ontology Concept Placement is distinct from entity linking to a concept, which alternatively links an out-of-KB mention to an edge (of subsumption relations) in the structure of an ontology.

2.2   Pre-trained Language Models for Ontology Concept Placement

We consider pre-trained language model as a neural, Transformer model [27] that can be pre-trained using corpora using masked modelling or by predicting future tokens, processing very large amounts of text [18]. A Large Language Model is a scaled PLM to a vastly higher degree which can result in improved performance and emergent capabilities [35].

A relevant line of work to ontology concept placement is Knowledge Graph Construction, where BERT is evaluated and shows promise to enhance several relations in WikiData [28]. Other studies focus on formal KBs which are usually expressed as OWL Ontology, e.g., by predicting the subsumption relations [4].

---

[1] Our implementation of the methods and experiments are available at https://github.com/KRR-Oxford/LM-ontology-concept-placement.

The work [17] predicts a wider range of inter-ontology relations (e.g., equivalence, subsumption, meronymy, etc.) using PLMs (e.g., DistillBERT, RoBERTa, etc.).

For ontology concept placement, PLMs have also been applied. The study [21] aims to place concepts to SNOMED CT by pre-training and fine-tuning BERT for subsumption prediction. The study [23] uses a similar BERT-based Bi-encoder architecture and experiments with more medical ontologies. However, both works always place a concept as a leaf node, instead of higher levels.

Another approach utilising LLMs is through a prompting-based approach. The idea is to formulate an ontology-related task using natural language input that leverages the generative capability of a language model. While the recent study [29] explored prompting-based approaches of LLMs for concept equivalence linking, few studies have explored them for ontology concept placement.

In this paper, we propose an LM-based framework that leverages embedding, fine-tuning, prompting, and instruction-tuning of PLMs and LLMs for ontology concept placement. The task also considers contexts in a mention and the logically complex concepts in ontologies that are not considered in previous work.

3   Problem Statement

We use the definition of an OWL ontology, a Description Logic KB that contains a set of axioms [2, 12]. We focus on the TBox (terminology) part of an ontology, containing General Concept Inclusion axioms, each as $A \sqsubseteq B$, where A (and B) are atomic or complex concepts [1]. *Complex concepts* mean concepts that involve at least one logical operator, e.g., negation ($\neg$), conjunction ($\sqcap$), disjunction ($\sqcup$), existential restriction ($\exists r.C$), universal restriction ($\forall r.C$), etc. [1].

An ontology $\mathcal{O}$ can be more simply defined as a set of concepts D (possibly complex) and directed edges E. A directed edge contains a direct parent and a direct child, where the parent or child can be complex concepts.[2]

Formally, the task is to place a new concept mention m (with surrounding contexts in a corpus) into edges in an ontology $\mathcal{O}$ so that $C \sqsubseteq m \sqsubseteq P$ for an edge $< P, C >$ (or as $P \rightarrow C$) that contains a parent concept P and a child concept C. The child concept C can be NULL when the mention is to be placed as a leaf node. Using SNOMED CT (version 1703) as an example, a mention "Psoriatic arthritis" (in a scientific paper) is to be placed as Psoriatic arthritis with distal interphalangeal joint involvement $\sqsubseteq$ Psoriatic arthritis $\sqsubseteq$ Psoriasis with arthropathy; and a mention "Neurocognitive Impairment" is to be placed as a leaf concept (so C is NULL), and the axioms include Neurocognitive Impairment $\sqsubseteq$ Cognitive disorder and Neurocognitive Impairment $\sqsubseteq \exists$ RoleGroup.($\exists$DueTo.Disease)[3].

Ontology concept placement can thus be considered matching from a textual mention (possibly surrounded by a context window) to edges in the structure of

---

[2] We focus on the common case that only the parent can be a complex concept, as in the explicit axioms in the SNOMED CT ontology.

[3] This means that Neurocognitive Impairment belongs to the role group [25] or a grouping of the characteristic that is caused by ("due to") a disease.

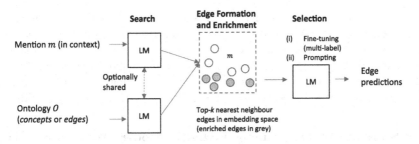

Fig. 1. An overall three-step framework for ontology concept placement with LMs.

an ontology. Given that a concept may have more than one parent and more than one child, it can be placed into many edges. Thus ontology concept placement can be formulated as a *multi-label learning* problem [10]. The task is to learn a mapping function f that can map the input (a textual mention possibly with contexts) to a set of labels (here as edges in E). Typically, a multi-label learning process can create a label *ranking* based on a metric score that orders the whole set of labels [10] or an ordered set without an explicit metric (e.g., by the order of text generation). This is distinct from the entity linking task which usually maps the input to only a single label (as an entity or a NIL entity) [8].

4   Methodology

Extending the general ideas in information retrieval and entity linking, we propose a three-step framework for ontology concept placement, as shown in Fig. 1 below. Usually, retrieving a set of correct items (e.g., edges) needs two steps, *search* (or candidate generation) and *selection* (or candidate ranking). The search step aims to find a set of seed concepts (to form edges) or a set of seed edges directly. The selection step finds (and also ranks, as in multi-label classification) the correct edges among the candidates. Considering the structural nature of the edge generation process, we add another step in between, *edge formation and enrichment*, which forms seed edges from a seed concept (optionally) and enriches seed edges to derive the full candidate edges. We employ LMs in both the search and the selection steps, and further leverage the ontological structure for the edge formation and enrichment step.

4.1   Edge Search: Searching Seed Concepts or Edges

The search step inputs a textual mention m (with a context window) and an ontology $\mathcal{O}$, both represented using LMs. For concept search, we encode a mention and the label of a concept using an LM with fixed parameters (or as two same LMs sharing parameters). For edge search, we encode a mention and an edge using two LMs with fine-tuning to align them into the same embedding space, given the distinct types of texts (in corpora and in ontologies) between them.

Concept Search with Fixed Embeddings. We search concepts by using the nearest neighbours of LM-based embeddings, i.e., ranking using the cosine similarity of the mention embedding and every concept embedding in the ontology. A domain-specific ontology-pre-trained BERT, SapBERT [20], is used to represent both a mention and a concept. Complex concepts, with logical operators, can be verbalised using a rule-based verbaliser (e.g., in [16]), before their embedding.

Edge Search with Fine-Tuning Edge-Bi-encoder. We use two LMs to encode the mention and the edge separately, using the representation of the [CLS] token in the last layer, adapting the Bi-encoder architecture [8,31]. A mention is represented as [CLS] ctxt$_l$ [M$_s$] mention [M$_e$] ctxt$_r$ [SEP], where ctxt$_l$ and ctxt$_r$ are the left and right contexts of the mention in the document, resp., and [M$_s$], [M$_e$] are the special tokens placed before and after the mention. In the setting without contexts, we set both ctxt$_l$ and ctxt$_r$ as empty strings. A directed edge (having a direct parent and a direct child) is represented as "[CLS] parent tokens [P-TAG] child tokens [C-TAG] [SEP]". We use a special token [NULL] to represent the child tokens of a leaf concept in the ontology.

The training follows a contrastive loss, more specifically, a max-margin triplet loss [22] described below, where α is a margin of small value (e.g., 0.2) and $[x]_+$ denotes $\max(x, 0)$, for each mention to its gold edge (the i-th) in a batch, $s(m, e)$ is the mention-edge similarity, calculated as the dot-product of the mention embedding and the concept embedding. The idea is to make each mention close to one of its edges in the embedding space, but far away from the other edges within the same batch. We use in-KB data for training and validation to form a model and then finally validate and test on out-of-KB data.

$$L_{m_i, e_i} = \sum_{j \neq i} [\alpha - s(m_i, e_i) + s(m_i, e_j)]_+; \quad s(m, e) = v_m \cdot v_e \qquad (1)$$

4.2 Edge Formation and Enrichment

The idea of edge formation and enrichment is to leverage the ontological structure together with the LM-based embedding for candidate retrieval. The detailed process with examples is presented in Fig. 2.

Edge Formation from Seed Concepts. When concept candidates are selected from entities, for each concept A, we traverse the ontology by one hop to find the parents $P_1, ..., P_n$ and children $C_1, ..., C_n$ of the concept, and then using the set $S = \bigcup_i \{P_i \rightarrow A\} \cup \bigcup_j \{A \rightarrow C_j\} \cup \bigcup_i \bigcup_j \{P_i \rightarrow C_j\}$ as the candidate edge set, which includes all one-hop edges containing A and all two-hop edges which traverse through A (see an example in the left part of Fig. 2). We further added leaf edges, $A \rightarrow$ NULL, to S.

Edge Ranking after Edge Formation. Then the edge set is ranked using the LM-based embedding w.r.t. the mention m, as the average cosine similarity of

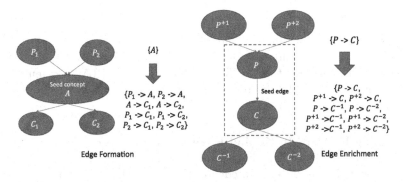

Fig. 2. An example of the edge formation and enrichment process using ontology structure. Edge formation transforms a seed concept into a set of edges, while edge enrichment augments the set of edges one by one. For methods that directly search edges (e.g., Edge-Bi-encoder), no edge formation is needed and only enrichment is applied.

m to the parent and m to the child in the embedding space (see Eq. 2 below). For the edge score of leaf edges (where $C =$ NULL), we first set a rule to deduce whether the mention is to be placed on a leaf edge by checking if the top ranked seed concept is a leaf concept, if so, we prioritise all enriched leaf edges of the mention with the highest edge score (i.e., better ranked than non-leaf edges).

$$\text{Edge\_score}_{fixed}(m, < P, C >) = \frac{\text{sim}(m, P) + \text{sim}(m, C)}{2}, \text{where } C \neq \text{NULL} \quad (2)$$

Edge Enrichment from Seed Edges. We further enrich the edges by traversing one-hop upper for parents and one-hop lower for children in the ontology. For each edge $P \rightarrow C$, we thus first find their one-hop upper parents $P^{+1},...,P^{+i}$ and one-hop lower children $C^{-1},...,C^{-j}$, and enrich the set to $\{P \rightarrow C\} \cup \bigcup_i \{P^{+i} \rightarrow C\} \cup \bigcup_j \{P \rightarrow C^{-j}\} \cup \bigcup_i \bigcup_j \{P^{+i} \rightarrow C^{-j}\}$. We combine the enriched edges from all seed edges and then remove the duplicated edges (given that some of the enriched edges can be the same for different but similar seed edges). This covers more related edges based on the ontological structure and LM-based similarity and can greatly improve the recall of the edge retrieval. We also enrich a "leaf" edge, i.e., $P \rightarrow$ NULL, when a parent P in a non-leaf edge is predicted.

Edge Ranking after Edge Enrichment. The enriched edges are then ranked with scores from different edge search methods. For the fixed embedding approach, edges are ranked based on the edge score in Eq. 2. For the fine-tuned embedding (Edge-Bi-encoder) approach, edges are ranked using the dot product scores, $s(m, e)$ (see Eq. 3 below and the right part of Eq. 1) for all edges (including both leaf and non-leaf edges) after the enrichment for the fine-tuned embedding (Edge-Bi-encoder) approach. The top-k candidate edges are then retrieved from the seed edges after this process.

$$\text{Edge\_score}_{fine-tuned}(m, < P, C >) = s(m, < P, C >) \quad (3)$$

4.3   Edge Selection

The edge selection step aims to find the correct edges to place the concept mention from the k candidate edges. We utilise LMs based on their distinct architectures, i.e., we fine-tune BERT-like, encoder-only PLMs for multi-label classification, and prompts and instruction-tunes LLMs, which have a decoder, for result generation.

Fine-Tuning PLMs: Multi-label Edge-Cross-encoder. We adapt an LM-based cross-encoder in [8,31], that encodes the interaction between sub-tokens in the contextual mention and an edge in the top-k edges, for multi-label classification. Specifically, for each of the k candidate edges the input is a concatenation of the contextual mention with the edge, i.e., [CLS] ctxt$_l$ [M$_s$] mention [M$_e$] ctxt$_r$ [SEP] parent tokens [P-TAG] child tokens [C-TAG] [SEP], and the output is a multi-label classification over all the inputs, i.e., the selection of the correct edges from the candidates. We use a special token [NULL] to represent the child tokens of a leaf edge. Each input is encoded with a BERT model into a vector v_{cross} (we use the representation of [CLS] in the last layer).

Therefore, the loss is a binary cross-entropy loss after a sigmoid activation of the score, linearly transformed from the representation vector, $s_{m,e}^{(cross)} = v_{cross}w$, of each input. All the inputs share the same BERT model for fine-tuning.

Zero-Shot Prompting LLMs. Alternatively, a recent paradigm is to prompt LLMs to generate answers directly. We formulate a prompt to allow LLMs to generate the indices of the options. The prompt provides contexts and all necessary information including the top-k candidate edge options to allow the LLMs to be conditioned and generate the answer. The prompt is structured as below, which contains an input (including task description, the mention in context, and the options of k edges) and a response headline. The sequence which is underlined (after ### Response) is expected to be generated by the LLM.

Input:
Can you identify the correct ontological edges for the given mention (marked with *) based on the context? The ontological edge consists of a pair where the left concept represents the parent of the mention, and the right concept represents the child of the mention. If the mention is a leaf node, the right side of the edges will be NULL. If the context is not relevant to the options, make your decision solely based on the mention itself. There may be multiple correct options. Please answer briefly using option numbers, separated by commas. If none of the options is correct, please answer None.

mention in context:
Our aim was to verify the occurrence of selected mutations of the EZH2 and ZFX genes in an Italian cohort of 23 sporadic *parathyroid carcinomas*, 12 atypical and 45 typical adenomas.

options:
0.primary malignant neoplasm → parathyroid carcinoma
1.malignant neoplastic disease → malignant tumor of parathyroid gland
2.malignant neoplastic disease → primary malignant neoplasm of parathyroid gland
...
8.primary malignant neoplasm of parathyroid gland → NULL

...
(till all the k candidates are listed)
Response:
2,8

Explainable Instruction-Tuning LLMs. We can observe that it would not be straightforward for an LLM to directly figure out the edges and output the option numbers (e.g., 2,8 in the example above). Fine-tuning with in-KB training data would be needed. To bridge the reasoning gap between the input and the response, we propose to add an explanation section that describes the reasoning steps in a narrative form.

Thus, we automatically synthesise explanations by steps to solve the new concept placement problem: (i) List all possible parents in the candidates; (ii) Find correct parents; (iii) Narrow the list of children based on the correct parents; (iv) Find correct children; (v) List the final answer based on the correct children. The explanation Expl_texts is a function of the k candidate edges and the gold edges of a mention, i.e., Expl_texts = Template(E_{cand}, E_{gold}). The template is below, where elements in the lists (in square brackets) are separated by comma.

Explanation:
From the parents in the options above, including [all candidate parents], the correct parents of the mention, [mention name], include [correct gold parents]. Thus the options are narrowed down to [option numbers having correct gold parents]. From the children in the narrowed options, including [children in the filtered options], the correct children of the mention, [mention name], include [correct gold children in the filtered options]. Thus, the final answers are [correct option numbers].

We place the explanation section (### Explanation) before the response section (### Response). During training, the whole explanation is fed into the LLM to allow it to be conditioned to generate the response. During inference, the instruction-tuned LLM is expected to generate an explanation of the same template structure, after the explanation section mark (### Explanation), with a response (as a part of the explanation and also in the response section).

An issue with current LLMs is the limited text window it can support. This long context issue however will be addressed with future LLMs. At this stage, we test the framework with an openly available LLM, Llama 2, which supports 4,096 tokens as input, sufficient for a low or medium top-k setting as 10 or 50.

5   Experiments

5.1   Data Construction

We adapt datasets MM-S14-Disease and MM-S14-CPP from the work in [7] for new concept placement in ontologies[4]. The datasets are constructed by using two versions of SNOMED CT (2014.09 and 2017.03) with a text corpus where mentions are linked to UMLS. Then mapping between UMLS and SNOMED CT is also available in the UMLS. New mentions are therefore synthesised by considering the gap between the two versions of SNOMED CT. The edges to be inserted into the ontology for each new mention are also created, by finding the nearest parents and children for the new mention in the old version of SNOMED CT. The statistics of the dataset are displayed in Table 1.

---

[4] https://zenodo.org/records/10432003.

Table 1. Statistics for datasets for Concept Placement, for SNOMED CT (ver 20140901, "S14") under different categories: "Disease" and "CPP", i.e., $\underline{C}$linical finding, $\underline{P}$rocedure, and $\underline{P}$harmaceutical/biologic product. A mention-edge pair or link (in L) denotes a mention (in M) and one of its directed edges in the KB. The mention-edge pair is complex (i.e. L_{comp}) when the edge involves a complex concept. Mentions are from the MedMentions dataset ("MM"). The numbers of edges are those having one hop (including leaf nodes to NULL) and two hops from any paths in the ontology. (Table adapted from the study [7].)

		MM-S14-Disease	MM-S14-CPP
Ontology: # all (# complex)	concepts	64,900 (824)	175,895 (2,718)
	edges	237,826 (4,997)	625,994 (19,401)
Corpus: # M/# L/# L_{comp}	train, in-KB	11,812/887,840/917	34,704/1,398,111/9,475
	valid, in-KB	4,248/383,457/203	11,707/548,295/4,305
	valid, out-of-KB	329/672/10	568/979/13
	test, out-of-KB	276/965/3	432/1,152/9

The number of edges (one-hop including leaf nodes and two-hop) is numerous, over 3.5 times of the number of concepts. This makes the task of placement into edges less tractable than entity linking into a concept for a mention.

We consider the unsupervised setting of concept placement common to the real-world scenario, which means that no mention-edge pairs for out-of-KB concepts are available for the training. This can, however, be approached using in-KB self-supervised data creation: we can see from Table 1 that it is possible to generate edges for in-KB concepts; this is simply by looking at the directed parents and children of a concept in the current ontology (i.e., the older version of SNOMED CT). Thus, we use in-KB data for training and validation, and then use out-of-KB data solely for external validation and testing.

5.2 Metrics

We present new metrics for new concept placement, as insertion rate for any edges (InR_{any}) and for all edges (InR_{all}) predicted for mentions. Here "any" means that one of the gold edges is predicted for a mention, whereas "all" means that all of the gold edges are predicted. The metrics can be defined as below in Eq. 4, where the value of $\mathbb{1}(x)$ is 1 where the statement x is true, otherwise 0, and Z_i and Y_i are the set of predicted edges and gold labels (or edges), resp. Also, we use the insertion rates at k (i.e., $InR_{any}@k$ and $InR_{all}@k$) to denote the performance after predicting the top-k edges, this measures whether the "positive" edges are ranked before the "negative" ones. We select k as 1, 5, and 10, considering that terminologists can select from a few edges (as few as 10 or less) suggested by a system for updating an ontology.

$$InR_{any} = \frac{1}{|M|} \sum_{m_i \in M} \mathbb{1}(Z_i \cap Y_i \neq \emptyset); \quad InR_{all} = \frac{1}{|M|} \sum_{m_i \in M} \mathbb{1}(Z_i \supseteq Y_i) \quad (4)$$

The proposed metrics can be considered a loose version ("any") and a strict version ("all") of the example-based metrics [10] for multi-label learning. The standard multi-label learning requires a complete set of gold labels, while ontologies that follow the open-world assumption are inherently incomplete (i.e., edges which are not in the gold standard may also be correct), thus the ranking-based metrics, $InR_{any}@k$ and $InR_{all}@k$, are more appropriate.

The insertion rate metrics can be used to evaluate both edge candidates and final edge selection. We also separately evaluate the insertion rate metrics for leaf edges (where the child edge is NULL) and non-leaf edges.

5.3 Experimental Settings and Baseline Methods

We select two representative top-k values after the edge enrichment step, $k = 10$ and $k = 50$, enriched from $\frac{k}{2}$ edges, or 5 and 25 seed edges resp., after an initial investigation of a range of k values[5]. Then for each of the top-k settings, the models select the final set of top 1, 5, and 10 edges after the edge selection step.

For edge search, the baselines include an inverted index based approach, fixed BERT embeddings, and fine-tuned BERT embeddings with contrastive learning (as Edge-Bi-encoder). For all methods, the sub-token length of contexts and concepts are 32 and 128 resp. We choose SapBERT [20] as the BERT model in edge search (fixed and fine-tuned embeddings). For the inverted index based approach, we create an inverted index from all SNOMED CT concepts, where a key is a sub-token from a concept and a corresponding value is all the concepts, and we use the index of sub-tokens created using the SentencePiece tokenizer (also used by FLAN-T5) [19]. The similarity score based on the inverted index between a mention and a concept is then calculated as the sum of inverse document frequency scores $(\text{sim}_{idf}(m, C, \mathcal{T}, I, |D|) = \sum_{t \in \mathcal{T}(C) \cap \mathcal{T}(m)} \log \frac{|D|}{|I[t]|})$ of all the common sub-tokens t that appear in both the mention m and a concept C in the set of all concepts D, and $\mathcal{T}$ is the tokenizer and I is the index from a sub-token t to the list of concepts. Then the edge score w.r.t. a mention is calculated similarly to Eq. 2, as the average of mention-parent similarity and mention-child similarity score using the inverted index. For inverted index and fixed embedding, we use the mention only without contexts, considering that methods do not learn the relation between the concept and the natural language context; for Edge-Bi-encoder we explored mentions with or without contexts.

We apply all baseline methods with the steps of edge formation and enrichment. Then, for edge selection, we choose PubMedBERT [13] as the model for fine-tuning cross-encoder-based method; we also choose GPT-3.5 ("gpt-3.5-turbo-0613")[6], and Llama 2 [26] for the zero-shot prompting of LLMs, both models allowing 4,096 sub-tokens as input. FLAN-T5 [5] has a limited input token length of 512, below the token usage of our prompts with the top-50 settings (between 1,556 and 3,014 sub-tokens for the datasets for top-50), thus we only use it for the top-10 setting. The model GPT-4 has a much higher cost, 30 folds of the price compared to GPT-3.5, and is slower and less stable in querying, and GPT-4 is also under updating, thus we only report results for GPT-3.5.

---

[5] We also investigated k up to 300, while the insertion rate at k improves, the overall results after edge selection are worse than smaller k values as 10 and 50. A larger k also leads to a substantially longer running time for edge enrichment and selection.

[6] https://platform.openai.com/docs/models/gpt-3-5.

Table 2. Results on edge search, formation and enrichment for MM-S14-Disease and MM-S14-CPP datasets. Each setting has validation and testing results, separated by a slash (/) sign. "*lf*" and "*nlf*" mean *leaf* and *non-leaf*, resp.

MM-S14-Disease	k	InR_{any}	InR_{all}	InR_{any}, *lf*	InR_{all}, *lf*	InR_{any}, *nlf*	InR_{all}, *nlf*
Inverted Index	10	10.0/12.0	9.1/10.1	9.5/14.2	9.2/11.9	14.3/3.4	8.6/3.4
	50	41.3/40.6	37.7/**38.8**	44.6/**50.0**	**41.2/48.2**	14.3/5.2	8.6/3.4
Fixed embs	10	16.1/13.0	7.0/12.3	18.0/16.1	7.8/16.0	0.0/1.7	0.0/0.0
	50	35.3/31.9	28.3/30.8	38.4/38.1	30.6/37.6	8.6/8.6	8.6/5.2
Fine-tuned embs	10	31.9/25.7	14.6/8.0	28.9/12.4	14.6/8.7	57.1/75.9	14.3/5.2
(Edge-Bi-enc)	50	**57.8/50.0**	40.1/38.0	**55.4/38.1**	38.4/33.5	**77.1/94.8**	**54.3/55.2**
MM-S14-CPP	k	InR_{any}	InR_{all}	InR_{any}, *lf*	InR_{all}, *lf*	InR_{any}, *nlf*	InR_{all}, *nlf*
Inverted Index	10	5.5/5.8	5.1/5.3	5.3/5.7	5.1/5.7	6.9/6.3	5.2/3.1
	50	23.1/23.4	21.0/22.5	24.9/26.9	22.8/25.8	6.9/3.1	5.2/3.1
Fixed embs	10	11.3/8.3	8.3/7.4	12.4/9.2	9.2/8.7	1.7/3.1	0.0/0.0
	50	28.4/26.9	25.9/25.2	30.4/30.2	28.8/29.6	10.3/7.8	0.0/0.0
Fine-tuned embs	10	32.0/27.8	19.7/14.4	31.2/19.3	21.4/16.0	39.7/76.6	5.2/4.7
(Edge-Bi-enc)	50	**50.9/48.4**	**36.8/34.5**	**50.4/42.4**	**38.8/38.0**	**55.2/82.8**	**19.0/14.1**

For LLM instruction-tuning, we use the Supervised Fine-tuning (SFT)[7] with 4-bit quantisation to fine-tune the Llama-2 model; the efficient instruction-tuning uses QLoRA, quantisation with Low Rank Adapters (LoRA) [6].

For all supervised models (fine-tuning and instruction-tuning), we use in-KB data for training. The best models were selected by using the validation set of the in-KB data. We then report results on the validation and the test sets for the out-of-KB data. Note that the out-of-KB validation set is not used for parameter tuning and is independent of model development.[8]

5.4   Results

We report results on the first two steps to determine the best edge search methods, followed by the overall results of the full framework, with edge selection. The metric results in all Tables are presented as percentage scores.

Results on Edge Search, Formation and Enrichment. Results are presented in Table 2. The "all" metrics are generally lower than the "any" metrics (also for results in the other tables) as the full completion for concept placement is more challenging than the placement into any correct edges.

It can be observed that the fine-tuned Edge-Bi-encoder achieves the best overall results under the settings. The inverted index approach has a higher coverage of leaf edges for Diseases (but not for the broader categories of CPP) - this may be because the parent disease names are likely to be lexically similar to the new mention, while for non-leaf edges, fixed and fine-tuned embedding-based methods achieve higher performance; also, the inverted index and fixed embeddings tend to prioritise leaf edges, based on the rule by checking whether the top seed concept is a leaf concept.

---

[7] https://huggingface.co/docs/trl/sft_trainer.

[8] More details on experimental settings and time usage are in Appendix 1.

Table 3. Overall results after edge selection for MM-S14-Disease and MM-S14-CPP datasets. Each setting has validation and testing results, separated by a slash (/) sign.

MM-S14-Disease	k	InR_{any}@1	InR_{all}@1	InR_{any}@5	InR_{all}@5	InR_{any}@10	InR_{all}@10
Inverted Index	10	0.6/0.0	0.0/0.0	1.8/2.5	0.9/0.7	10.0/12.0	9.1/10.1
	50	0.6/0.0	0.0/0.0	0.9/1.8	0.0/0.0	3.3/4.0	0.9/1.5
Fixed embs	10	4.0/1.4	0.9/0.0	6.7/2.2	1.5/0.7	16.1/13.0	7.0/**12.3**
	50	4.0/1.4	0.9/0.0	6.7/2.2	1.5/0.7	13.4/4.4	3.0/2.5
Edge-Bi-enc	10	4.0/11.6	0.3/0.0	9.7/**17.4**	2.7/1.4	**31.9/25.7**	**14.6/8.0**
	50	4.0/11.6	0.3/0.0	9.7/**17.4**	2.7/1.4	13.7/20.3	4.3/2.5
+ Edge-Cross-enc	10	0.6/2.2	0.0/0.4	12.2/14.1	1.5/3.6	**31.9/25.7**	**14.6/8.0**
	50	**7.3/7.6**	**1.8/1.5**	**17.9/15.6**	**7.3/4.7**	25.8/**26.5**	10.6/8.7
+ GPT-3.5	10	4.0/4.0	0.0/0.0	5.5/4.3	2.4/1.4	5.5/4.3	2.4/1.4
	50	3.3/1.5	0.0/0.0	4.6/3.6	1.5/0.4	4.6/3.6	1.5/0.4
+ FLAN-T5-XL	10	2.7/1.8	0.6/0.0	2.7/1.8	0.6/0.0	2.7/1.8	0.6/0.0
+ Llama-2-7B	10	2.7/4.3	0.3/0.0	5.8/6.2	2.1/0.0	8.8/7.2	3.3/1.1
	50	1.8/3.3	0.0/0.0	3.7/5.8	1.2/0.7	4.0/6.9	1.2/0.7
+ Llama-2-7B-tuned	10	5.2/**13.8**	0.0/0.0	7.6/16.3	1.5/1.8	7.6/16.3	1.5/1.8
	50	6.1/13.0	0.0/0.0	8.5/15.2	1.5/1.1	8.5/15.6	1.5/1.5
MM-S14-CPP	k	InR_{any}@1	InR_{all}@1	InR_{any}@5	InR_{all}@5	InR_{any}@10	InR_{all}@10
Inverted Index	10	0.4/0.0	0.0/0.0	0.9/0.5	0.0/0.0	5.5/5.8	5.1/5.3
	50	0.4/0.0	0.0/0.0	0.4/0.0	0.0/0.0	0.9/1.4	0.0/0.5
Fixed embs	10	2.8/1.2	0.7/0.2	6.3/3.7	2.3/2.1	11.3/8.3	8.3/7.4
	50	2.8/1.2	0.7/0.2	6.3/3.7	2.3/2.1	7.9/6.0	3.9/4.6
Edge-Bi-enc	10	2.5/6.3	0.0/0.2	6.2/11.8	1.2/1.9	**32.0/27.8**	**19.7/14.4**
	50	2.5/6.3	0.0/0.2	6.2/11.8	1.2/1.9	8.6/14.4	3.0/3.5
+ Edge-Cross-enc	10	3.4/**9.3**	0.2/0.0	7.8/13.7	2.1/2.3	**32.0/27.8**	**19.7/14.4**
	50	4.9/3.9	**2.1**/0.2	**15.3/17.6**	**6.3/6.9**	24.8/26.6	13.2/14.4
+ GPT-3.5	10	**5.1**/3.9	0.0/0.0	7.9/6.0	3.3/3.5	7.9/6.0	3.3/3.4
	50	1.8/1.9	0.0/0.0	3.9/2.8	0.9/0.7	4.0/2.8	0.9/0.7
+ FLAN-T5-XL	10	2.6/1.9	0.5/**0.7**	2.6/1.9	0.5/0.7	2.6/1.9	0.5/0.7
+ Llama-2-7B	10	1.8/4.6	0.0/0.2	4.8/7.2	0.7/1.9	8.8/10.4	3.9/3.9
	50	1.2/3.5	0.0/0.0	2.5/5.1	0.7/0.9	3.0/6.3	1.1/1.2
+ Llama-2-7B-tuned	10	2.6/7.2	0.0/0.0	6.5/10.6	1.9/1.2	7.6/12.7	2.5/3.2
	50	2.5/4.6	0.0/0.0	3.3/6.7	0.5/0.6	4.0/8.1	0.9/1.4

Overall Results After Edge Selection. We then add the edge selection steps mainly on the candidates from the fine-tuned embedding (Edge-Bi-encoder) approach, given its best overall performance in generating edge candidates.

As shown in Table 3, the multi-label Edge-Cross-encoder achieves the best performance in most experimental settings. Edge-Cross-encoder further reranks the edge candidates and helps substantially improve the performance over Edge-Bi-encoder (edge search only), e.g., by around 8–9% absolute scores for InR_{any}@5 for the datasets and around 12–16% for InR_{any}@10 (except for the same @10 results for top-10 setting, where re-ranking does not make a difference).

We also test the LLMs, it can be seen that the tested medium scale LLMs (GPT-3.5, FLAN-XL, and Llama-2-7B), especially not instruction-tuned for the task, can still not be directly used for concept placement, although GPT-3.5 has notably better results on top edge suggestion ($InR_{any}@1$) from the top-10 setting. The explainable instruction tuning approach greatly improves the performance of Llama-2-7B. This shows that training on in-KB data by generating an automated explanation before generating the results is practically useful to enhance the capability of LLMs on ontology reasoning tasks. Most results from LLMs, except for the top edge suggestion ($InR_{any}@1$ and $InR_{all}@1$), are still below the original candidates from Edge-Bi-encoder. Nevertheless, the results are encouraging and can motivate future studies using LLMs for concept placement.

We also notice a performance gap between the validation set and the test set on the two datasets in Tables 2–3, which may be due to the high variance caused by the small number of mentions in the sets (between 200 and 600, see Table 1) and the distinct data distribution based on concept drift (e.g., different lexical mentions between the sets), showing the challenge to generalise to new concepts.

Discussion on Model Applicability. The overall performance of the models is not high, especially for LLMs, as shown in Table 3. The best InAny@10 is around 30% with Edge-Bi-encoder and Edge-Cross-encoder. This shows that the models cannot support an automated application, but still, they may potentially be applicable to suggest a ranking of the edges for human terminologists to add a new concept to an ontology. In practice, having a larger k can help improve the metrics of $InR_{any}@k$ and $InR_{all}@k$, but can also increase the effort of manual selection, thus a balance needs to be achieved and warrants future studies.

Case Study. In Appendix 3, we select a few test mentions and display the 5 top edge suggestions, under the top-50 setting from Edge-Bi-encoder, Edge-Cross-encoder, Llama-2-7B, and Llama-2-7B fine-tuned models. For Llama-2 models, we display the generated answers. Without instruction tuning, Llama-2-7B sometimes generates answers in an incorrect format or generates irrelevant outputs. With explainable instruction-tuning, Llama-2-7B generates explanations that follow a natural language reasoning path to lead to the correct edge option. We also note that merging with existing concepts is needed as a further step after the placement of the mention, e.g., for Chronic kidney disorder.

We also note that many edge predictions are not completely wrong, for example for the first case, the predicted parent (e.g., kidney disease) in the methods is more general than the gold direct parent (e.g., renal impairment), and the predicted child (e.g., hypertensive heart and renal disease with renal failure) is not far from the gold children in the ontology structure. However, calculating a lenient, soft score (e.g., with Wu & Palmer similarity [32]) between every prediction and the set of gold edges instead of a binary evaluation is not time efficient in our experiments. We leave an efficient, lenient evaluation for future studies.

5.5 Ablation Studies

Our ablation studies aim to investigate how edge enrichment and automated explainable instruction tuning can enhance the performance of concept place-

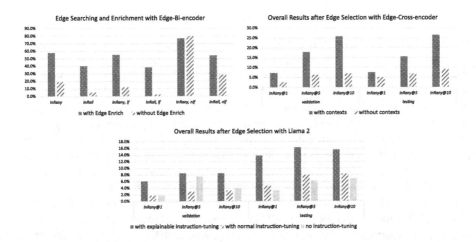

Fig. 3. Ablation results on top-50 edge candidates with MM-S14-Disease dataset: **(a)** Top-left: results with Edge-Bi-encoder, with or without the edge enrichment step, on validation set; **(b)** Top-right: overall results with Edge-Cross-encoder, with or without contexts; **(c)** Bottom: overall results after Edge Selection with Llama-2-7B, with explainable instruction-tuning, normal instruction-tuning, or without instruction-tuning.

ment. We use top-50 setting and MM-S14-Disease dataset as an example, and other top-k settings and MM-S14-CPP dataset follow a similar pattern of results.

Edge Enrichment. Edge enrichment has greatly improved the overall results for Edge-Bi-encoder, about above 30% absolute improvement of InR_{any} and InR_{all} as shown in Fig. 3 part (a). Results with inverted index and fixed embeddings are in Appendix 2, showing a general improvement with edge enrichment.

Contextual Information of Mention. The left and right contexts of a mention are useful in edge search with Edge-Bi-encoder to learn the similarity between a contextual mention and a concept, even though the type of texts is distinct from ontology concept labels, as shown in Fig. 3 part (b).

Explainable Instruction Tuning. Explainable instruction tuning helps improve the performance of the LLM, Llama-2-7B, under the top-50 setting, especially on the test set of MM-S14-Disease, as displayed in Fig. 3 part (c). In contrast, normal instruction tuning that directly generates the edge option number does not always improve over the case without instruction tuning.

5.6   Conclusion and Future Studies

We propose an LM-based framework for new concept placement in ontologies. The framework uses a three-step approach, that enhances the two-step information retrieval with edge formation and enrichment leveraging the ontological structure. The results overall show that methods that fine-tune PLMs perform the best, while there is an encouraging performance for the recent LLMs, especially with explainable instruction tuning. Our case study shows that explanations can be generated to detail the steps for concept placement. We focused on placing directly the mentions into edges in this work, a following step is to group or merge mentions of the same new concept, and also with existing concepts if they have the same meaning, when they are placed into the same edges. Future studies will further explore LLM-generated explanations and leverage advanced Retrieval Augmented Generation [9] and prompting strategies. Future studies also need to investigate how to use the methods to assist human terminologists.

Acknowledgements. This work is supported by EPSRC projects, including ConCur (EP/V050869/1), OASIS (EP/S032347/1), UK FIRES (EP/S019111/1); and Samsung Research UK (SRUK).

Appendix 1: Detailed model settings and time usage

The approaches are implemented using PyTorch and Huggingface Transformers. Edge-Bi-encoder and Edge-Cross-encoder are originally based on the architectures of BLINKout [8] (based on BLINK [31]). Inverted index with ontology concepts is based on DeepOnto Library [15]. The batch sizes for Edge-Bi-encoder and Edge-Cross-encoder are 16 and 1, resp. The fine-tuning of Edge-Bi-encoder and Edge-Cross-encoder takes 1 and 4 epochs, resp. We limit the rows to 200,000 for training the Edge-Cross-encoder models given the sufficient amount the data for model convergence and the long time of training. The instruction tuning of Llama-2-7B uses a 4-bit quantisation and takes 3 epochs with a batch size of 4.

Time Usage. We run all models using an NVIDIA Quadro RTX 8000 GPU card (48GB GPU). We report the time usage estimate for MM-S14-Disease under the top-50 setting. Training bi-encoder took around 29 h. Training cross-encoder took around 4 h. Instruction tuning of Llama-2-7B took around 16 h. Inferencing with fixed embeddings and inverted index with edge enrichment is within around 0.5 and 1 s per mention, resp. Inferencing with Edge-Bi-encoder only takes around

0.2 s per mention. The whole inferencing with both Edge-Bi-encoder and Edge-Cross-encoder takes around 2.3 s per mention. The prompting of an explainable instruction-tuned Llama-2-7B model takes around 78 s per mention to output natural language explanations.

Appendix 2: Detailed results on edge enrichment

We applied edge formation enrichment over inverted index and fixed embedding approach. Results in Table 4 show a substantial improvement for InR_{any} and InR_{all}. We see that the mentions to be placed to non-leaf edges are not improved with inverted index and fixed embeddings, but are improved with the fine-tuned, Edge-Bi-encoder, this is because the latter places a more lenient score for the leaf edges that do not always rank them before the non-leaf edges.

Table 4. Results on edge search and enrichment (vs. not using edge enrichment) for MM-S14-Disease, under the top-50 setting. Each setting has validation and testing results, separated by a slash (/) sign. "*lf*" and "*nlf*" mean *leaf* and *non-leaf*, resp.

MM-S14-Disease	k	InR_{any}	InR_{all}	InR_{any}, lf	InR_{all}, lf	InR_{any}, nlf	InR_{all}, nlf
Inverted Index	50	**41.3/40.6**	**37.7/38.8**	**44.6/50.0**	**41.2/48.2**	14.3/5.2	8.6/3.5
w/o Edge Enrich	50	7.3/6.5	4.6/3.6	2.7/3.7	0.7/0.5	**45.7/17.2**	**37.1/15.5**
Fixed embs	50	**35.3/31.9**	**28.3/30.8**	**38.4/38.1**	**30.6/37.6**	8.6/8.6	**8.6/5.2**
w/o Edge Enrich	50	19.1/17.3	4.3/4.3	16.0/6.4	3.7/4.1	**45.7/58.6**	**8.6/5.2**
Edge-Bi-Enc	50	**57.8/50.0**	**40.1/38.0**	**55.4/38.1**	**38.4/33.5**	77.1/**94.8**	**54.3/55.2**
w/o Edge Enrich	50	19.1/23.9	5.2/5.8	11.9/5.0	2.4/2.8	**80.0/94.8**	28.6/17.2

Appendix 3: Qualitative examples

Examples of a non-leaf and a leaf concept placement, with prompt options, model predictions, and instruction-tuned Llama-2-7B's explanations, are in Table 5.

Table 5. Examples of two mentions in the out-of-KB test set of MM-S14-Disease to enrich SNOMED CT 2014.09. The correct predictions are in **bold**. (Note: while the concept Chronic kidney disease in SNOMED CT ver 2017.03 is not in ver 2014.09, it is modified from Chronic renal impairment, ID 236425005, in the older ontology.)

	Test, out-of-KB, 13	Test, out-of-KB, 138
Mention in contexts	...Since no one had *CKD* in partial nephrectomized patients, we determined risk factors for CKD in radical nephrectomized patients...	Development of a novel near-infrared fluorescent theranostic combretastain A-4 analogue, YK-5-252, to target triple negative breast cancer. The treatment of triple negative breast cancer (*TNBC*) is a significant challenge to cancer research...
Gold Concept	http://snomed.info/id/709044004 Chronic kidney disease (not available in SNOMED CT 2014.09)	http://snomed.info/id/706970001 Triple negative malignant neoplasm of breast (not available in SNOMED CT 2014.09)
Gold Edges	**Parents:** (i) Renal impairment → **Children:** (i) Chronic renal impairment associated with type II diabetes mellitus; (ii) Hypertensive heart and chronic kidney disease; (iii) Chronic kidney disease stage 1; (iv) Chronic kidney disease stage 2; (v) Chronic kidney disease stage 3; (vi) Chronic kidney disease stage 4; (vii) Chronic kidney disease stage 5; (viii) Chronic renal failure syndrome; (viiii) Hypertensive heart AND chronic kidney disease on dialysis; (x) Chronic kidney disease due to hypertension; Malignant hypertensive chronic kidney disease	**Parents:** (i) Human epidermal growth factor 2 negative carcinoma of breast; (ii) Malignant tumor of breast; (iii) Hormone receptor negative neoplasm → **Children:** (i) NULL
Edge-Bi-enc	(i) renal impairment → end stage renal disease (ii) renal impairment → renal failure following molar and/or ectopic pregnancy (iii) renal impairment → renal failure syndrome (vi) **renal impairment → chronic kidney disease due to hypertension** (v) kidney disease → impaired renal function disorder	(i) malignant tumor of breast → lobular carcinoma of breast (ii) carcinoma of breast → lobular carcinoma of breast (iii) malignant tumor of breast → mucinous carcinoma of breast (iv) carcinoma of breast → cancer en cuirasse (v) malignant tumor of breast → malignant phyllodes tumor of breast
Edge-Cross-enc	(i) kidney disease → renal function impairment with growth failure (ii) kidney disease → impaired renal function disorder (iii) disorder of the genitourinary system → impaired renal function disorder (iv) **renal impairment → chronic kidney disease due to hypertension** (v) kidney disease → renal failure syndrome	(i) neoplasm of breast → malignant phyllodes tumor of breast (ii) carcinoma of breast → cancer en cuirasse (iii) **malignant tumor of breast → NULL** (iv) carcinoma of breast → NULL (v) neoplasm of breast → NULL
Prompt, only options are displayed (correct options in bold), the full prompt template is in Sect 4.3	options: 0.renal impairment → end stage renal disease 1.renal impairment → renal failure following molar and/or ectopic pregnancy 2.renal impairment → renal failure syndrome 3.**renal impairment → chronic kidney disease due to hypertension** 4.kidney disease → impaired renal function disorder 5.kidney disease → renal function impairment with growth failure 6.kidney disease → renal failure syndrome 7.**renal impairment → chronic renal failure syndrome** 8.disorder of the genitourinary system → impaired renal function disorder 9.renal impairment → hypertensive heart and renal disease with renal failure	options: 0.malignant tumor of breast → lobular carcinoma of breast 1.carcinoma of breast → lobular carcinoma of breast 2.malignant tumor of breast → mucinous carcinoma of breast 3.carcinoma of breast → cancer en cuirasse 4.malignant tumor of breast → malignant phyllodes tumor of breast 5.malignant tumor of breast → cancer en cuirasse 6.neoplasm of breast → malignant phyllodes tumor of breast 7.**malignant tumor of breast → NULL** 8.carcinoma of breast → NULL 9.neoplasm of breast → NULL
Gold answer	3,7 (option incomplete)	7 (option incomplete)
Llama-2-7B	1	Fail to generate answer with the relevant content
Llama-2-7B instruction-tuned	### Explanation: From the parents in the options above, including renal impairment , kidney disease , disorder of the genitourinary system , the correct parents of the mention, CKD, include **renal impairment** , kidney disease , disorder of the genitourinary system . Thus the options are narrowed down to 0, 1, 2, 3, 7, 9, 4, 5, 6, 8. From the children in the narrowed options, including end stage renal disease , renal failure following molar and/or ectopic pregnancy , renal failure syndrome , chronic kidney disease due to hypertension , chronic renal failure syndrome , impaired renal function disorder , renal function impairment with growth failure , the correct children of the mention, CKD, include **chronic kidney disease due to hypertension** , **chronic renal failure syndrome** , impaired renal function disorder , renal function impairment with growth failure . Thus, the final answers are **3**, **7**, 5, 6, 8. ### Response: 3,7,5,6,8	### Explanation: From the parents in the options above, including malignant tumor of breast , carcinoma of breast , neoplasm of breast , the correct parents of the mention, TNBC, include **malignant tumor of breast** . Thus the options are narrowed down to 0, 2, 4, 5, 7. From the children in the narrowed options, including lobular carcinoma of breast , mucinous carcinoma of breast , malignant phyllodes tumor of breast , cancer en cuirasse , NULL, the correct children of the mention, TNBC, include NULL. Thus, the final answers are **7**. ### Response: 7

References

1. Baader, F., Horrocks, I., Lutz, C., Sattler, U.: A Basic Description Logic, pp. 10–49. Cambridge University Press, Cambridge (2017). https://doi.org/10.1017/9781139025355.002
2. Baader, F., Horrocks, I., Lutz, C., Sattler, U.: Ontology Languages and Applications, pp. 205–227. Cambridge University Press, Cambridge (2017). https://doi.org/10.1017/9781139025355.008
3. Chen, J., et al.: Knowledge graphs for the life sciences: recent developments, challenges and opportunities. arXiv preprint arXiv:2309.17255 (2023)
4. Chen, J., He, Y., Geng, Y., Jiménez-Ruiz, E., Dong, H., Horrocks, I.: Contextual semantic embeddings for ontology subsumption prediction. World Wide Web, pp. 1–23 (2023)
5. Chung, H.W., et al.: Scaling instruction-finetuned language models. arXiv preprint arXiv:2210.11416 (2022)
6. Dettmers, T., Pagnoni, A., Holtzman, A., Zettlemoyer, L.: Qlora: efficient finetuning of quantized llms. arXiv preprint arXiv:2305.14314 (2023)
7. Dong, H., Chen, J., He, Y., Horrocks, I.: Ontology enrichment from texts: a biomedical dataset for concept discovery and placement. In: Proceedings of the 32nd ACM International Conference on Information & Knowledge Management. Association for Computing Machinery, New York, NY, USA (2023). https://doi.org/10.1145/3583780.3615126
8. Dong, H., Chen, J., He, Y., Liu, Y., Horrocks, I.: Reveal the unknown: out-of-knowledge-base mention discovery with entity linking. In: Proceedings of the 32nd ACM International Conference on Information and Knowledge Management, pp. 452–462. CIKM '23, Association for Computing Machinery, New York, NY, USA (2023). https://doi.org/10.1145/3583780.3615036
9. Gao, Y., et al.: Retrieval-augmented generation for large language models: a survey. arXiv preprint arXiv:2312.10997 (2023)
10. Gibaja, E., Ventura, S.: A tutorial on multilabel learning. ACM Comput. Surv. **47**(3) (2015). https://doi.org/10.1145/2716262
11. Glauer, M., Memariani, A., Neuhaus, F., Mossakowski, T., Hastings, J.: Interpretable ontology extension in chemistry. Semantic Web **Pre-press**(Pre-press), 1–22 (2023)
12. Grau, B.C., Horrocks, I., Motik, B., Parsia, B., Patel-Schneider, P., Sattler, U.: Owl 2: the next step for owl. J. Web Semant. **6**(4), 309–322 (2008). semantic Web Challenge 2006/2007
13. Gu, Y., et al.: Domain-specific language model pretraining for biomedical natural language processing. ACM Trans. Comput. Healthc. **3**(1) (2021). https://doi.org/10.1145/3458754
14. He, Y., Chen, J., Dong, H., Horrocks, I.: Exploring large language models for ontology alignment. arXiv preprint arXiv:2309.07172 (2023)
15. He, Y., et al.: Deeponto: a python package for ontology engineering with deep learning. arXiv preprint arXiv:2307.03067 (2023)
16. He, Y., Chen, J., Jimenez-Ruiz, E., Dong, H., Horrocks, I.: Language model analysis for ontology subsumption inference. In: Rogers, A., Boyd-Graber, J., Okazaki, N. (eds.) Findings of the Association for Computational Linguistics: ACL 2023, pp. 3439–3453. Association for Computational Linguistics, Toronto, Canada, July 2023. https://doi.org/10.18653/v1/2023.findings-acl.213, https://aclanthology.org/2023.findings-acl.213

17. Hertling, S., Paulheim, H.: Transformer based semantic relation typing for knowledge graph integration. In: Pesquita, C., et al. (eds.) The Semantic Web. ESWC 2023. LNCS, vol. 13870, pp. 105–121. Springer, Cham (2023). https://doi.org/10.1007/978-3-031-33455-9_7

18. Jurafsky, D., Martin, J.H.: Speech and Language Processing (3rd Edition) (2023). Online

19. Kudo, T., Richardson, J.: SentencePiece: a simple and language independent subword tokenizer and detokenizer for neural text processing. In: Blanco, E., Lu, W. (eds.) Proceedings of the 2018 Conference on Empirical Methods in Natural Language Processing: System Demonstrations, pp. 66–71. Association for Computational Linguistics, Brussels, Belgium, November 2018.https://doi.org/10.18653/v1/D18-2012, https://aclanthology.org/D18-2012

20. Liu, F., Shareghi, E., Meng, Z., Basaldella, M., Collier, N.: Self-alignment pretraining for biomedical entity representations. In: Proceedings of the 2021 Conference of the North American Chapter of the Association for Computational Linguistics: Human Language Technologies, pp. 4228–4238. Association for Computational Linguistics, Online, June 2021. https://doi.org/10.18653/v1/2021.naacl-main.334

21. Liu, H., Perl, Y., Geller, J.: Concept placement using BERT trained by transforming and summarizing biomedical ontology structure. J. Biomed. Inform. **112**(C) (2020)

22. Reimers, N., Gurevych, I.: Sentence-BERT: sentence embeddings using Siamese BERT-networks. In: Proceedings of the 2019 Conference on Empirical Methods in Natural Language Processing and the 9th International Joint Conference on Natural Language Processing (EMNLP-IJCNLP), pp. 3982–3992. Association for Computational Linguistics, Hong Kong, China, November 2019. https://doi.org/10.18653/v1/D19-1410

23. Ruas, P., Couto, F.M.: Nilinker: attention-based approach to nil entity linking. J. Biomed. Inform. 104137 (2022). https://doi.org/10.1016/j.jbi.2022.104137, https://www.sciencedirect.com/science/article/pii/S1532046422001526

24. Shen, W., Wang, J., Han, J.: Entity linking with a knowledge base: Issues, techniques, and solutions. IEEE Trans. Knowl. Data Eng. **27**(2), 443–460 (2014)

25. Spackman, K.A., Dionne, R., Mays, E., Weis, J.: Role grouping as an extension to the description logic of ontylog, motivated by concept modeling in snomed. In: Proceedings of the AMIA Symposium, p. 712. American Medical Informatics Association (2002)

26. Touvron, H., et al.: Llama 2: open foundation and fine-tuned chat models. arXiv preprint arXiv:2307.09288 (2023)

27. Vaswani, A., et al.: Attention is all you need. In: Guyon, I., Luxburg, U.V., Bengio, S., Wallach, H., Fergus, R., Vishwanathan, S., Garnett, R. (eds.) Advances in Neural Information Processing Systems, vol. 30. Curran Associates, Inc. (2017)

28. Veseli, B., Singhania, S., Razniewski, S., Weikum, G.: Evaluating language models for knowledge base completion. In: Pesquita, C., et al. (eds.) The Semantic Web. ESWC 2023. LNCS, vol. 13870, pp. 227–243. Springer, Cham (2023). https://doi.org/10.1007/978-3-031-33455-9_14

29. Wang, Q., Gao, Z., Xu, R.: Exploring the in-context learning ability of large language model for biomedical concept linking. arXiv preprint arXiv:2307.01137 (2023)

30. Wang, S., Zhao, R., Zheng, Y., Liu, B.: Qen: applicable taxonomy completion via evaluating full taxonomic relations. In: Proceedings of the ACM Web Conference 2022, pp. 1008–1017. WWW '22, Association for Computing Machinery, New York, NY, USA (2022). https://github.com/sheryc/QEN

31. Wu, L., Petroni, F., Josifoski, M., Riedel, S., Zettlemoyer, L.: Scalable zero-shot entity linking with dense entity retrieval. In: Proceedings of the 2020 Conference on Empirical Methods in Natural Language Processing (EMNLP), pp. 6397–6407. Association for Computational Linguistics, Online, November 2020.https://doi.org/10.18653/v1/2020.emnlp-main.519

32. Wu, Z., Palmer, M.: Verb semantics and lexical selection. In: 32nd Annual Meeting of the Association for Computational Linguistics, pp. 133–138. Association for Computational Linguistics, Las Cruces, New Mexico, USA, June 1994. https://doi.org/10.3115/981732.981751, https://aclanthology.org/P94-1019

33. Zeng, Q., Lin, J., Yu, W., Cleland-Huang, J., Jiang, M.: Enhancing taxonomy completion with concept generation via fusing relational representations. In: Proceedings of the 27th ACM SIGKDD Conference on Knowledge Discovery & Data Mining, pp. 2104–2113. Association for Computing Machinery, New York, NY, USA (2021). https://doi.org/10.1145/3447548.3467308, https://github.com/DM2-ND/GenTaxo

34. Zhang, J., Song, X., Zeng, Y., Chen, J., Shen, J., Mao, Y., Li, L.: Taxonomy completion via triplet matching network. In: Proceedings of the AAAI Conference on Artificial Intelligence, pp. 4662–4670. AAAI Press, Palo Alto, California, USA (2021). https://github.com/JieyuZ2/TMN

35. Zhao, W.X., et al.: A survey of large language models. arXiv preprint arXiv:2303.18223 (2023)

Low-Dimensional Hyperbolic Knowledge Graph Embedding for Better Extrapolation to Under-Represented Data

Zhuoxun Zheng[1,2]([✉]), Baifan Zhou[2,3], Hui Yang[4], Zhipeng Tan[1],
Arild Waaler[2], Evgeny Kharlamov[1,2], and Ahmet Soylu[2,3]

[1] Bosch Center for AI, Renningen, Germany
zhengzx712@gmail.com
[2] University of Oslo, Oslo, Norway
[3] Oslo Metropolitan University, Oslo, Norway
[4] Laboratoire Interdisciplinaire des Sciences du Numérique, Paris, France

Abstract. Past works have shown knowledge graph embedding (KGE) methods learn from facts in the form of triples and extrapolate to unseen triples. KGE in hyperbolic space can achieve impressive performance even in low-dimensional embedding space. However, existing work limitedly studied extrapolation to under-represented data, including under-represented entities and relations. To this end, we propose HolmE, a general form of KGE method on hyperbolic manifolds. HolmE addresses extrapolation to under-represented entities through a special treatment of the bias term, and extrapolation to under-represented relations by supporting strong composition. We provide empirical evidence that HolmE achieves promising performance in modelling unseen triples, under-represented entities, and under-represented relations. We prove that mainstream KGE methods either: (1) are special cases of HolmE and thus support strong composition; (2) do not support strong composition. The code and data are open-sourced at https://github.com/nsai-uio/HolmE-KGE.

Keywords: Knowledge graph embedding · Knowledge graph

1 Introduction

Knowledge graphs (KGs) refer to multi-relational graphs that represent facts in the form of subject-relation-object triples. Studies in KG embedding (KGE) strive to represent KGs numerically via link prediction, enabling the possibility to leverage powerful machine learning (ML) in many practical applications, such as predicting interactions between drugs and targets [12] or diseases [16].

The extrapolation of KGE models to unseen triples (i.e., predicting unseen links) is the focus of current studies. Many past works have achieved promising results. Translational models have good performance as well as transparent geometrical explanations [3]. KGE models in hyperbolic space achieve impressive performance even in low-dimensional space, due to their expressiveness for hierarchical structures [6]. However, *extrapolation* is multi-faceted; past work limitedly discussed the extrapolation to *under-represented data*, i.e., long-tail data, which includes *under-represented entities*, and *relations* [33,35].

We argue that for good extrapolation to under represented entities, a special treatment to the bias terms in KGE models is needed. Many past models adopt scoring functions with entity-specific bias terms to evaluate the probability of predicted entities (higher score, higher chance to be predicted) [1,5,6]. We observe that the bias terms learned from the training set are highly correlated with the entity distribution in the training set (Fig. 1b). In benchmark datasets such as WN18RR and FB15k-237, the latter is also highly correlated with the entity distribution in the test set (Fig. 1a). This makes exploitation possible: to simply assign higher scores to entities that are more frequent via giving higher biases to these entities. In this regard, the bias terms essentially serves as prior probabilities of the entities in the training set. Adopting the bias terms assumes that the entity distribution of the training set is similar to that in the test set, which is true for benchmark datasets , but not necessarily true for many real-world cases, where most entities in test data are under-represented. In these cases, the use of bias terms can lead to performance deterioration.

On extrapolation to *under-represented relations*, we propose the notion of *strong composition*, The entire embedding space for relations of *strong composition* KGE model supports composition (Fig. 2), contrasting to *weak composition* KGE, whose embedding space only has a sub-space that supports composition. The key difference is that, *weak composition* KGE would assume relations to be not compositional and tend to put them into the sub-space not supporting com-

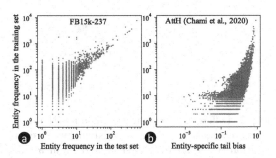

Fig. 1. (a) High correlations of the entity frequency (occurrence of triples) in the training set with that in the test set, and (b) with the scoring bias terms b_t of a KGE model [6].

position, when the compositional relations are under-represented in the training set. Practically, *strong composition* KGEs can better model under-represented relations, as these relations are not embedded in sub-spaces where composition patterns are not supported.

To these ends, we propose Holme that addresses multi-fold extrapolation: to unseen triples, to under-represented entities, and to under-represented relations. Our contributions are summarised as below:

– We give a detailed analysis of *extrapolation* in multiple aspects (Sect. 3), including extrapolation to: (1) unseen triples; (2) under-represeted entities; (3) under-represented relations;

– We propose a general form of KGE method HolmE, consisting of rotation and translation in product space of manifolds (Sect. 4). We provide extensive empirical evidence (Sect. 5) that HolmE outperforms SotA in low-dimensional space in multiple aspects of extrapolation.
– We give in-depth analysis of the influence of biases of KGE models (Sect. 5.3) and argue that the bias term should be treated with special care, depending on the assumption whether the test data has an entity distribution similar to that of the training data.
– We provide theoretical proof that HolmE supports strong composition, and that main stream KGE methods are either special cases of HolmE and thus support strong composition or they do not satisfy strong composition (Sect. 4).

2   Related Work

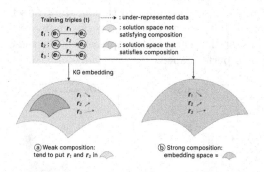

Fig. 2. Difference between weak composition and strong composition: models satisfying weak composition will tend to embed r_1 and r_2 in a subspace that does not satisfy composition if r_3 (or the composition pattern) is under-represented in the training set; models satisfying strong composition do not have this problem, because their whole embedding space $\mathcal{G}$ satisfies composition.

In general, KGE studies have explored in two directions: exploration of expressive spaces for entity embeddings, and appropriate mathematical transformation that model relations as mappings.

Geometric Space. The first group of work has focused on Euclidean spaces, including TransE [3], TransR [21]. Another group of work has explored 2D complex space, e.g., ComplEx [27], RotatE [25]. Hyperbolic space refers Riemannian distance spaces with negative curvature. Hyperbolic embedding achieves good performance even in low-dimensional embedding due to its high expressivity and suitability for hierarchical structure [6], which is common in KGs. Past work explored various curvature spaces, such as MurP [1], AttH [6], TorusE [14], and GIE [5].

Mapping Modelling. Translational or rotational models have clear geometrical explanations, and support many relation patterns, e.g., TransE [3], RotatE [25], and some variants such as TransH [30], TransD [17]. Bilinear models [2,34] as another group of KGE models treat both entities and relations as embedding vectors/matrices and perform multiplication scoring.

Neural Networks. These works, including ConvE [11], ConvKB [10] and CompGCN [28] use graph neural networks to encode KGs [9]. Recently transformer, which exhibits strong modelling power in natural language models has already been used for learning embeddings for various KG-related link prediction tasks, such as RETRA [31]. These methods can also have good performance, although they usually lack explicitly geometrical explanation on relation patterns and do not explicitly support relation patterns [5, 20, 24].

Extrapolation to Under-represented Data. The fairness by the view of the representation bias is becoming a hot topic in research [4, 8, 22]. In KGE-related fields, this issue refers to the modelling of entities or relations that are long-tail, i.e. under-represented in KGs. The existence of those long-tail data deteriorate not only the modelling capability for those specific items, but also the general performance of the mothods [9, 13]. KGEs for long-tail entities are studied in [19, 33]. The work [33] uses neural networks to learn KG structure and attribute information, and tested on datasets with attribute information, falling in a rather different framework than KGE. Another work [19] tested their models on long-tail entities of FB15k-237, but insufficiently explained what factors are important. KGE for under-represented relations is partially covered in few-shot relation KG completion. A series of work [29, 35] trains neural networks or bilinear models on triples with few-shot relations tested on datasets such as NELL, Wiki. This also falls in a rather different framework. In summary, exploring fairness from the view of representation bias, which focus on extrapolation to under-represented data in the KGE framework still remains to be explored. Our work also falls within this category.

3   Problem Setup

3.1   Knowledge Graph Embedding (KGE)

A knowledge graph (KG) is a multi-relational graph, denoted as $G = (\mathcal{E}, \mathcal{R}, \mathcal{F})$, where $\mathcal{E}$ is a set of entities (nodes), $\mathcal{R}$ is a set of binary relations (edge types) between entities, and $\mathcal{F}$ is a set of facts (edges) given in the triple form of $(h, r, t) \in \mathcal{F} \subseteq \mathcal{E} \times \mathcal{R} \times \mathcal{E}$, with h, t and r denoting the head entity, tail entity, and the relation in between respectively.

A common setting of KGE problem seeks to solve the problem of link prediction: $(h, r, ?)$ (or $(?, r, t)$), namely given the *query* of the head entity (or tail entity) and the relation, to find the most probable tail entity (or head entity). For simplicity, we denote the query in both two directions as $(h, r) \to t$.

Let $(\mathcal{P}, \mathcal{G})$ be an embedding space, where $\mathcal{P}$ is a distance space, $\mathcal{G}$ is a set of mappings that have domain and range defined on $\mathcal{P}$, and s is a scoring function: $s : \mathcal{P} \times \mathcal{G} \times \mathcal{P} \to \mathbb{R}$. The KGE problem aims to find an embedding from G to $(\mathcal{P}, \mathcal{G})$ that (i) maps entities $h, t \in \mathcal{E}$ to points (vectors) $\mathbf{e}_h, \mathbf{e}_t \in \mathcal{P}$; (ii) maps each relation $r \in \mathcal{R}$ to a map $g_r \in \mathcal{G}$ such that $s(\mathbf{e}_h, g_r, \mathbf{e}_t)$ ranks how probable that $(h, r, t) \in \mathcal{F}$.

3.2   Riemannian Geometry

We briefly introduce Riemannian geometry and refer the readers to text-books [32]. A Riemannian manifold is a distance space where the distance between two points are characterised by a Riemannian metric. A hyperbolic space is a Riemannian manifold with constant negative curvature $-\kappa$ and dimension d as $\mathbb{H}^{d,\kappa} := \{\mathbf{x} \in \mathbb{R}^d \mid ||\mathbf{x}||^2 < \frac{1}{\kappa}\}$, where $|| \cdot ||$ denotes the L2 norm.

For each point on a non-Euclidean manifold $\mathbf{x} \in \mathbb{H}^{d,\kappa}$, the tangent space $\mathcal{T}_{\mathbf{x}}^c$ is a d-dimensional vector space containing all possible directions of paths in $\mathbb{H}^{d,\kappa}$ leaving from $\mathbf{x}$. The transformation of $\mathbf{x}$ on $\mathbb{H}^{d,\kappa}$ to the $\mathcal{T}_{\mathbf{x}}^c$ is referred to as the *logarithmic map*, and from $\mathcal{T}_{\mathbf{x}}^c$ to $\mathbb{H}^{d,\kappa}$ the *exponential map*. Their closed-form expressions at the origin:

$$exp_0^\kappa(\mathbf{e}) = tanh(\sqrt{\kappa}||\mathbf{e}||)\frac{\mathbf{e}}{\sqrt{\kappa}||\mathbf{e}||} \tag{1}$$

$$log_0^\kappa(\mathbf{e}) = arctanh(\sqrt{\kappa}||\mathbf{e}||)\frac{\mathbf{e}}{\sqrt{\kappa}||\mathbf{e}||} \tag{2}$$

Following previous work, the translation in hyperbolic space is defined as the Möbius addition [15], denoted as $\oplus^\kappa$, which provides an analogue to Euclidean addition for hyperbolic space:

$$\mathbf{x} \oplus^\kappa \mathbf{y} = \frac{(1 + 2\kappa \langle \mathbf{x}, \mathbf{y} \rangle + \kappa||\mathbf{y}||^2)\mathbf{x} + (1 - \kappa||\mathbf{x}||^2)\mathbf{y}}{1 + 2\kappa \langle \mathbf{x}, \mathbf{y} \rangle + \kappa^2||\mathbf{x}||^2||\mathbf{y}||^2} \tag{3}$$

The distance on $\mathbb{H}^{d,\kappa}$ is defined as:

$$d^\kappa(\mathbf{x}, \mathbf{y}) = \frac{2}{\sqrt{\kappa}}arctanh(\sqrt{\kappa}|| - \mathbf{x} \oplus^\kappa \mathbf{y}||) \tag{4}$$

Product Space of Hyperbolic Space: $\mathbb{P}_{m,n,\kappa}$ of dimension $d = m \cdot n$ consists of m component spaces $\mathbb{H}^{n,\kappa}$ of dimension n: $\mathbb{P}_{m,n,\kappa} = \overbrace{\mathbb{H}^{n,\kappa} \times \ldots \times \mathbb{H}^{n,\kappa}}^{m \text{ times}}$.

A vector on $\mathbb{P}_{m,n,\kappa}$ can be decomposed into m sub-vectors in $\mathbb{H}^{n,\kappa}$ of dimension n. For any $\mathbf{x}, \mathbf{y} \in \mathbb{P}_{m,n,\kappa}$, where $\mathbf{x} = (\mathbf{x}^1, \ldots, \mathbf{x}^m)$, $\mathbf{y} = (\mathbf{y}^1, \ldots, \mathbf{y}^m)$, the Möbius addition $\oplus^\kappa$ can be extended to $\oplus_{\mathcal{P}}^\kappa$ on $\mathbb{P}_{m,n,\kappa}$:

$$\mathbf{x} \oplus_{\mathcal{P}}^\kappa \mathbf{y} = (\mathbf{x}^1 \oplus^\kappa \mathbf{y}^1, \ldots, \mathbf{x}^m \oplus^\kappa \mathbf{y}^m). \tag{5}$$

The distance between $\mathbf{x}, \mathbf{y}$ is calculated as the sum of all Riemannian distances between the sub-vectors $\mathbf{x}^i, \mathbf{y}^i$ (Eq. 6).

$$d_{\mathcal{P}}^\kappa(\mathbf{x}, \mathbf{y})^2 = \sum_{i=1}^m \left(\frac{2}{\sqrt{\kappa}}arctanh(\sqrt{\kappa}|| - \mathbf{x}^i \oplus^\kappa \mathbf{y}^i||)\right)^2. \tag{6}$$

3.3   Extrapolation

Extrapolation to Unseen Triples. This is the most studied part in the past work, which refers to the *ability of models that can extrapolate from seen facts to unseen facts*. Commonly, a KG is split into the training set $\mathcal{F}_{tr}$, validation set $\mathcal{F}_{val}$, and test set $\mathcal{F}_{tst}$. The $\mathcal{F}_{tr}$ is used to train the KGE model, the $\mathcal{F}_{val}$ is to select the hyper-parameters of the KGE model, and $\mathcal{F}_{tst}$ is to test the extrapolation, where the three sets do not share any triples.

Extrapolation to Under-represented Entities. Under-represented entities are those entities that have a limited number of occurrence in the training set $\mathcal{F}_{tr}$. We formulate it as entities with frequency less than a threshold ϵ_e: $e \in \{e|f_e < \epsilon_e\}$, where ϵ_e should be a reasonable number depending on actual situations. A good KGE model should still perform well on triples that contain under-represented entities. Past work rely on learning biases for different entities, which essentially service as a term that adjusts the scores to match the prior probability of the entities. This approach has a strong drawback: it assumes the prior probabilities of the entities are similar across the $\mathcal{F}_{tr}$ and $\mathcal{F}_{tst}$ as in Fig. 1. This is not guaranteed to be true in real-world situations. For example in industry the KGE models can trained on large general dataset but applied on specific test sets. This significantly limits the extrapolation of KGE models to datasets whose prior probabilities are different from the training set. A better solution would be to decide the usage of the prior probabilities of entities depending on the actual situations.

Extrapolation to Under-represented Relations. Under-represented relations are relations that have a limited presence in the training data $\mathcal{F}_{tr}$. We formulate it as relations with a frequency less than a threshold ϵ_r: $r \in \{r|f_r < \epsilon_r\}$.

3.4   Composition Patterns

We differentiate between *weak composition patterns* and *strong composition patterns*. An intuitive understanding of both is provided in Fig. 2, with formal definitions presented below.

Weak Composition. A KGE model $(\mathcal{P}, \mathcal{G}, s)$ satisfies *weak composition* if there exist $g_1, g_2, g_3 \in \mathcal{G}$ such that $\forall \mathbf{e}, \mathbf{e}', \mathbf{e}'' \in \mathcal{M}$, we have $g_1(\mathbf{e}) = \mathbf{e}' \wedge g_2(\mathbf{e}') = \mathbf{e}'' \Rightarrow g_3(\mathbf{e}) = \mathbf{e}''$.

Strong Composition. A KGE model $(\mathcal{P}, \mathcal{G}, s)$ satisfies *strong composition* if for any $g_1, g_2 \in \mathcal{G}$, there exists $g \in \mathcal{G}$ such that $\forall \mathbf{e}, \mathbf{e}', \mathbf{e}'' \in \mathcal{M}_{ent}$, we have $g_1(\mathbf{e}) = \mathbf{e}' \wedge g_2(\mathbf{e}') = \mathbf{e}'' \Rightarrow g(\mathbf{e}) = \mathbf{e}''$.

4   Method: HolmE

We aim to design a KGE model that can (1) encode strong relation patterns such as symmetry, inversion and strong composition; (2) learns KG embeddings in Riemannian space that allows better expressivity in low-dimensional space; (3) possesses good extrapolation to unseen triples, under-represented entities, and relations. We elaborate on the formulas of HolmE and the rationale behind them; then prove HolmE supports strong relation patterns; and then compare HolmE to representative KGE methods.

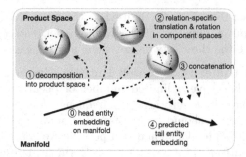

Fig. 3. HolmE performs relation-specific translation and rotation in component spaces of the product space (illustrated with 3D balls, with the shape $\mathbb{P}_{m,n,\kappa}, m, n, \kappa$ indicate the number of component spaces, the component space ($\mathbb{H}^{n,\kappa}$) dimension and the curvature of the hyperbolic space respectively.)

4.1 HolmE: Holomorphic KG Embedding

Intuition. HolmE decomposes the embeddings of the head entity in the manifold $\mathbb{P}_{m,n,\kappa}$ (Fig. 3.0) to sub-vectors in the component spaces of the product space $\mathbb{H}^{n,\kappa}$ (Fig. 3.1), and performs relation-specific translation and rotation to the sub-vectors (Fig. 3.2); then, the sub-vectors are concatenated (Fig. 3.3) to the predicted embedding of the tail entity (Fig. 3.4). We give the general form of HolmE as follows:

Definition 1. *Let $(\mathcal{P}, \mathcal{G})$ be the embedding space, where $\mathcal{P} = \mathbb{P}_{m,n,\kappa}$, and $\mathcal{G}$ is a group of mappings:*

$$g_{(a,A)} : \mathbb{P}_{m,n,\kappa} \rightarrow \mathbb{P}_{m,n,\kappa} \tag{7}$$
$$e \mapsto a \oplus_{\mathcal{P}}^{\kappa} (A \cdot e). \tag{8}$$

and the scoring function of a tuple (h, r, t) is defined by

$$s(e_h, g_r, e_t) = -d_{\mathcal{P}}^{\kappa} (g_r(e_h), e_t) \, [+b_h + b_t], \tag{9}$$
$$g_r(e_h) = a_r \oplus_{\mathcal{P}}^{\kappa} (A_r \cdot e_h), \tag{10}$$

where $e_h, e_t \in \mathcal{P}$ are embeddings of the head entity and tail entity on the manifold; $g_r = g_{(a_r, A_r)} \in \mathcal{G}$ is the relation-specific mapping, with parameters of translation $a_r \in \mathcal{P}_{m,n,\kappa}$ and rotation $A_r \in SO(d)$ (special orthogonal matrix); $\oplus_{\mathcal{P}}^{\kappa}$ is the extended Möbius addition on product space $\mathcal{P}$ with curvature κ; b_h and b_t are biases of the head entity and the tail entity (they are optional and thus in brackets, details see the bias paragraph and Sect. 5.3); the distance $d_{\mathcal{P}}^{\kappa}$ is calculated on the product space $\mathcal{P}$.

We now elaborate on each part of HolmE and the rationales behind them in the following paragraphs.

Entity Embedding with Learned Curvature. The entities are embedded as vectors in a hyperbolic manifold $\mathcal{M}$ with a learned constant curvature κ (Fig. 3.0). They are decomposed into sub-vectors in a product space $\mathcal{P}$ with a series of n dimensional component spaces (Fig. 3.1). These component spaces also share the same κ. The curvature is learned because it adjusts the embedding space to better distribute points throughout the space [6]. We adopt a constant curvature because it is required to support strong composition[1].

Relation Embedding. The relation mapping g_r consists of two components, introduced as below:

Rotation in Product Space. A tempting choice is to model rotation simply as a high dimensional rotation matrix, which contains a high number of parameters and is very difficult to learn. HolmE decomposes a high dimensional rotation into the product space with a series of n dimensional component spaces of curvature κ (Fig. 3.2). The rotation matrix $\mathbf{A}_r$ is a special orthogonal matrix (namely $|\mathbf{A}_r| = 1$), in the form of a diagonal rotation matrix consisting of a series of rotation matrices of n dimensions: $\mathbf{A}_r = diag[\mathbf{R}^n(\theta_{r,1}), \ldots, \mathbf{R}^n(\theta_{r,d/n})]$, where $\mathbf{R}^n(\theta_r)$ is a n dimensional rotation matrix, $(n = 2, 3, \ldots)$. When $n = 2$, $\mathbf{R}^n(\theta_r)$ is the special case of rotation in complex space (Eq. 11), and HolmE becomes a holomorphic function (this is where the name HolmE comes from).

$$\mathbf{R}^n(\theta_r) := \begin{bmatrix} cos(\theta_r) & -sin(\theta_r) \\ sin(\theta_r) & cos(\theta_r) \end{bmatrix}. \tag{11}$$

Translation in Product Space. The translation of HolmE is performed with extended *Möbius addition* (Eq. 5) in the component spaces with the curvature κ (Fig. 3.2). This means the Möbius addition is performed between the sub-vectors of $\mathbf{a}_r$ and the resulting vectors of the rotation $(\mathbf{A}_r \cdot \mathbf{e}_h)$, where each sub-vector has the shape $\mathcal{P}_{d,n}$. We adopt the translation in the product space instead of high dimensional translation, different from past work [6], because translation in the product space matches the rotation in the product space and thus geometrically makes more sense, and this matching is required by strong composition. Important to note is that the Möbius addition here is the *left addition*, namely the translation vector $\mathbf{a}_r$ must be on the left hand side, for ensuring strong composition (See footnote 1).

Scoring Function and Biases. HolmE has two forms of scoring functions: (1) one is simply the distance $s = d_{\mathcal{P}}^\kappa(g_r(\mathbf{e}_h), \mathbf{e}_t)$, which gives HolmE; (2) the other one is the distance adding the biases of head entity and tail entity $s = d_{\mathcal{P}}^\kappa(g_r(\mathbf{e}_h), \mathbf{e}_t) + b_h + b_t$, which results in HolmE-b. The two forms of HolmE should be used in different scenarios. HolmE-b should be used when the test set is assumed to have an entity distribution similar to the training set; HolmE should be used in other cases. Details see Sect. 5.3.

---

[1] Proof see https://github.com/nsai-uio/HolmE-KGE/blob/main/Proof.pdf.

Loss and Training of HolmE. We adopt cross-entropy loss with uniform negative sampling as past works:

$$\mathcal{L} = \sum_{(h,r,t)\in\mathcal{F}\cup\mathcal{F}'} log(1 + exp(-ys(\mathbf{e}_h, g_r, \mathbf{e}_t))),$$

where $\mathcal{F}$ and $\mathcal{F}'$ denote training examples and negative examples respectively. The negative examples are sampled uniformly from all possible triplets obtained by perturbing the tail entities in triplets (h, r, t). The label y is 1 if the triplet is training example, else is -1. The training setting follows the initialisation in tangent space as [7]. In particular, all parameters are initialised and optimised in the tangent space of the manifold using standard Euclidean techniques. These parameters are mapped to the manifold with exponential map, which ensures the embeddings are in the desired space [7].

4.2   Model Analysis

Benefits. HolmE has two major benefits and several supporting benefits: (1) The design of HolmE gives clear geometric meaning and relatively good transparency (illustrated in Fig. 3). (1.1) HolmE decomposes high dimensional computation in product space and saves parameters. (1.2) HolmE has a matching rotation, translation and distance function on the product spaces. (2) HolmE has good extrapolation to unseen triples, under-represented entities and relations (Sect. 5). (2.1) HolmE provides optional biases applied according to actual scenarios. (2.2) HolmE supports strong composition (Theorem 1).

Theorem 1. *The KGE defined by Definition 1 satisfies strong composition[2].*

Complexity Analysis. The time complexity of HolmE mainly comes from scoring calculation and cross entropy loss calculation. While the operation for calculating loss is consistent across various KGE models, the complexity of the score function in d dimentional HolmE is linear, i.e., $O(d)$ for each triple. This complexity is comparable to that of traditional KGE models such as TransE and lower than that of neural network-based models such as ConvE, as no additional graph aggregation operations are incorporated, making it more efficient in scenarios where computational simplicity and speed are critical. In terms of space complexity, for a KG with $|\mathcal{E}|$ entities and $|\mathcal{R}|$ relations, HolmE need $d\,|\mathcal{E}|+2d\,|\mathcal{R}|$ parameters, while HolmE-b need more d parameters for entity-specific bias term. Compared with other Riemannian KGE, such as AttH and GIE, (both need at least $(d+1)\,|\mathcal{E}| + 3d\,|\mathcal{R}|$ parameters), HolmE need fewer parameters. Compared with other Euclidean KGE methods, HolmE achieves similar representation performance in a lower embedding dimension [1,6].

---

[2] Proof see https://github.com/nsai-uio/HolmE-KGE/blob/main/Proof.pdf..

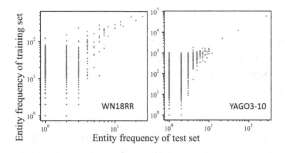

Fig. 4. Entity frequency in training and test set.

Limitations. HolmE is limited in providing theoretically strict mappings satisfying the *transitivity* pattern and mapping patterns of *one-to-many* (1-to-N), *many-to-one* (N-to-1), and *many-to-many* (N-to-N). However, empirical studies show that HolmE still has good performance on these relation patterns or mapping patterns, due to the mechanism in KGE that selects the best scored predicted tail embedding instead of the exact tail embedding (see Appendix).

Comparison to TransE/RotatE. HolmE adopted similar components of translation and rotation as in other translational KGEs. HolmE is a general form that unites KGEs supporting strong composition, including TransE and RotatE. KGEs like neural networks, bilinear models, mixture models cannot guarantee to satisfy strong composition because they lack explicit geometric explanation and the notion of *relation embedding as mapping* is not clearly defined.

Comparison to Hyperbolic KGE. MurP [1] adopts translation on high dimensional hyperbolic space and does not adopt rotation. AttH [6] adopts rotation in product space, but translation and distance function on high dimensional manifolds, with the result that AttH does not support strong composition. GIE [5] is a mixture model of different curvature spaces and not pure hyperbolic, and also does not support strong composition. In contrast, HolmE has rotation, translation, distance on product space and supports strong composition; HolmE also has a special treatment to the bias terms.

5   Experiments

We evaluates these hypotheses: (H1) HolmE outperforms SotA on extrapolation to unseen triples in low-dimensional space; (H2) HolmE outperforms SotA on extrapolation to under-represented data; (H2.1) in scenarios where entity's frequency distribution between the training and testing sets are sufficiently different, HolmE outperforms others; (H2.2) in scenarios where relation's distribution between the training and testing sets are sufficiently different, HolmE outperforms others.

Table 1. Dataset statistics. $\bar{f}_e$, $\bar{f}_r$: average of entity and relation frequency.

Dataset	#ent	#rel	#tri	$\bar{f}_e$	$\bar{f}_r$
WN18RR	41k	11	93k	4.3	7.9k
FB15k-237	15k	237	310k	37.5	1.1k
YAGO3-10	123k	37	1M	17.5	29.1k

5.1   Overall Experiment Design

To test H1, we conduct experiments (Sect. 5.2) to train HolmE and baseline KGE models on $\mathcal{F}_{tr}$ and test on unseen triples $\mathcal{F}_{tst}$. This follows the standard setting of past work. Moreover, we design experiments evaluating the influence of strong composition. To test H2 (H2.1 in Sect. 5.3, H2.2 in Sect. 5.4), we plot the MRR along entity (Sect. 5.3) or relation frequency to evaluate performance of various models. In addition, we split the benchmark data to create scenarios where the frequency distribution of entities and relations are sufficiently different, we test the model performance on these scenarios. In all of our experiments, we conducted a hyperparameter search for the batch size b (chosen from {500, 1000, 2000}); optimizer op (selected from Adam or Adagrad); the learning rate lr (for Adam ranging from [0.0001, 0.005], for Adagrad ranging from [0.01, 0.1]); negative sampling size neg (chosen from {50, 150, 200}). For the high-dimensional embedding, the embedding size is searched in the range of {200, 400, 500}. All the experiments are done using GPU Nvidia A100-80GB (training and testing models) and CPU Core(TM) i7-11850H (other scripts).

5.2   Extrapolation to Unseen Triples

Benchmark Datasets. We use 3 standard datasets: WN18RR [11], FB15k-237 [26], YAGO3-10 [23]. The statistics of these datasets are reported in Table 1. Figure 1a and Fig. 4 indicate in all three benchmark datasets, there is high correlation between entity frequency in the training and test set.

Datasets of under Represented Entities. The datasets WN18RR-under-e and FB15k-237-under-e are generated by selecting all triples with under-represented target entities, where under-represented entities are entities with frequency less than a threshold and stand for about 80% of total entities. Table 2 reports the statistics of these new test data. Appendix shows that the different values of threshold do not influence the validity of claims in the paper.

The series of test sets with different percentage of under-represented entities shown in Fig. 7 is generated by random sampling. We denote the set of triples of under-represented entities in the WN18RR or FB15k-237 as $\mathcal{F}_{ue}$ (with cardinality f_{ue}) and the set of other triples as $\mathcal{F}_{we}$ (with cardinality f_{we}). We want to generate a dataset with the percentage of triples with under-represented entities of p_{tst}. then the percentage of other triples is $1 - p_{tst}$. For each p_{tst}, we would like to generate the test set repeatedly in five different scales $sc \in \{20\%, 40\%, 60\%, 80\%, 100\%\}$. Then the random sampling is performed as: For a given p_{tst}, for a scale $sc \in \{20\%, 40\%, 60\%, 80\%, 100\%\}$, we uniformly random sample $f_{ue} \times po$ triples from $\mathcal{F}_{ue}$ and uniformly random sample $(f_{ue} \times po)/p_{tst} * (1 - p_{tst})$ triples from $\mathcal{F}_{we}$.

Datasets of Under-represented Relations. Similarly, the datasets WN18RR-under-r and FB15k-237-under-r are generated by selecting all triples with under-represented relations, where under-represented relation are relations

Table 2. Dataset statistics for test sets with under-represented entities and relations. f_u/f_{total} indicates the number of test queries in the dataset defined by the number of test queries in their original test sets of WN18RR/FB15k-237. Threshold is the frequency to determine under-represented entities/relations such that under-represented entities/relations always take about 80% of the total entities/relations.

Dataset	#entities	entity percentage	#test queries	f_u/f_{total}	threshold
WN18RR-under-e	33k	81.0%	3487	55.6%	5
FB15k-237-under-e	11k	80.4%	18k	44.9%	47
Dataset	#relations	relation percentage	#test queries	f_u/f_{total}	threshold
WN18RR-under-r	9	81.8%	1618	25.8%	7402.0
FB15k-237-under-r	189	79.7%	10k	26.5%	1282.4

Table 3. HolmE outperforms SotA in link prediction for extrapolation to unseen triples in low-dimensional space (d = 32). Best score in bold and second best underlined. Sources are indicated by citations, or generated/reproduced by us with open source code.

$\mathcal{M}$	Model	WN18RR				FB15k-237				YAGO3-10			
		MRR	H@1	H@3	H@10	MRR	H@1	H@3	H@10	MRR	H@1	H@3	H@10
$\mathbb{R}$	TransE	.177	.031	.269	.447	.110	.066	.114	.196	–	–	–	–
$\mathbb{C}$	RotatE [6]	.387	.330	.417	.491	.290	.208	.316	.458	–	–	–	–
$\mathbb{C}$	ComplEx-N3 [6]	.420	.390	.420	.460	.294	.211	.322	.463	.336	.259	.367	.484
$\mathbb{R}$	ConvE	.438	.410	.440	.520	.325	.237	.356	.501	.430	.335	.479	.580
$\mathbb{R}$	CompGCN	.467	.433	.481	.532	.323	.237	.351	.500	.438	.329	.502	.586
$\mathbb{H}$	MurP [6]	.465	.420	.484	.544	.323	.235	.353	.501	.230	.150	.247	.392
$\mathbb{H}$	AttH [6]	.466	.419	.484	.551	.324	.236	.354	.501	.397	.310	.437	.566
M	GIE	<u>.472</u>	<u>.424</u>	<u>.492</u>	.558	.320	.230	.350	.503	–	–	–	
$\mathbb{P}$	HolmE-b	**.486**	**.440**	**.502**	**.575**	<u>.329</u>	**.238**	<u>.358</u>	<u>.511</u>	**.454**	**.349**	**.518**	**.655**
$\mathbb{P}$	HolmE	.466	.415	.489	<u>.561</u>	**.331**	<u>.237</u>	**.366**	**.517**	<u>.441</u>	<u>.333</u>	<u>.507</u>	<u>.641</u>

with frequency less than a threshold and stand for about 80% of total relations. Table 2 reports the statistics of these new test data.

Baselines. We compare HolmE and the variant with bias (HolmE-b) with three types of baselines: (1) representative works that provide insights in this field, including TransE [3], RotatE [25], and ComplEx-N3 [18]; (2) graph neural network based models, including ConvE [11] and CompGCN [28] (2) Riemannian KGE, including MurP [1], AttH [6], and a mixture model of Euclidean, sphere, and hyperbolic space, GIE [5].

Evaluation Metrics. We adopt *mean reciprocal rank* (MRR), calculated as the mean of reciprocal ranks of the predicted entity; hits at k (H@K, $k \in \{1, 3, 10\}$), calculated as the percentage of the correct triples among the top k predictions.

Results Analysis. The results (Table 3) show that the HolmE models can outperform baselines in low-dimensional embedding space. On WN18RR, FB15k-237, YAGO3-10, the best HolmE model outperform the best published baseline

Table 4. Verifying the influence of strong composition with pretraining and fine-tuning

Training set:	Pretrained on train-remain			Fine tuned on train-com		
Test set	test-remain	test-com	total-test	test-remain	test-com	total-test
AttH	0.324	0.064	0.294	0.227	0.272	0.228
AttH-ub	0.318	**0.105**	0.294	0.228	0.248	0.229
HolmE-b	<u>0.337</u>	<u>0.090</u>	**0.342**	**0.246**	**0.285**	**0.247**
HolmE	**0.331**	0.050	<u>0.299</u>	<u>0.236</u>	<u>0.276</u>	<u>0.237</u>

by 3%, 2.2%, and 14.4%, respectively. HolmE is sometimes worse than models with bias (GIE, HolmE-b), because the three datasets have similar entity distribution in the training set and test set (Fig. 1a and Fig. 4). Though, HolmE can still perform impressively well, especially on FB15k-237. We postulate in this case, the gain of supporting strong composition sometimes outmatches the loss of dropping the bias terms.

Link Prediction Performance along Embedding Dimension. Following [6], we test HolmE with different embedding sizes and compare with representative baselines. The results (Fig. 5) reveal that HolmE achieves impressive performance in low dimensions; and it consistently outperform these baselines in a wide range of dimensions.

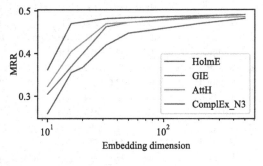

Fig. 5. MRR for KGE models with $d \in \{10, 16, 20, 32, 50, 200, 500\}$ on WN18RR.

Influence of Strong Composition. To verify that strong composition can really improve model performance for under-represented relations, we design experiments of fine tuning. The rationale behind is that strong composition has entire relation embedding space supporting composition, thus under-represented relations will also be embedded to support composition. To verify this, we remove a subset of triples from the training set of FB15k-237, such that some relations (denoted as r^{com}) that support composition pattern become under-represented in the remaining training set (train-remain). The removed subset is denoted as train-com. Correspondingly, we also remove all triples that contain r^{com} from the test set of FB15k-237 (test-total), the remaining test set is denoted as test-remain, and the removed subset is denoted as test-com. KGE models will be first pre-trained on FB15k-237, and tested on test-remain and test-com. Then these pre-trained models will be fine-tuned with train-com and again tested on test-remain and test-com. The results (Table 4) shows after fine-tuning, HolmE always outperform their counterpart in the case of with bias or

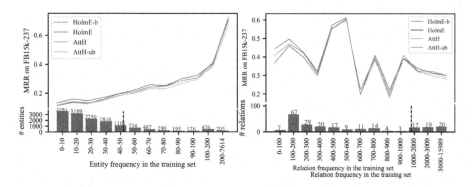

Fig. 6. MRR of four models (Table 3) along entity and relation frequency for FB15k-237. Left: for triples with under-represented entities, unbiased model (HolmE, AttH-ub) outperform biased model (HolmE-b, AttH), and HolmE outperform other models. Right: HolmE achieves better performance for triples with under-represented relations.

without bias: HolmE-b > AttH, HolmE > AttH-ub. The performance of all model deteriorate on test-remain and increases significantly on test-com. We postulate that HolmE models can easily adapt the embeddings learned on *train-remain* for under-represented relations to new embeddings when the composition patterns are revealed by *train-com*. While AttH models are worse in this regard, we postulate the reason is the relation embeddings learned by AttH are in the sub-space that does not support composition after pretraining on *train-remain*, and need more change to be adapted to the sub-space that supports composition when the composition patterns are revealed by *test-com*. HolmE models are easier to adapt the relation embeddings because the its entire relation embeddings space supports strong composition.

5.3   Extrapolation to Under-Represented Entities

Correlations between Entity Distribution and Bias Terms. To verify H2.1, we test the correlations between entity frequency in the training set and test set $(F_{tr} - F_{tst})$, between the training set and the bias terms $(F_{tr} - b_t)$ for the WN18RR and FB15k-237. The results show that these distributions are highly correlated (Fig. 1 and Appendix). Especially FB15k-237 exhibits a Pearson correlation coefficient of 0.91 for $F_{tr} - F_{tst}$. The other correlations are also evident, and sometimes of higher orders than simple linear correlation.

Model Performance Influenced by Entity Frequency. To verify H2.1, we plot the number of entities along the entity frequency (occurrence of triples) in the training set of FB15k-237 (Fig. 6.left), which has a reasonably rich set of relations and entities. We also plot performance of four KGE models in Table 3 along the entity frequency in Fig. 6.left. We choose to compare AttH with HolmE because it is a representative hyperbolic KGE model and it is the best SotA model in low-dimensional space for FB15k-237. In addition to the original AttH

Table 5. Testing extrapolation to under-represented entities and relations. Models without bias (HolmE, AttH-ub) outperform models with bias (HolmE-b, AttH) on test set of triples only with under-represented entities. HolmE outperforms SotA models.

Model	WN18RR-under-e				FB15k-237-under-e				WN18RR-under-r				FB15k-237-under-r			
	MRR	H@1	H@3	H@10	MRR	H@1	H@3	H@10	MRR	H@1	H@3	H@10	MRR	H@1	H@3	H@10
AttH	.371	.335	.383	.437	.142	.078	.148	.269	.312	.242	.338	.434	.390	.301	.426	.566
AttH-ub	.391	.355	.402	.463	.148	.086	.154	.271	.276	.209	.300	.408	.392	.304	.425	.568
HolmE-b	.391	.350	.403	.472	.148	.081	.157	.280	.341	.265	.374	.494	.403	.311	.443	.579
HolmE	.401	.355	.419	.483	.162	.093	.175	.297	.308	.227	.342	.461	.413	.322	.452	.590

version with bias, we add an extra version of AttH without bias (denoted as AttH-ub). Figure 6.left reveals several observations: (1) Most entities are under-represented and have very low frequency compared to the mean frequency of 37.5 and the maximum frequency of 7614; over 10k entities have samples below 50. (2) All model performance is relatively poor for triples with under-represented entities, and increases as the entity frequency increases; (3) Models without bias (HolmE, AttH-ub) outperform models with bias (HolmE-b, AttH) for triples with under-represented entities; and the other way around for other triples. This is further studied in the next paragraph.

Scenarios Where Entity Frequency Differs between the Training and the Test Set. To better understand the influence of bias and under-represented entities on model performance and further verify H2.1, we create subsets of test data of the two benchmark data sets, which include only triples with under-represented entities (WN18RR-test-under-e and FB15k-237-test-under-e). To quantify under-represented data, we consider the Pareto principle (80%-20% principle) and choose a frequency threshold (Fig. 6.left) such that 80% entities are regarded as under-presented entities (the choice of threshold does not influence the validity of the observations and our claims, see influence of threshold in Appendix). We test the same models in Fig. 6.left on these two test sets. The results (Table 5) shows that models without bias (HolmE, AttH-ub) are persistantly better than models with bias (HolmE-b, AttH), and HolmE outperforms other models.

To ensure the created test sets reveal systematic effect, we additionally create a series of test sets with different percentage of triples with under-represented entities via random sampling from the test set of FB15k-237 (Table 2). The same four models are tested on these test sets. Figure 7 reveals the tendency of model performance along the percentage of triples with under-represented entities in the test set. including these observations: (1) As the percentage of triples with under-represented entities (p_{tst}) increases in the test sets, all model performance deteriorates; (2) As p_{tst} increases, AttH-ub (without bias) gradually outmatches AttH (with bias), and the degree that HolmE (no bias) outmatches HolmE-b increases; (3) HolmE always outperforms other models.

5.4   Extrapolation to Under-Represented Relations

Model Performance Influenced by Relation Frequency. To verify H2.2, we plot the performance of KGE model trained on FB15k-237-train along the relation frequency (Fig. 6.right). We can see for under-represented relations (rare relations) HolmE models have better performance than AttH. For well-represented relations (the very right bar), the performance of models is very close. The model performance does not change monotonously along the relation frequency. This is likely because the results are aggregated from both under-represented entities and well-represented entities, leading to a complex trend.

Scenarios Where Relation Frequency Differs between the Training and the Test Set. To verify H2.2, we create test sets of triples only with under represented relations (WN18RR-test-under-r and FB15k-237-test-under-r). The under-represented relations are relations with frequency lower than a threshold. Similar to Sect. 5.3 we adopt a threshold such that 80% relations are under-represented (the choice of threshold does not influence the validity of the observations and our claims, see influence of threshold in Appendix). We test the same four models as in Fig. 6.left on these two datasets. The results (Table 5) shows that sometimes models with bias are

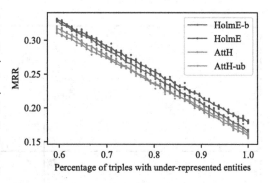

Fig. 7.   Each dot is an experiment on test sets created based on FB15k-237. The lines are average performance of five repeated experiments. Performance of unbiased model (HolmE) increases as percentage of triples with under-represented entities increases. The results on the very left hand side are namely in Table 3. The results on the very right hand side are namely in Table 5.

better than without bias (for WN18RR). We postulate it is because the results are aggregated from both under-represented entities and well-represented entities, and sometimes well-represented entities takes a major effect. HolmE models consistently outperform AttH models for under-represented relations for both the cases of with bias or without bias. This confirms H2.2.

6   Conclusion

This work proposes HolmE, a general form of hyperbolic KG embedding method that addresses multi-fold extrapolation: to unseen triples, under-represented entities, and relations (the latter two limitedly discussed in past work). We give in-depth analysis of the influence of bias terms on model performance on under-represented entities. We also show that HolmE supports strong composition, with the entire relation embedding space that support composition. We prove that main stream KGE methods are either special cases of HolmE and thus support strong composition, or they do not support strong composition.

A   Result Details

Correlation between Entity Frequency and Bias Term. Fig. 8 shows that in different data sets and models, there is strong correlation between bias term and entity frequency.

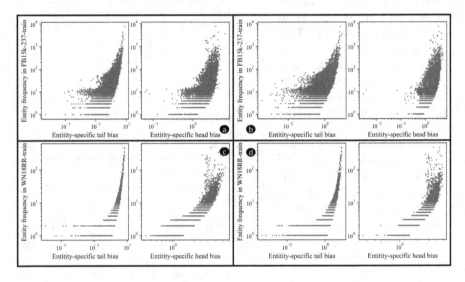

Fig. 8. Correlation between bias term and entity frequency. a,b: Atth and HolmE in FB15k-237 respectively; c,d: Atth and HolmE in WN18RR respectively.

Influence of Entity and Relation Frequency Threshold. Fig. 9 shows that results for different choice of entity and relation frequency threshold does not influence the validity of the observations and claims: (Fig. 9a) models without bias (HolmE, AttH-ub) outperform their counterparts with bias (HolmE-b, AttH), and HolmE outperform other models for triples with under-represented entities; (Fig. 9b) the majority of entities are under-represented entities as long as the threshold is not chosen to be extremely small, considering that the mean of entity frequency is 37.5 and the maximum is 7614. The results are obtained on FB15k-237. HolmE outperforms other models for triples with under-represented relations (Fig. 9c); the majority of relations are under-represented relations as long as the threshold is not chosen to be extremely small (Fig. 9d), considering that the mean of relation frequency is 1148.2 and the maximum is 15989. The results are obtained on FB15k-237.

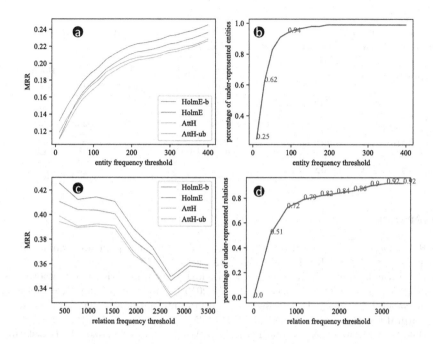

Fig. 9. The choice of threshold does not influence the observations or claims

Link Prediction Results in High Dimension. As expected, embeddings in different spaces achieve similar results in high dimensional space (Table 6), because both Euclidean and hyperbolic spaces become expressive enough to represent complex hierarchies in KGs [6]. Similar to the results in low-dimensional space, HolmE (without bias) is slightly worse than the models with bias (GIE, HolmE-b), although HolmE does not use the advantage of prior probabilities provided by the bias term (Tables 7 and 8).

Table 6. HolmE achieves comparable results as SotA hyperbolic KGE in high dimensional embedding space. Best score in bold and second best underlined. Sources are indicated by citations, or reproduced by us with open source code.

$\mathcal{M}$	Model	WN18RR				FB15k-237				YAGO3-10			
		MRR	H@1	H@3	H@10	MRR	H@1	H@3	H@10	MRR	H@1	H@3	H@10
ℝ	TransE [3]	.226	–	–	.501	.294	–	–	.465	–	–	–	–
ℂ	RotatE [25]	.476	.428	.492	.571	.338	.241	.375	.533	.495	.402	.550	.670
ℂ	ComplEx-N3 [18]	.480	.435	.495	.572	<u>.357</u>	<u>.264</u>	<u>.392</u>	<u>.547</u>	.569	<u>.498</u>	.609	.701
ℝ	ConvE	.43	.40	.44	.52	.325	.237	.356	.501	–	–	–	–
ℝ	CompGCN	.479	.443	.494	.546	.355	.264	.390	.535	–	–	–	–
ℍ	MurP [1]	.481	.440	.495	.566	.335	.243	.367	.518	.354	.249	.400	.567
ℍ	AttH [6]	.486	.443	.499	.573	.348	.252	.384	.540	.568	.493	<u>.612</u>	<u>.702</u>
M	GIE [5]	**.491**	**.452**	<u>.505</u>	<u>.575</u>	**.362**	**.271**	**.401**	**.552**	**.579**	**.505**	**.618**	**.709**
ℍ	HolmE-b	**.491**	<u>.445</u>	**.509**	**.584**	.352	.260	.383	.542	<u>.570</u>	.497	<u>.612</u>	.701
ℍ	HolmE	.479	.433	.497	.567	.352	.259	.389	.545	.546	.469	.597	.694

Table 7. Detailed analysis for relation patterns on FB15k-237. Sym. : Symmetry, Asym.: ASymmetry, Inv.: Inversion, Tran.: Transitivity, Comp.: Composition.

	AttH	AttH-ub	HolmE	HolmE-b
Sym	0.322	0.312	0.330	**0.336**
Asym	0.326	0.319	**0.338**	0.336
Inv	0.319	0.312	**0.331**	0.329
Tran	0.314	0.296	0.317	**0.320**
Comp	0.279	0.273	**0.292**	**0.292**

Table 8. Detailed analysis for relation mapping properties in low-dimensional space on FB15k-237.

Task	RMPs	AttH	AttH-ub	HolmE	HolmE-b
Predicting Head (MRR)	1-to-1	.460	.458	**.485**	.478
	1-to-N	.430	.411	.456	**.459**
	N-to-1	.079	.077	**.099**	.085
	N-to-N	.243	.235	**.253**	**.253**
Predicting Tail (MRR)	1-to-1	.452	.467	.475	**.477**
	1-to-N	.060	.059	**.077**	.072
	N-to-1	.728	.725	**.743**	.742
	N-to-N	.350	.343	**.360**	**.360**

References

1. Balažević, I., Allen, C., Hospedales, T.: Multi-relational poincaré graph embeddings. Adv. Neural Inf. Process. Syst. **32** (2019)
2. Balažević, I., Allen, C., Hospedales, T.: TuckER: tensor factorization for knowledge graph completion. In: Proceedings of the 2019 Conference on Empirical Methods in Natural Language Processing and the 9th International Joint Conference on Natural Language Processing (EMNLP-IJCNLP), pp. 5185–5194 (2019)
3. Bordes, A., Usunier, N., Garcia-Duran, A., Weston, J., Yakhnenko, O.: Translating embeddings for modeling multi-relational data, vol. 26 (2013)
4. Cao, E., Wang, D., Huang, J., Hu, W.: Open knowledge enrichment for long-tail entities. In: Proceedings of The Web Conference 2020, pp. 384–394 (2020)
5. Cao, Z., Xu, Q., Yang, Z., Cao, X., Huang, Q.: Geometry interaction knowledge graph embeddings. In: AAAI Conference on Artificial Intelligence (2022)
6. Chami, I., Wolf, A., Juan, D.C., Sala, F., Ravi, S., Ré, C.: Low-dimensional hyperbolic knowledge graph embeddings. In: Proceedings of the 58th Annual Meeting of the Association for Computational Linguistics, pp. 6901–6914 (2020)
7. Chami, I., Ying, Z., Ré, C., Leskovec, J.: Hyperbolic graph convolutional neural networks. Adv. Neural Inf. Process. Syst. **32** (2019)
8. Chu, P., Bian, X., Liu, S., Ling, H.: Feature space augmentation for long-tailed data. In: Vedaldi, A., Bischof, H., Brox, T., Frahm, J.-M. (eds.) ECCV 2020. LNCS, vol. 12374, pp. 694–710. Springer, Cham (2020). https://doi.org/10.1007/978-3-030-58526-6_41
9. Dai, E., et al.: A comprehensive survey on trustworthy graph neural networks: privacy, robustness, fairness, and explainability. arXiv preprint arXiv:2204.08570 (2022)
10. Dai Quoc Nguyen, T.D.N., Nguyen, D.Q., Phung, D.: A novel embedding model for knowledge base completion based on convolutional neural network. In: Proceedings of NAACL-HLT, pp. 327–333 (2018)
11. Dettmers, T., Minervini, P., Stenetorp, P., Riedel, S.: Convolutional 2D knowledge graph embeddings. In: Proceedings of the AAAI Conference on Artificial Intelligence, vol. 32 (2018)
12. Djeddi, W.E., Hermi, K., Ben Yahia, S., Diallo, G.: Advancing drug-target interaction prediction: a comprehensive graph-based approach integrating knowledge graph embedding and protbert pretraining. BMC Bioinformatics **24**(1), 488 (2023)
13. Dong, Y., Ma, J., Wang, S., Chen, C., Li, J.: Fairness in graph mining: a survey. IEEE Transactions on Knowledge and Data Engineering (2023)

14. Ebisu, T., Ichise, R.: Toruse: knowledge graph embedding on a lie group. In: Thirty-Second AAAI Conference on Artificial Intelligence (2018)
15. Ganea, O., Bécigneul, G., Hofmann, T.: Hyperbolic neural networks. Adv. Neural Inf. Process. Syst. **31** (2018)
16. Islam, M.K., Amaya-Ramirez, D., Maigret, B., Devignes, M.D., Aridhi, S., Smaïl-Tabbone, M.: Molecular-evaluated and explainable drug repurposing for COVID-19 using ensemble knowledge graph embedding. Sci. Rep. **13**(1), 3643 (2023)
17. Ji, G., He, S., Xu, L., Liu, K., Zhao, J.: Knowledge graph embedding via dynamic mapping matrix. In: Proceedings of the 53rd Annual Meeting of the Association for Computational Linguistics and the 7th International Joint Conference on Natural Language Processing (volume 1: Long papers), pp. 687–696 (2015)
18. Lacroix, T., Usunier, N., Obozinski, G.: Canonical tensor decomposition for knowledge base completion. In: International Conference on Machine Learning, pp. 2863–2872. PMLR (2018)
19. Li, M., Sun, Z., Zhang, S., Zhang, W.: Enhancing knowledge graph embedding with relational constraints. Neurocomputing **429**, 77–88 (2021)
20. Li, R., et al.: House: knowledge graph embedding with householder parameterization. arXiv preprint arXiv:2202.07919 (2022)
21. Lin, Y., Liu, Z., Sun, M., Liu, Y., Zhu, X.: Learning entity and relation embeddings for knowledge graph completion. In: Twenty-Ninth AAAI Conference on Artificial Intelligence (2015)
22. Liu, X., Zhao, F., Gui, X., Jin, H.: LeKAN: extracting long-tail relations via layer-enhanced knowledge-aggregation networks. In: Bhattacharya, A., et al. (eds.) Database Systems for Advanced Applications. DASFAA 2022. LNCS, vol. 13245, pp. 122–136. Springer, Cham (2022). https://doi.org/10.1007/978-3-031-00123-9_9
23. Mahdisoltani, F., Biega, J., Suchanek, F.: YAGO3: a knowledge base from multilingual wikipedias. In: 7th Biennial Conference on Innovative Data Systems Research. CIDR Conference (2014)
24. Pavlović, A., Sallinger, E.: Expressive: a spatio-functional embedding for knowledge graph completion. arXiv preprint arXiv:2206.04192 (2022)
25. Sun, Z., Deng, Z.H., Nie, J.Y., Tang, J.: RotatE: knowledge graph embedding by relational rotation in complex space. In: International Conference on Learning Representations (2018)
26. Toutanova, K., Chen, D.: Observed versus latent features for knowledge base and text inference. In: Proceedings of the 3rd Workshop on Continuous Vector Space Models and their Compositionality, pp. 57–66 (2015)
27. Trouillon, T., Welbl, J., Riedel, S., Gaussier, É., Bouchard, G.: Complex embeddings for simple link prediction. In: International Conference on Machine Learning, pp. 2071–2080. PMLR (2016)
28. Vashishth, S., Sanyal, S., Nitin, V., Talukdar, P.: Composition-based multi-relational graph convolutional networks. arXiv preprint arXiv:1911.03082 (2019)
29. Wang, S., Huang, X., Chen, C., Wu, L., Li, J.: Reform: error-aware few-shot knowledge graph completion. In: Proceedings of the 30th ACM International Conference on Information & Knowledge Management, pp. 1979–1988 (2021)
30. Wang, Z., Zhang, J., Feng, J., Chen, Z.: Knowledge graph embedding by translating on hyperplanes. In: Proceedings of the AAAI Conference on Artificial Intelligence, vol. 28 (2014)
31. Werner, S., Rettinger, A., Halilaj, L., Lüttin, J.: RETRA: recurrent transformers for learning temporally contextualized knowledge graph embeddings. In: Verborgh, R., et al. (eds.) ESWC 2021. LNCS, vol. 12731, pp. 425–440. Springer, Cham (2021). https://doi.org/10.1007/978-3-030-77385-4_25

32. Willmore, T.J.: An Introduction to Differential Geometry. Courier Corporation, Honolulu (2013)
33. Xu, Y.W., Zhang, H.J., Cheng, K., Liao, X.L., Zhang, Z.X., Li, Y.B.: Knowledge graph embedding with entity attributes using hypergraph neural networks. Intell. Data Anal. **26**(4), 959–975 (2022)
34. Yang, B., Yih, S.W.t., He, X., Gao, J., Deng, L.: Embedding entities and relations for learning and inference in knowledge bases. In: Proceedings of the International Conference on Learning Representations (ICLR) 2015 (2015)
35. Zhang, C., Yao, H., Huang, C., Jiang, M., Li, Z., Chawla, N.V.: Few-shot knowledge graph completion. In: Proceedings of the AAAI Conference on Artificial Intelligence, vol. 34, pp. 3041–3048 (2020)

SC-Block: Supervised Contrastive Blocking Within Entity Resolution Pipelines

Alexander Brinkmann[1]([✉]) [ID], Roee Shraga[2] [ID], and Christina Bizer[1] [ID]

[1] University of Mannheim, 68131 Mannheim, Germany
{alexander.brinkmann,christian.bizer}@uni-mannheim.de
[2] Worcester Polytechnic Institute, Worcester, USA
rshraga@wpi.edu

Abstract. Millions of websites use the schema.org vocabulary to annotate structured data describing products, local businesses, or events within their HTML pages. Integrating schema.org data from the Semantic Web poses distinct requirements to entity resolution methods: (1) the methods must scale to millions of entity descriptions and (2) the methods must be able to deal with the heterogeneity that results from a large number of data sources. In order to scale to numerous entity descriptions, entity resolution methods combine a blocker for candidate pair selection and a matcher for the fine-grained comparison of the pairs in the candidate set. This paper introduces SC-Block, a blocking method that uses supervised contrastive learning to cluster entity descriptions in an embedding space. The embedding enables SC-Block to generate small candidate sets even for use cases that involve a large number of unique tokens within entity descriptions. To measure the effectiveness of blocking methods for Semantic Web use cases, we present a new benchmark, WDC-Block. WDC-Block requires blocking product offers from 3,259 e-shops that use the schema.org vocabulary. The benchmark has a maximum Cartesian product of 200 billion pairs of offers and a vocabulary size of 7 million unique tokens. Our experiments using WDC-Block and other blocking benchmarks demonstrate that SC-Block produces candidate sets that are on average 50% smaller than the candidate sets generated by competing blocking methods. Entity resolution pipelines that combine SC-Block with state-of-the-art matchers finish 1.5 to 4 times faster than pipelines using other blockers, without any loss in F1 score.

Keywords: Identity Resolution · Blocking · schema.org · Benchmarking · Supervised Contrastive Learning

© The Author(s), under exclusive license to Springer Nature Switzerland AG 2024
A. Meroño Peñuela et al. (Eds.): ESWC 2024, LNCS 14664, pp. 121–142, 2024.
https://doi.org/10.1007/978-3-031-60626-7_7

1 Introduction

The Web Data Commons (WDC) project regularly extracts schema.org data from the Common Crawl[1] [5]. The extraction[2] from the October 2022 version of the Common Crawl has shown that 2.5 million websites (hosts) use the schema.org vocabulary to annotate product offers, 1.2 million websites annotate information about local businesses such as addresses and opening hours, and 50,000 websites annotate job postings. In total, 502 million records describing products and 55 million records describing local businesses were extracted from the Common Crawl. Due to the shallow coverage of the Common Crawl, the extracted data only represents a fraction of the schema.org data available on the Semantic Web. Applications seeking to integrate schema.org data from the Semantic Web for use cases such as product recommendation, price comparison, or complementing knowledge graphs face two challenges in the link discovery [2,11,35] step of their data integration and cleansing pipeline: (1) they require entity resolution methods that can scale to millions of entity descriptions and (2) these methods must be able to handle the heterogeneity that arises from using data from multiple sources. A scalable blocking method is crucial for these use cases as the pairwise comparison of all records is infeasible. Blocking [9,40,49] employs a computationally inexpensive method to generate a set of candidate pairs which likely contains most matches while being as small as possible. Afterwards, a matcher derives the final set of matching pairs using a computationally more expensive and more precise matching method [30,31,33,36].

In this paper, we propose SC-Block, a blocking method that applies supervised contrastive learning to cluster records that likely describe the same real-world entity in an embedding space. In this blocking-only scenario, SC-Block is compared to different state-of-the-art blocking methods [34,39,49,51]. Additionally, SC-Block and the two most competitive blockers are combined with different state-of-the-art matching methods [6,30,31,44] to evaluate the F1 performance and runtime of complete entity resolution pipelines. The experiments are conducted using existing entity resolution benchmarks as well as WDC-Block, a large new benchmark which is introduced in this paper. Existing benchmarks that are used for evaluating blocking methods are either rather small [33], mostly Wikipedia-related [17] or rely on synthetic data [41]. None of the benchmarks uses large amounts of real-world e-commerce data originating from numerous data sources. WDC-Block fills this gap by requiring the blocking of product offers from 3,259 e-shops. In summary, this paper makes the following contributions:

1. We propose SC-Block a blocking method which applies supervised contrastive learning to position records likely describing the same real-world entity close to each other in an embedding space.

---

[1] https://commoncrawl.org/.
[2] https://webdatacommons.org/structureddata/2022-12/stats/schema_org_subsets.html.

2. We introduce WDC-Block a new large blocking benchmark that requires blocking schema.org product offers from the Semantic Web and features a maximal Cartesian product of 200 billion record pairs and almost 7 million unique tokens within the entity descriptions.
3. We show that SC-Block creates smaller candidate sets than state-of-the-art blocking methods leading to pipelines that execute 1.5 to 2 times faster on the smaller benchmark datasets and 4 times faster on the largest product matching task of WDC-block without negatively affecting F1 scores.
4. We relate the training time of SC-Block to the overall runtime reduction of entity resolution pipelines and show that the runtime reduction overcompensates the training time of SC-Block for larger datasets.

The paper is structured as follows. Section 2 presents the WDC-Block benchmark. Section 3 introduces SC-Block and discusses its supervised contrastive training. SC-Block and the impact of SC-Block on complete entity resolution pipelines are evaluated in Sect. 4 and in Sect. 5, respectively. Related work is discussed in Sect. 6. The new benchmark[3] and the code[4] for replicating all experiments are available online.

2   WDC-Block: A Large Product Blocking Benchmark

This section introduces WDC-Block and compares it to blocking benchmarks from the related work. WDC-Block is built by extending the WDC Products entity matching benchmark[5] [45] with additional product offers from the WDC Product Data Corpus V2020[6]. WDC Products is a multi-dimensional benchmark which supports the evaluation of matching systems along combinations of three dimensions: (1) amount of corner cases, (2) training set size, and (3) amount of unseen entities in the test set. The product offers in WDC Products and the WDC Product Data Corpus have been extracted from the Common Crawl using schema.org annotations. The product offers are clustered by product identifiers such as MPN and GTIN [45]. All offers in the same cluster describe the same real-world entity. Record pairs from the clusters are classified as corner cases if they are difficult to match for a range of baseline matchers [45]. 500 clusters are selected as *seen*. The record pairs in the *seen* clusters are split into train, validation and test sets. Afterwards, record pairs in the test set are replaced with record pairs from *unseen* clusters until the desired percentage of unseen record pairs in the test set is reached. For WDC-Block, we select the dataset containing 80% corner cases, because corner cases are challenging for entity resolution pipelines. We chose the largest training set of WDC Products as our seed training set. We choose the test set with 50% unseen record pairs. This ensures a balance between products that were part of the training set and those that were not.

---

[3] https://webdatacommons.org/largescaleproductcorpus/wdc-block/.

[4] https://github.com/wbsg-uni-mannheim/SC-Block/.

[5] https://webdatacommons.org/largescaleproductcorpus/wdc-products/.

[6] https://webdatacommons.org/largescaleproductcorpus/v2/index.html.

Three Sizes of WDC-Block. To match the setup of previous blocking benchmarks, we divided the WDC Products data into two separate datasets, A and B. The maximum cardinality of matching records across datasets A and B is 15. We then extended these datasets with additional randomly selected offers from the WDC Product Data Corpus V2020 to create three versions of WDC-Block: a small version (WB_{small}), a medium version (WB_{medium}), and a large version (WB_{large}). We ensure that the randomly selected records do not match existing records in the datasets to avoid introducing additional matching pairs. Through these additional records and identical train, validation and test pairs in WB_{small}, WB_{medium} and WB_{large}, we can measure the effect of significantly increased vocabulary sizes (67K to 6.9M tokens) and large Cartesian products ($A \times B$ between $2.5 \cdot 10^7$ and $2.0 \cdot 10^{11}$ pairs) on the performance of blockers. Table 1 provides statistics about the different versions of the WDC-Block benchmark and other blocking benchmarks from related work. The vocabulary size is defined as the number of unique tokens in all datasets that belong to the benchmark task. Tokens are derived from the records by serializing them into entity descriptions (see Sect. 3) and splitting the entity descriptions by whitespace.

Table 1. Statistics of the benchmark tasks.

Benchmark	Dataset A	Dataset B	Pos Train	Neg Train	Pos Val.	Neg Val.	Pos Test	Neg Test	Vocab Size	Cartesian Product
WDC-Block$_{small}$	5.0k	5.0k	6.5k	10.5k	3.2k	5.3k	0.5k	4.0k	67k	$2.5 \cdot 10^7$
WDC-Block$_{medium}$	5.0k	0.2M	6.5k	10.5k	3.2k	5.3k	0.5k	4.0k	1.2M	$1.0 \cdot 10^9$
WDC-Block$_{large}$	0.1M	2.0M	6.5k	10.5k	3.2k	5.3k	0.5k	4.0k	6.9M	$2.0 \cdot 10^{11}$
Abt-Buy	1.1k	1.1k	0.6k	5.1k	0.2k	1.7k	0.2k	1.7k	7.9k	$1.2 \cdot 10^6$
Amazon-Google	1.4k	3.3k	0.7k	6.2k	0.2k	2.1k	0.2k	2.1k	7.4k	$4.5 \cdot 10^6$
Walmart-Amazon	2.6k	22.1k	0.6k	5.6k	0.2k	0.9k	0.2k	0.9k	49.2k	$5.6 \cdot 10^6$
DM$_{2M}$	2.0M	–	–	–	–	–	1.7M	–	3.8M	$2.0 \cdot 10^{12}$
BTC12-Infoboxes	1.6M	8.9M	–	–	–	–	1.5M	–	4.9M	$1.5 \cdot 10^{13}$

Comparison to other Blocking Benchmarks. Similar to WDC-Block, the Abt-Buy (A-B), Amazon-Google (A-G) and Walmart-Amazon (W-A)[7] benchmarks, which are widely used in the related work [31,33,37,43,44,49], cover e-commerce use cases. The statistics in Table 1 show that these benchmarks have much smaller Cartesian products $A \times B$ and vocabulary sizes compared to WDC-Block. Each of these benchmarks only requires blocking data from two e-shops while WDC-Block contains data from 3,259 sources. The DM$_{2M}$ benchmark [37] offers a large Cartesian product but consists of synthetic data. The benchmark is thus biased by its data generation process. Similar to WDC-Block, the blocking benchmarks introduced in [17] are based on data from the Semantic

---

[7] https://github.com/anhaidgroup/deepmatcher/blob/master/Datasets.md.

Web. The largest benchmark from this paper, BTC12-Infoboxes, requires blocking data from the Billion Triples Challenge 2012 dataset which has been crawled from the LOD Cloud and infobox data from DBpedia [17]. Although the dataset BTC12-Infoboxes is larger than WDC-Block, the matches in BTC12-Infoboxes all involve entities that appear in Wikipedia. In contrast, WDC-Block contains data from a different topical domain: e-commerce data from many e-shops. In addition, the e-commerce data in WDC-Block is more recent than the Wikipedia data in BTC12-Infoboxes (2022 versus 2012).

3   SC-Block: Supervised Contrastive Blocking

There are unsupervised [16,42], self-supervised [34,49,51,53], and supervised blocking methods [4,18,39]. The supervised methods use a training set containing matching and non-matching record pairs. Most state-of-the-art matching methods require training sets, which are assembled by many entity resolution projects. The motivation behind supervised blocking is to use the available training data not only for matching but also for blocking. In the context of the Semantic Web, the necessary training data can be derived from schema.org annotations that contain identifiers like GTINs, MPNs or ISBNs. SC-Block utilizes supervised contrastive learning to position record embeddings likely describing the same real-world entity close to each other in an embedding space. Figure 1 gives an overview of SC-Block while the different steps of the method are described below.

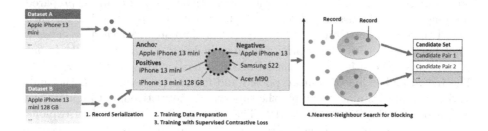

Fig. 1. Overview of the SC-Block method

(1) Record Serialization. Following recent works [31,49,51], records in datasets A and B are serialized into textual entity descriptions. Each attribute of a record is serialized as follows, "[Col] attribute_name [Val] actual_attribute_value". For the entity descriptions, all attribute serializations are concatenated. In the appendix, we evaluate the impact of adding the attribute names to the serialization.

(2) Training Data Preparation. This section explains how the training data is prepared for supervised contrastive learning and how source-aware sampling is used to reduce inter-label noise. The supervised contrastive loss requires records to share the same label if they describe the same real-world entity. The benchmark datasets provide pairs of records instead of clusters of matching records that share the same label. To identify records that refer to the same real-world entity, a correspondence graph is constructed, following the approach of Peeters and Bizer [44]. The records serve as vertices in the graph, and an edge between two vertices indicates a match between the corresponding records. A unique label is assigned to each connected component in the graph, ensuring that records describing the same real-world entity receive the same label. The labelling procedure may introduce inter-label noise as only a subset of all matches is known. Consequently, some matching records may not share the same label. During training, these matching records are treated as non-matches and are not embedded in a nearby location in the embedding space. This inter-label noise reduces the effectiveness of the embedding [12]. We apply source-aware sampling to reduce inter-label noise [44]. To achieve source-aware sampling, we create two training sets: A and B. Training set A contains all records from dataset A and the records from dataset B that share a label with a record from dataset A. Training set B contains all records from dataset B and the records from dataset A that share a unique label with a record from dataset B. During training, batches of offers are sampled from either training dataset A or B. This sampling strategy reduces the noise resulting from missing matches [44].

(3) Training with Supervised Contrastive Loss. In this section, we introduce the supervised contrastive loss and how it is applied to train effective embeddings to cluster matching records in an embedding space. The training procedure starts with a batch of N entity descriptions sampled from the prepared training data as discussed in Sect. 3. We duplicate all records so that for each record there is at least one matching record in the batch. An *encoder network* $Enc(\cdot)$ maps each entity description t to an embedding, $z = Enc(t) \in R^D$. In our experiments, $Enc(\cdot)$ is a pre-trained RoBERTa-base model [32] with $D = 768$. The record embedding z is mean pooled and normalized using the L_2 normalization. During training, we apply the supervised contrastive loss to update the parameters of the RoBERTa model. The supervised contrastive loss exploits label information of matching records by maximizing the agreement of records with the same label (Positives) and minimizing the agreement of records from different labels (Negatives) to train effective embeddings for SC-Block. Formally, supervised contrastive loss is defined as follows [29]: Given a batch of $2N$ embedded records $z = Enc(t) \in R^D$:

$$\mathcal{L} = \sum_{i \in I} \mathcal{L}_i = \sum_{i \in I} \frac{-1}{|P(i)|} \sum_{p \in P(i)} \log \frac{exp(z_i \cdot z_p / \mathcal{T})}{\sum_{a \in A(i)} exp(z_i \cdot z_a / \mathcal{T})} \tag{1}$$

Here, $i \in I \equiv 1...N$ is the index of an embedded record z. The index i is called an anchor embedding z_i, $P(i) \equiv \{p \in A(i) : y_p = y_i\}$ is the set of indices of all records with the same label in the batch distinct from i, and $|P(i)|$ is its cardinality. Also recall that the dot $(\cdot)$ symbol denotes the inner product,

$A(i) \equiv I \backslash i$, and $\mathcal{T} \in R^+$ is a scalar temperature parameter. Within a batch, each embedding becomes an anchor embedding z_i that is pulled close to all record embeddings from the same unique label while it is pushed away from all record embeddings with different labels. By setting the batch size to 1024 we ensure that each embedding is compared to many other embeddings, which is beneficial for the supervised contrastive loss [29]. Additionally, in each training epoch, different records are sampled into a batch leading to a variation in the record comparison during training. The large number of comparisons and the variation of the batches enable supervised contrastive training to learn better embeddings than pair-wise training which only uses the record pairs from the training set. In our experiments, $\mathcal{T}$ is set to 0.07 and the encoder network is trained for 20 epochs with a learning rate of $5e-5$. After training the encoder network embeds the entity descriptions from the input datasets A and B.

(4) Nearest-Neighbour Search for Blocking. SC-Block uses nearest neighbour search to create blocks of similar records, which is a common approach for blocking [49,51,53]. The nearest neighbour search exploits that similar records are close in the embedding space due to supervised contrastive training. For the search, we define dataset A, which contains fewer records than dataset B, to serve as the query table Q_A and dataset B to serve as an index table I_B. In the appendix, we evaluate the impact of switching the query and the index table. We index the embeddings z_B of I_B using FAISS [26] to allow an efficient search. For each query record $z_A \in Q_A$, the search ranks candidate records $z_B \in I_B$ based on their cosine similarity with the query record $z_A \in Q_A$. A hyperparameter k determines how many nearest neighbours are retrieved for each query record. A record $r_A \in A$ and its k corresponding neighbouring records $r_1, r_2 \ldots, r_k, r_i \in B$ build the set of pairs $(r_A, r_1), (z_A, r_2) \ldots (r_A, r_k)$ and are added to the candidate set C. For the nearest neighbour search, FAISS performs a merge-sort on I_B, which has a time complexity of $O(n \log_2 n)$ [26].

4 Blocking-Only Evaluation

This section evaluates the candidate sets generated by SC-Block concerning recall and candidate set size and compares them to the sets generated by seven other blocking methods from related work. The evaluation of SC-Block's candidate sets conducted in two scenarios: (1) with a fixed value of k=5 to draw general conclusions about the recall and precision of the candidate sets, and (2) with k tuned to minimize the number of missed matching pairs while keeping the candidate set size as small as possible.

4.1 Baseline Blocking Methods

This section describes the blocking methods from related work that we compare to SC-Block. Further details on the baseline blocker configurations are available online[8]. Table 2 provides an overview of the methods based on the criteria

---

[8] https://webdatacommons.org/largescaleproductcorpus/wdc-block/.

blocking technique, training procedure, supervision, and encoder network. If a criterion is not applicable, it is marked with a '–'.

Table 2. Blocking techniques.

	Learning	Blocking technique	Training	Encoder
JedAI [38]	unsupervised	key-based blocking	–	–
BM25 [42]	unsupervised	nearest-neighbour search	–	–
Auto [49]	self-supervised	nearest-neighbour search	autoencoder	fasttext
CTT [49]	self-supervised	nearest-neighbour search	pair-wise	fasttext
BT [51]	self-supervised	nearest-neighbour search	contrastive	RoBERTa
SimCLR [51]	self-supervised	nearest-neighbour search	contrastive	RoBERTa
SBERT [46]	supervised	nearest-neighbour search	pair-wise	RoBERTa
SC-Block	supervised	nearest-neighbour search	contrastive	RoBERTa

JedAI. The JedAI framework implements symbolic techniques for entity resolution pipelines. For our experiments, we run the default configuration of the block building, block cleaning & filling steps of the linked tutorial[9] [16].

BM25. BM25 is an unsupervised blocker [42] that uses a vector space model [48] and the BM25 term weighting scheme [47] to compute a similarity score for the nearest neighbour search. BM25 is evaluated using whitespace tokenization (referred to as BM25) and tri-grams (referred to as $BM25_3$).

Autoencoder (Auto) and Cross Tuple Training (CTT). Auto and CTT use fasttext to embed the tokens of the entity descriptions and average the token embeddings to obtain a record embedding [49]. For Auto, the record embeddings are sent through an autoencoder. Auto is self-supervised and thus requires no labelled training data. For CTT, the record embeddings are sent through a Siamese summarizer and a classifier learns to detect matches based on the element-wise difference of the created embeddings. CTT is trained on synthetically produced training data derived from the two blocked datasets.

Barlow Twins (BT) and SimCLR. BT and SimCLR, two self-supervised blockers, duplicate and augment entity descriptions in training batches by dropping random tokens [51]. The entity descriptions are embedded using a RoBERTa model to be comparable to SC-Block. During training, BT aligns the cross-correlation of augmented and original entity descriptions with the identity matrix [52], while SimCLR aims to maximize the similarity between embeddings of the same records and minimize it for different records within a batch [8].

Sentence-Bert (SBERT). SBERT requires labelled pairs of matching and non-matching records for training [46]. The entity descriptions of a labelled pair

---

[9] https://github.com/AI-team-UoA/pyJedAI/blob/main/docs/tutorials/CleanCleanER.ipynb.

are embedded using a pre-trained RoBERTa language model to ensure that the encoder network of SBERT and SC-Block is the same. During training, the cosine similarity of both embeddings is calculated, and the weights of the language model are updated using mean-squared error loss.

4.2 Implementation

We used a shared server with 96×3.6 GHz CPU cores, 1024 GB RAM and an NVIDIA RTX A6000 GPU for the experiments. We use Elasticsearch[10] to implement BM25 and $BM25_3$. The Elasticsearch instance runs on a virtual machine with 4×2.1 GHz CPU cores, 32 GB RAM and 512 GB storage. If an execution exceeds 48 h or if the shared server's memory is insufficient, the experiment is labelled as timed-out or as an out-of-memory error, respectively.

4.3 Results for Fixed K

We now analyze the candidate sets of the nearest neighbour blockers with a fixed number of nearest neighbours $k = 5$. By fixing the hyperparameter k, differences in recall and precision become visible that are not visible when k is tuned. $k = 5$ is chosen because it allows the blockers to score high recall, especially on the datasets A-B, A-G and W-A, which exhibit a maximum number of matching neighbours smaller than 5. At the same time, blockers miss matching pairs, because $k = 5$ is not sufficiently large, making it possible to see differences. In Sect. 4.4, we adjust the value of k to ensure that the candidate sets of the nearest neighbour blockers exceed a threshold of 99.5% on the validation set, to minimize the number of missed matches. Table 3 shows runtime, recall and precision of the candidate sets generated by the nearest neighbour blockers. The candidate sets are evaluated based on the record pairs of the respective test set. The lowest runtime and the highest score per column are marked in bold.

Table 3. Runtime (RT) in seconds, recall (R) in %, and precision (P) in % of the candidate sets generated by all nearest neighbour blockers with fixed $k = 5$

	A-B			A-G			W-A			WB$_{small}$			WB$_{medium}$			WB$_{large}$		
	RT	R	P	RT	R	P	RT	R	P	RT	R	P	RT	R	P	RT	R	P
SC-Block	175	**100**	36	290	**97**	36	588	96	**33**	402	**72**	**57**	0.7k	**66**	**64**	18.1k	**57**	74
BM25$_3$	18	98	27	36	94	31	290	**97**	21	300	53	39	1.0k	47	44	172.8k	timed-out	
BM25	**9**	92	27	**12**	94	32	**45**	96	22	**44**	59	40	**0.2k**	54	45	**4.1k**	42	54
Auto	20	75	31	27	83	36	353	82	23	127	43	40	0.7k	36	41	out-of-memory		
CTT	95	77	32	174	82	36	745	81	24	448	43	38	9.8k	35	41	out-of-memory		
SimCLR	178	90	**44**	41	92	**37**	534	91	32	728	35	39	1.1k	21	39	11.4k	3	33
BT	185	95	29	266	90	34	210	93	25	838	32	35	1.2k	21	36	9.6k	13	33
SBERT	498	74	75	307	29	41	441	31	17	1.9k	45	49	1.5k	35	55	8.4k	24	57

---

[10] https://www.elastic.co/what-is/elasticsearch.

On average, SC-Block has the highest recall and precision scores compared to the other blockers. It is also evident that the recall of all generated candidate sets decreases from WB_{small} to WB_{large}, indicating that a larger vocabulary and Cartesian product make the dataset more challenging. In the following paragraphs, we compare the performance of SC-Block to the performance of the other blockers.

Unsupervised Blockers. On datasets A-B, A-G, and W-A, BM25 and $BM25_3$ exhibit similar performance to SC-Block. However, on WDC-Block, BM25 and $BM25_3$ miss an average of 16.3% more pairs than SC-Block. This decrease in performance can be attributed to the larger token vocabulary of WDC-Block, which complicates the identification of matching pairs through token overlap by both BM25 blockers.

Self-supervised Blockers. Auto and CTT have the poorest performance among dense nearest neighbour-based blockers in datasets A-B, A-G, and W-A. This is due to the pre-trained fasttext embeddings used in Auto and CTT being less potent compared to the robust RoBERTa embeddings utilized in other blockers. SimCLR and BT lose between 3% and 9% more pairs than SC-Block on the datasets A-B, A-G and W-A with BT performing marginally better than SimCLR. On the W-A dataset, the slight difference between SC-Block, SimCLR, and BT can be attributed to the relatively low number of positive training pairs available. When supervision is absent, the supervised contrastive loss algorithm degrades to the loss of SimCLR [29]. On WDC-Block SimCLR and BT miss on average 45% and 43% more pairs than SC-Block. SC-Block holds an advantage as it leverages the guidance of matching and non-matching training pairs present in the datasets. Auto and CTT outperform BT and SimCLR on WDC-Block. During training, Auto and CTT require exposure to all tokens in a dataset, which necessitates training on all records from the initial datasets. However, BT and SimCLR are solely trained on the records mentioned in the training and validation set. This difference in training accounts for the performance gap and highlights the susceptibility of self-supervised blockers to unseen out-of-distribution records. SC-Block uses the same records for training as BT and SimCLR. However, its recall and precision scores demonstrate increased robustness against noise. The nearest neighbour search on WB_{large} of Auto and CTT results in an out-of-memory error, revealing a limitation of this implementation for large blocked datasets.

Supervised Blockers. The supervised blocker SBERT performs poorly. This indicates that SC-Block's supervised contrastive loss utilizes supervision more effectively than SBERT's pair-wise cosine similarity loss.

4.4   Results for 99.5% Recall on Validation Set

This section analyses the impact of tuning the hyperparameter k of the nearest neighbour blockers. Increasing k raises the likelihood of including all matches in the candidate set. However, higher k values generate larger candidate sets, which increase the runtime of the entity resolution pipeline since the matcher must compare more candidate pairs. To analyse how the blockers handle this trade-off, we evaluate each nearest neighbour blocker with increasing values of k, starting from $k = 1$. Once the recall of the candidate set exceeds 99.5% on the validation set, k is fixed and the recall is evaluated on the test sets. To limit the search space, we cap k at a maximum of 50 on A-B, A-G, W-A and WB_{small}, 100 on WB_{medium} and 200 on WB_{large}. Table 4 presents the values of k, the runtime, the recall achieved on the test set, and the size of the candidate sets. The highest recall and lowest k, runtime and candidate set size per dataset are highlighted.

Table 4. k, runtime (RT) in seconds, recall (R) in % and candidate set size ($|C|$) of the candidate sets on A-B, A-G and W-A. k is tuned on the validation set.

	A-B				A-G				W-A			
	k	RT	R	$|C|$	k	RT	R	$|C|$	k	RT	R	$|C|$
SC-Block	5	175	**100**	**5k**	8	293	**100**	11k	12	558	97	**31k**
BM25₃	13	18	**100**	14k	27	43	**100**	37k	12	290	99	**31k**
BM25	7	**9**	95	8k	29	**15**	99	40k	21	**45**	99	54k
JedAI	–	**9**	97	13k	–	19	98	20k	–	340	99	172k
Auto	50	20	97	54k	50	27	95	68k	50	353	93	128k
CTT	50	95	97	54k	50	174	95	68k	50	745	92	128k
BT	20	182	99	22k	30	49	98	41k	26	545	99	66k
SimCLR	29	186	96	31k	30	276	98	41k	23	221	96	59k
SBERT	26	502	86	28k	30	314	51	41k	50	441	49	128k
	WB_{small}				WB_{medium}				WB_{large}			
	k	RT	R	$|C|$	k	RT	R	$|C|$	k	RT	R	$|C|$
SC-Block	**14**	406	94	70k	**20**	426	92	**100k**	50	**8.9k**	90	**5M**
BM25₃	50	679	94	250k	100	779	94	500k	200	173.0k	timed-out	
BM25	50	86	**97**	250k	100	**186**	98	500k	200	14.2k	**96**	20M
JedAI	–	**50**	55	**51k**	–	1.6k	81	561k	–	173k	timed-out	
Auto	50	127	85	250k	100	450	80	500k	out-of-memory			
CTT	50	448	85	250k	100	672	80	500k	out-of-memory			
BT	50	770	67	250k	100	870	43	500k	200	10.7k	34	20M
SimCLR	50	923	70	250k	100	1.0k	46	500k	200	12.9k	36	20M
SBERT	50	1567	59	250k	100	2.1k	78	500k	200	13.5k	59	20M

Overall, the candidate sets of the blockers differ mainly in size, while most of the nearest neighbour search blocking techniques have a recall close to 1. SC-Block creates relatively small candidate sets. The candidate sets are now compared in detail.

Unsupervised Blockers. The BM25 weighting schema is insufficient to achieve competitive candidate set sizes on the WDC-Block benchmark, as evident from the large candidate sets of unsupervised blockers. This is due to the considerable number of corner cases in the validation sets and the vast vocabulary of the datasets. SC-Block is fine-tuned and can exploit matching information from the training set to acquire knowledge about the corner cases. JedAi's pruning of blocks negatively affects the recall score on WB_{small} and WB_{medium} and causes a timeout on WB_{large}.

Self-supervised Blockers. The self-supervised blockers BT and SimCLR require two to six times higher values of k than SC-Block to produce candidate sets that meet the recall threshold of 99.5% on the validation set. Auto and CTT fail to meet the recall threshold for any of the datasets. On the WDC-Block benchmark, Auto and CTT again outperform BT and SimCLR on the datasets WB_{small} and WB_{medium} by finding on average 25% more pairs. The out-of-memory error on WB_{large} arises from the nearest neighbour search implementation of Auto and CTT.

Supervised Blockers. SBERT overfits the training and validation data because it reaches a high recall score on the validation set but reaches a much lower recall on the test set. SC-Block's supervised contrastive loss better utilizes the training data than SBERT's mean-squared error loss on the cosine similarity of the training pairs.

The best-performing blockers SC-Block, $BM25_3$ and BT produce candidate sets with a recall close to 1, yet vary in their sizes. In Sect. 5, we combine these blockers with different matchers and measure the impact of the blockers on the F1 score and runtime of the entire entity resolution pipeline.

5   Evaluation Within Entity Resolution Pipelines

To evaluate the impact of the SC-Block, $BM25_3$, and BT blockers on entity resolution pipelines, we assess (1) the F1 score and runtime of the pipelines with these blockers and (2) whether the reduced runtime of the pipeline with SC-Block compensates for the training time of the blocker. The candidate sets, generated by the blockers, are processed by the matchers Magellan [30], RoBERTa Cross Encoder (CE) [6], Ditto [31], and SupCon-Match [44] to produce final sets of matching pairs. Each blocker uses the tuned k from Sect. 4.4. The matchers are fine-tuned on the same training sets as the blockers.

5.1   F1 Score and Runtime

This section analyses the impact of SC-Block, $BM25_3$, and BT on the F1 score and runtime of entity resolution pipelines. The runtime refers to the execution time of blocking and matching, excluding training times, which is discussed in Sect. 5.2. Table 5 shows the F1 score and the runtime in seconds, with the highest F1 score and lowest runtime highlighted.

Table 5. F1 score in % and runtime in seconds (RT) of state-of-the-art entity resolution pipelines

Blocker	Matcher	A-B		A-G		W-A		WB$_{small}$		WB$_{medium}$		WB$_{large}$	
		F1	RT	F1	RT	F1	RT	F1	RT	F1	RT	F1	RT
SC-Block	SupCon	**93**	71	**80**	133	81	355	71	**383**	72	**742**	72	**30.7k**
	Ditto	91	185	76	336	**86**	754	**77**	918	**78**	1.5k	**78**	66.5k
	CE	80	**57**	64	101	**86**	303	**77**	351	77	606	76	27.9k
	Magellan	52	335	58	511	68	1.5k	59	2.1k	61	2198	61	46.2k
BM25$_3$	SupCon	**93**	88	**80**	467	82	490	69	2.0k	70	4.2k	timed-out	
	Ditto	90	228	76	941	**86**	1.0k	74	3.6k	75	7.9k	timed-out	
	CE	79	70	64	353	85	437	73	1.8k	74	4k	timed-out	
	Magellan	51	216	58	555	67	1.9k	44	2.4k	43	3.7k	timed-out	
BT	SupCon	**93**	125	**80**	157	81	548	58	1.5k	45	2.9k	40	119.6k
	Ditto	90	296	75	442	**86**	1.0k	64	3.0k	50	6.3k	44	246.1k
	CE	79	75	64	**93**	85	**188**	63	1.4k	49	2.6k	43	109.8k
	Magellan	51	212	57	346	66	1.2k	42	2.4k	36	2.7k	32	72.1k

F1 Score. Table 5 shows that the F1 scores of the pipeline depend on the matcher for candidate sets with high recall. Pipelines with BT on WDC-Block are an exception because of the low recall of the respective candidate set, which harms the F1 scores as known from Sect. 4.4. On WDC-Block, pipelines consisting of SC-Block and Ditto or CE yield better results than pipelines with SC-Block and SupCon. This is because Ditto and CE learn distinctive patterns that are not acquired by SC-Block and SupCon. By combining SC-Block with Ditto or CE, these varied patterns are effectively exploited.

Runtime. In general, it can be concluded that using an effective blocker such as SC-Block can reduce the runtime of a pipeline. When comparing SC-Block to BM25$_3$ and BT, we can observe that the smaller candidate sets of SC-Block result in pipelines that run 4 to 7 times faster. For example, the runtime of the pipeline consisting of BT and CE on the WB$_{large}$ dataset is 30 h, whereas SC-Block reduces the workload of the CE matcher and reduces the pipeline runtime to 8 h. On the WB$_{large}$, BM25$_3$ requires a significant amount of time to generate large candidate sets, leading to pipeline timeouts after a two-day runtime. However, on datasets A-B, A-G, and W-A, using the smaller candidate sets of SC-Block results in pipelines that finish 1.5 to 2 times faster compared to pipelines with BM25$_3$ and BT. To see if training SC-Block is reasonable, we set the reduction in runtime into context with the training time in Sect. 5.2.

5.2   Impact of Training Time

If online training of a blocker is necessary, it is only reasonable to do so if the runtime of the pipeline, including training time, is shorter than the runtime of a pipeline with a blocker that does not require further training. Therefore, we consider the training time of the SC-Block and BT blockers in relation to the overall runtime of the entity resolution pipelines. Table 6 displays the blocker training time (BTT) and the complete runtimes (CT), which include the blocker training time, blocking time, and matching time of all pipelines.

Table 6. Blocker Training Time (BTT) and complete runtime of entity resolution pipelines (CT) in seconds

Blocker	Matcher	A-B		A-G		W-A		WB_{small}		WB_{medium}		WB_{large}	
		BTT	CT	BTT	CT	BTT	CT	BTT	CT	BTT	RT	BTT	RT
SC-Block	SupCon	227	298	259	392	430	0.8k	**343**	**0.7k**	**343**	1.1k	**343**	31.0k
	Ditto	227	412	259	595	430	1.2k	**343**	1.3k	**343**	1.8k	**343**	66.9k
	CE	227	284	259	360	430	0.7k	**343**	**0.7k**	**343**	**0.9k**	**343**	**28.2k**
	Magellan	227	562	259	770	430	1.9k	**343**	2.4k	**343**	2.5k	**343**	46.6k
$BM25_3$	SupCon	–	88	–	467	–	0.5k	–	1.9k	–	4.2k	timed-out	
	Ditto	–	228	–	941	–	1.1k	–	3.6k	–	7.9k	timed-out	
	CE	–	70	–	353	–	0.4k	–	1.8k	–	4.0k	timed-out	
	Magellan	–	216	–	555	–	1.9k	–	2.4k	–	3.7k	timed-out	
BT	SupCon	**158**	283	**236**	393	**130**	0.7k	779	2.3k	779	3.7k	779	120.4k
	Ditto	**158**	454	**236**	678	**130**	1.1k	779	3.9k	779	7.1k	779	246.9k
	CE	**158**	233	**236**	**329**	**130**	**0.3k**	779	2.2k	779	3.4k	779	110.6k
	Magellan	**158**	370	**236**	582	**130**	1.3k	779	3.2k	779	3.5k	779	72.9k

Small Benchmark Datasets. For the small datasets A-B, A-G and W-A, the unsupervised $BM25_3$ blocker is as efficient as the supervised and self-supervised blocking methods. Although SC-Block and BT show a small improvement in runtime on A-B, A-G, and W-A, their advantage is counterpoised by the training time of the blockers. Training SC-Block for these small datasets is not practical since entity resolution pipelines with the unsupervised $BM25_3$ blocker and a tuned k value generate recall scores near 1, equivalent F1 scores, require no time for training and runtime is faster than SC-Block's runtime and training time.

WDC-Block. The runtime of pipelines using $BM25_3$ on WDC-Block increases with a larger vocabulary and a larger cartesian product of candidate pairs. In this scenario, training SC-Block is a reasonable option since the training data is available. The five minutes of training time reduce the pipeline's runtime compared to $BM25_3$ even on WB_{small} by 20 min for the matchers SupCon and CE. Due to our computational restrictions, it was only possible to run competitive pipelines on WB_{large} if SC-Block was applied for blocking.

6   Related Work

Entity resolution [11], also known as Link Discovery [35], is a crucial task in the process of integrating data from the Web. The entity resolution pipelines consist of two main steps: blocking and matching, which both have been studied for decades [2,11,35,36,40].

Blocking. Blocking is traditionally tackled as an unsupervised task by generating a blocking key value from each record [1,10,50]. Records with the same blocking key value are assigned to the same block. Instead of one blocking key value, some works use multiple blocking keys [13,14,22], or multi-dimensional blocking [3,14,24]. Meta-blocking, implemented in the JedAI toolkit [38], extends blocking by blocking key values with an additional pruning step that first weights candidate record pairs by their matching likelihood and discards pairs with the lowest scores [16]. This pruning step has also been implemented using supervision [18,39]. Other works add a clustering step to exploit the transitivity of candidate record pairs [3,50]. The blocking key value has also been used as a sorting key for sorted neighbourhood blockers [23,27]. Related works also utilized supervision [4,53] or self-supervision [28] to learn a blocking strategy. Recently, deep learning for blocking has become popular [25,53]. DeepBlocker [49] explored the use of deep self-supervised learning for blocking, introducing two blockers based on auto-encoding (Auto) and cross-tuple training (CTT). Sudowoodo [51] applies self-supervised learning in combination with a transformer model and the two loss functions SimCLR and Barlow Twins (BT). The supervised contrastive loss of SC-Block is an extension of the SimCLR loss [8,29]. We compare SC-Block to the blockers JedAI, Auto, CTT, SimCLR and BT from the related work. Except for JedAI all benchmarked blockers apply the nearest neighbour search for candidate set generation. Mugeni et al. use self-supervised contrastive learning to position embeddings in the embeddings space and apply an unsupervised community detection technique from graph structure mining to block record pairs [34]. Sparkly has recently demonstrated that TF-IDF and BM25 are strong baseline blockers [42].

Entity Matching. Most current entity matching methods rely on deep learning techniques [2]. This trend was initialized by Ebraheem et al. [15] and Mudgal et al. [33]. Recently, several transformer-based matching methods have achieved state-of-the-art performance [6,20,21,31,43]. Among them, Peeters and Bizer [44] and Wang, Li and Wang [51] use contrastive learning, similar to our work.

Contrastive Learning. SC-Block is inspired by ideas from computer vision [29], information retrieval [19], and entity matching [44] where contrastive learning has shown to be more effective than the traditional cross-entropy-based learning. Specifically, Gao, Yao and Chen [19] use contrastive learning to learn sentence embeddings without any supervision. Supervised contrastive learning still suffers from challenges, including the robustness of learned representations [12] and class collapse, i.e., all samples from a cluster are mapped to the same representation [7]. Similar to SupCon [44], SC-Block applies source-aware sampling to increase the robustness of the learned embeddings. The main difference to SupCon is that SC-Block's embeddings are optimized for blocking and not for matching, which requires a longer pre-training of the embeddings.

7 Conclusion

The paper introduced the WDC-Block blocking benchmark, which employs data from a large number of Web data sources. The benchmark offers a large maximal Cartesian product and vocabulary size. The paper further proposes the SC-Block blocking method, which employs supervised contrastive learning to position records in an embedding space. The evaluation of SC-Block using WDC-Block and three smaller benchmark datasets showed that SC-Block generates smaller candidate sets than other state-of-the-art blockers. When using SC-Block together with the best-performing matcher on WDC-Block, the runtime for pipelines decreased from 30 to 8 h, performing 1.5 to 4 times faster than pipelines utilizing other state-of-the-art blockers and the same matcher. The reduced runtime overcompensates the time that is required to train SC-Block.

A Appendix - Additional Experiments

This appendix presents additional experiments that empirically justify several low-level design decisions made for SC-Block. We evaluate (1) the impact of another record serialization and (2) the impact of switching query and index table. Furthermore, WDC-Block is available with three training set sizes. We evaluate (3) how the different training set sizes impact SC-Block's performance.

Serialization. The default serialization method adds the attribute name to entity descriptions when serializing records. It is unclear whether the attribute name affects how SC-Block places embeddings in the embedding space. In the 'No attribute name' experiment, we only concatenate attribute values to derive entity descriptions. A fixed value of k=5 is used in these experiments to observe differences in precision and recall. We use a fixed $k = 5$ for these experiments to see differences in precision and recall. The results in Table 7 show that for the benchmark datasets A-B, A-G and W-A the serialization has only a marginal impact on SC-Block's recall and precision. On WDC-Block we see that the serialization with attribute names achieves better recall and precision scores than the serialization without attribute names. We can conclude that the attribute names are not harmful and structure the entity descriptions, which makes the entity descriptions better readable.

Table 7. Recall (R) in % and Precision (P) in % for different Record Serialization

Dataset	A-B		A-G		W-A		WB_{small}		WB_{medium}		WB_{large}	
Serialization	R	P	R	P	R	P	R	P	R	P	R	P
Attribute Names	99	36	97	36	96	33	72	57	66	64	57	74
No attribute name	100	39	98	34	96	33	69	54	62	62	51	69

Switch Query Table and Index Table. By default, the table with fewer records is selected as the query table and the larger table is selected as the index table. In this study, we examine the effects of swapping the query and index tables. The baseline for these experiments is the corresponding SC-Block run with an optimized k from Sect. 4.4. The value of k for the experiments involving switched tables is selected to ensure that the candidate sets have the same size, denoted by $|C|$, as the candidate sets of the baseline experiment where tables are not switched. If this is not possible, the smallest possible value of k is chosen that generates a candidate set larger than the candidate set of the baseline experiment where the query table and the index table are not switched.

Table 8. Recall (R) in %, Query Table (QT) Size and k for switching query and index table

QT	DS	R	QT Size	k	$|C|$
A	A-B	99	1,081	5	5,405
B	A-B	99.0	1,092	5	5,460
A	A-G	99.6	1,363	8	10,904
B	A-G	99.1	3,266	4	13,064
A	W-A	96.9	2,554	12	30,648
B	W-A	96.9	22,074	2	44,148
A	WB_{small}	93.5	5,000	14	70,000
B	WB_{small}	91.7	5,000	14	70,000
A	WB_{medium}	91.9	5,000	20	100,000
B	WB_{medium}	0.0	200,000	1	200,000
A	WB_{large}	89.5	100,000	50	5,000,000
B	WB_{large}	51.6	2,000,000	3	6,000,000

The results in Table 8 show that for the benchmarks A-B, A-G, W-A and WD_{small} switching query and index table while keeping the candidate set size stable results in a comparable recall. The advantage for A-G and W-A is that k is much smaller, meaning fewer k values need to be tested during the hyperparameter search. On WB_{medium} and WB_{large} table B is 20 to 40 times larger. If the query table and the index table are switched and k is chosen such that the candidate set size is similar, the recall is much lower.

Different Training Set Sizes. WDC-Block is available with three training set sizes. To analyze how the different training set sizes impact the performance of SC-Block, we train SC-Block on the different training sets (SC-Block$_{small}$, SC-Block$_{medium}$ and SC-Block$_{large}$) and search for k as described in Sect. 4.4. We compared SC-Block to SimCLR, which was trained on records from the same three training sets. Unlike SC-Block, SimCLR does not consider the information about matching records in the training set. If only datasets without matching information are available, SC-Block's loss is equivalent to SimCLR's loss.

Table 9. k and Recall (R) in % for different Training Set Sizes

	Pos.	Neg.	WDC-B$_{small}$			WDC-B$_{medium}$			WDC-B$_{large}$		
	Dev.	Dev.	k	R	$|C|$	k	R	$|C|$	k	R	$|C|$
SC-Block$_{large}$	9,680	15,752	14	93.5	70k	20	91.9	100k	50	89.5	5M
SC-Block$_{medium}$	2,338	2,819	50	92.6	250k	100	86.6	500k	200	77.8	20M
SC-Block$_{small}$	399	611	50	71.8	250k	100	52.0	500k	200	42.2	20M
SimCLR	0	0	50	69.5	250k	100	46.0	500k	200	36.1	20M

The results in Table 9 demonstrate that more training data improves recall and decreases the candidate set size of SC-Block. Furthermore, the small training set alone is adequate to enhance the results beyond those of the SimCLR baseline. These findings confirm the usefulness of training data for SC-Block.

References

1. Aizawa, A., Oyama, K.: A fast linkage detection scheme for multi-source information integration. In: Proceedings of the Sixth IEEE International Conference on Data Mining, pp. 30–39. IEEE, Tokyo, Japan (2005). https://doi.org/10.1109/WIRI.2005.2
2. Barlaug, N., Gulla, J.A.: Neural networks for entity matching: a survey. ACM Trans. Knowl. Discov. Data **15**(3), 1–37 (2021). https://doi.org/10.1145/3442200
3. van Bezu, R., Borst, S., Rijkse, R., Verhagen, J., Vandic, D., Frasincar, F.: Multi-component similarity method for web product duplicate detection. In: Proceedings of the 30th Annual ACM Symposium on Applied Computing, pp. 761–768. Association for Computing Machinery, New York, NY, USA (2015). https://doi.org/10.1145/2695664.2695818
4. Bilenko, M., Kamath, B., Mooney, R.J.: Adaptive blocking: learning to scale up record linkage. In: Proceedings of the Sixth IEEE International Conference on Data Mining, pp. 87–96. IEEE, Hong Kong, China (2006). https://doi.org/10.1109/ICDM.2006.13
5. Brinkmann, A., Primpeli, A., Bizer, C.: The web data commons Schema.org data set series. In: Companion Proceedings of the ACM Web Conference 2023, pp. 136–139. Association for Computing Machinery, New York, NY, USA (2023). https://doi.org/10.1145/3543873.3587331
6. Brunner, U., Stockinger, K.: Entity matching with transformer architectures - a step forward in data integration. In: Proceedings of the 23rd International Conference on Extending Database Technology, pp. 463–473. OpenProceedings.org, Copenhagen, Denmark (2020). https://doi.org/10.5441/002/edbt.2020.58
7. Chen, M., Fu, D.Y., Narayan, A., Zhang, M., Song, Z., Fatahalian, K., et al.: Perfectly balanced: improving transfer and robustness of supervised contrastive learning. In: Proceedings of the 39th International Conference on Machine Learning, vol. 162, pp. 3090–3122. PMLR, Baltimore, Maryland, USA (2022)
8. Chen, T., Kornblith, S., Norouzi, M., Hinton, G.: A simple framework for contrastive learning of visual representations. In: Proceedings of the 37th International Conference on Machine Learning, vol. 119, pp. 1597–1607. PMLR, Virtual (2020)

9. Christen, P.: Data Matching: Concepts and Techniques for Record Linkage, Entity Resolution, and Duplicate Detection, 1st edn. Springer, Berlin, Heidelberg (2012). https://doi.org/10.1007/978-3-642-31164-2
10. Christen, P.: A survey of indexing techniques for scalable record linkage and deduplication. IEEE Trans. Knowl. Data Eng. **24**(9), 1537–1555 (2012). https://doi.org/10.1109/TKDE.2011.127
11. Christophides, V., Efthymiou, V., Palpanas, T., Papadakis, G., Stefanidis, K.: An overview of end-to-end entity resolution for big data. ACM Comput. Surv. **53**(6), 1–42 (2021). https://doi.org/10.1145/3418896
12. Chuang, C.Y., Robinson, J., Lin, Y.C., Torralba, A., Jegelka, S.: Debiased contrastive learning. In: Larochelle, H., Ranzato, M., Hadsell, R., Balcan, M.F., Lin, H. (eds.) Proceedings of the 33rd Annual Conference on Neural Information Processing Systems, vol. 33, pp. 8765–8775. Curran Associates, Inc., Virtual (2020)
13. van Dam, I., van Ginkel, G., Kuipers, W., Nijenhuis, N., Vandic, D., Frasincar, F.: Duplicate detection in web shops using LSH to reduce the number of computations. In: Proceedings of the 31st Annual ACM Symposium on Applied Computing, pp. 772–779. Association for Computing Machinery, New York, NY, USA (2016). https://doi.org/10.1145/2851613.2851861
14. De Assis Costa, G., Parente De Oliveira, J.M.: A blocking scheme for entity resolution in the semantic web. In: Proceedings of the 30th International Conference on Advanced Information Networking and Applications, pp. 1138–1145. IEEE, Crans-Montana (2016). https://doi.org/10.1109/AINA.2016.23
15. Ebraheem, M., Thirumuruganathan, S., Joty, S., Ouzzani, M., Tang, N.: Distributed representations of tuples for entity resolution. Proc. VLDB Endow. **11**(11), 1454–1467 (2018). https://doi.org/10.14778/3236187.3236198
16. Efthymiou, V., Papadakis, G., Papastefanatos, G., Stefanidis, K., Palpanas, T.: Parallel meta-blocking for scaling entity resolution over big heterogeneous data. Inf. Syst. **65**, 137–157 (2017). https://doi.org/10.1016/j.is.2016.12.001
17. Efthymiou, V., Stefanidis, K., Christophides, V.: Benchmarking blocking algorithms for web entities. IEEE Trans. Big Data **6**(2), 382–395 (2020). https://doi.org/10.1109/TBDATA.2016.2576463
18. Gagliardelli, L., Papadakis, G., Simonini, G., Bergamaschi, S., Palpanas, T.: Generalized supervised meta-blocking. Proc. VLDB Endow. **15**(9), 1902–1910 (2022). https://doi.org/10.14778/3538598.3538611
19. Gao, T., Yao, X., Chen, D.: SimCSE: simple contrastive learning of sentence embeddings. In: Moens, M.F., Huang, X., Specia, L., Yih, S.W.T. (eds.) Proceedings of the 2021 Conference on Empirical Methods in Natural Language Processing, pp. 6894–6910. Association for Computational Linguistics, Virtual and Punta Cana, Dominican Republic (2021). https://doi.org/10.18653/v1/2021.emnlp-main.552
20. Genossar, B., Gal, A., Shraga, R.: The battleship approach to the low resource entity matching problem. Proc. ACM Manag. Data **1**(4), 1–25 (2023). https://doi.org/10.1145/3626711
21. Genossar, B., Shraga, R., Gal, A.: FlexER: flexible entity resolution for multiple intents. Proc. ACM Manag. Data **1**(1), 1–27 (2023). https://doi.org/10.1145/3588722
22. Hartveld, A., et al.: An LSH-based model-words-driven product duplicate detection method. In: Krogstie, J., Reijers, H.A. (eds.) CAiSE 2018. LNCS, vol. 10816, pp. 409–423. Springer, Cham (2018). https://doi.org/10.1007/978-3-319-91563-0_25
23. Hernández, M.A., Stolfo, S.J.: The merge/purge problem for large databases. ACM SIGMOD Rec. **24**(2), 127–138 (1995). https://doi.org/10.1145/568271.223807

24. Isele, R., Jentzsch, A., Bizer, C.: Efficient multidimensional blocking for link discovery without losing recall. In: Marian, A., Vassalos, V. (eds.) Proceedings of the 14th International Workshop on the Web and Databases, pp. 1–6. Association for Computing Machinery, New York, NY, USA (2011)

25. Javdani, D., Rahmani, H., Allahgholi, M., Karimkhani, F.: DeepBlock: a novel blocking approach for entity resolution using deep learning. In: Proceedings of the 5th International Conference on Web Research, pp. 41–44. IEEE, Tehran, Iran (2019). https://doi.org/10.1109/ICWR.2019.8765267

26. Johnson, J., Douze, M., Jégou, H.: Billion-scale similarity search with GPUs. IEEE Trans. Big Data 7(3), 535–547 (2021). https://doi.org/10.1109/TBDATA.2019. 2921572

27. Kejriwal, M., Miranker, D.P.: Sorted neighborhood for schema-free RDF data. In: Gandon, F., Guéret, C., Villata, S., Breslin, J., Faron-Zucker, C., Zimmermann, A. (eds.) ESWC 2015. LNCS, vol. 9341, pp. 217–229. Springer, Cham (2015). https:// doi.org/10.1007/978-3-319-25639-9_38

28. Kejriwal, M., Miranker, D.P.: An unsupervised instance matcher for schema-free RDF data. J. Web Semant. 35(2), 102–123 (2015). https://doi.org/10.1016/j. websem.2015.07.002

29. Khosla, P., Teterwak, P., Wang, C., Sarna, A., Tian, Y., Isola, P., et al.: Supervised contrastive learning. In: Larochelle, H., Ranzato, M., Hadsell, R., Balcan, M.F., Lin, H. (eds.) Proceedings of the 33rd Annual Conference on Neural Information Processing Systems, vol. 33, pp. 18661–18673. Curran Associates, Inc. (2020)

30. Konda, P., Das, S., Doan, A., Ardalan, A., Ballard, J.R., et al.: Magellan: toward building entity matching management systems over data science stacks. Proc. VLDB Endow. 9(13), 1581–1584 (2016). https://doi.org/10.14778/3007263. 3007314

31. Li, Y., Li, J., Suhara, Y., Doan, A., Tan, W.C.: Deep entity matching with pre-trained language models. Proc. VLDB Endow. 14(1), 50–60 (2020). https://doi. org/10.14778/3421424.3421431

32. Liu, Y., Ott, M., Goyal, N., Du, J., Joshi, M., Chen, D., et al.: RoBERTa: a robustly optimized bert pretraining approach (2019). arXiv:1907.11692 [cs]

33. Mudgal, S., Li, H., Rekatsinas, T., Doan, A., Park, Y., Krishnan, G., et al.: Deep learning for entity matching: a design space exploration. In: Proceedings of the 2018 International Conference on Management of Data, pp. 19–34. Association for Computing Machinery, New York, NY, USA (2018). https://doi.org/10.1145/ 3183713.3196926

34. Mugeni, J.B., Amagasa, T.: A graph-based blocking approach for entity matching using contrastively learned embeddings. ACM SIGAPP Appl. Comput. Rev. 22(4), 37–46 (2023). https://doi.org/10.1145/3584014.3584017

35. Nentwig, M., Hartung, M., Ngonga Ngomo, A.C., Rahm, E.: A survey of current link discovery frameworks. Semant. Web 8(3), 419–436 (2016). https://doi.org/10. 3233/SW-150210

36. Ngomo, A.C.N., Auer, S.: LIMES: a time-efficient approach for large-scale link discovery on the web of data. In: Proceedings of the 22nd international joint conference on Artificial Intelligence, vol. 3, pp. 2312–2317. AAAI Press, Barcelona, Catalonia, Spain (2011)

37. Papadakis, G., Fisichella, M., Schoger, F., Mandilaras, G., Augsten, N., Nejdl, W.: Benchmarking filtering techniques for entity resolution. In: Proceedings of the IEEE 39th International Conference on Data Engineering, pp. 653–666. IEEE, Anaheim, CA, USA (2023). https://doi.org/10.1109/ICDE55515.2023.00389

38. Papadakis, G., Mandilaras, G., Gagliardelli, L., Simonini, G., Thanos, E., Giannakopoulos, G., et al.: Three-dimensional entity resolution with JedAI. Inf. Syst. **93**, 101565 (2020). https://doi.org/10.1016/j.is.2020.101565

39. Papadakis, G., Papastefanatos, G., Koutrika, G.: Supervised meta-blocking. Proc. VLDB Endow. **7**(14), 1929–1940 (2014). https://doi.org/10.14778/2733085.2733098

40. Papadakis, G., Skoutas, D., Thanos, E., Palpanas, T.: Blocking and filtering techniques for entity resolution: a survey. ACM Comput. Surv. **53**(2), 1–42 (2020). https://doi.org/10.1145/3377455

41. Papadakis, G., Svirsky, J., Gal, A., Palpanas, T.: Comparative analysis of approximate blocking techniques for entity resolution. Proc. VLDB Endow. **9**(9), 684–695 (2016). https://doi.org/10.14778/2947618.2947624

42. Paulsen, D., Govind, Y., Doan, A.: Sparkly: a simple yet surprisingly strong TF/IDF blocker for entity matching. Proc. VLDB Endow. **16**(6), 1507–1519 (2023). https://doi.org/10.14778/3583140.3583163

43. Peeters, R., Bizer, C.: Dual-objective fine-tuning of BERT for entity matching. Proc. VLDB Endow. **14**(10), 1913–1921 (2021). https://doi.org/10.14778/3467861.3467878

44. Peeters, R., Bizer, C.: Supervised contrastive learning for product matching. In: Companion Proceedings of the Web Conference 2022, pp. 248–251. Association for Computing Machinery, New York, NY, USA (2022). https://doi.org/10.1145/3487553.3524254

45. Peeters, R., Der, R.C., Bizer, C.: WDC products: a multi-dimensional entity matching benchmark. In: Proceedings of the 27th International Conference on Extending Database Technology, vol. 27, pp. 22–33. OpenProceedings.org, Konstanz (2023). https://doi.org/10.48786/edbt.2024.03

46. Reimers, N., Gurevych, I.: Sentence-BERT: sentence embeddings using Siamese BERT-networks. In: Inui, K., Jiang, J., Ng, V., Wan, X. (eds.) Proceedings of the 2019 Conference on Empirical Methods in Natural Language Processing and the 9th International Joint Conference on Natural Language Processing, pp. 3982–3992. Association for Computational Linguistics, Hong Kong, China (2019). https://doi.org/10.18653/v1/D19-1410

47. Robertson, S., Zaragoza, H.: The probabilistic relevance framework: BM25 and beyond. Found. Trends® Inf. Retr. **3**(4), 333–389 (2009). https://doi.org/10.1561/1500000019

48. Salton, G., Wong, A., Yang, C.S.: A vector space model for automatic indexing. Commun. ACM **18**(11), 613–620 (1975). https://doi.org/10.1145/361219.361220

49. Thirumuruganathan, S., et al.: Deep learning for blocking in entity matching: a design space exploration. Proc. VLDB Endow. **14**(11), 2459–2472 (2021). https://doi.org/10.14778/3476249.3476294

50. Vandic, D., Frasincar, F., Kaymak, U., Riezebos, M.: Scalable entity resolution for web product descriptions. Inf. Fusion **53**, 103–111 (2020). https://doi.org/10.1016/j.inffus.2019.06.002

51. Wang, R., Li, Y., Wang, J.: Sudowoodo: contrastive self-supervised learning for multi-purpose data integration and preparation. In: Proceedings of the IEEE 39th International Conference on Data Engineering, pp. 1502–1515. IEEE, Anaheim, CA, USA (2023). https://doi.org/10.1109/ICDE55515.2023.00391
52. Zbontar, J., Jing, L., Misra, I., LeCun, Y., Deny, S.: Barlow twins: self-supervised learning via redundancy reduction. In: Meila, M., Zhang, T. (eds.) Proceedings of the 38th International Conference on Machine Learning, vol. 139, pp. 12310–12320. PMLR, Virtual (2021)
53. Zhang, W., Wei, H., Sisman, B., Dong, X.L., Faloutsos, C., Page, D.: AutoBlock: a hands-off blocking framework for entity matching. In: Proceedings of the 13th International Conference on Web Search and Data Mining, pp. 744–752. Association for Computing Machinery, New York, NY, USA (2020). https://doi.org/10.1145/3336191.3371813

Navigating Ontology Development with Large Language Models

Mohammad Javad Saeedizade[(⊠)] and Eva Blomqvist[iD]

Linköping University, Linköping, Sweden
{javad.saeedizade,eva.blomqvist}@liu.se

Abstract. Ontology engineering is a complex and time-consuming task, even with the help of current modelling environments. Often the result is error-prone unless developed by experienced ontology engineers. However, with the emergence of new tools, such as generative AI, inexperienced modellers might receive assistance. This study investigates the capability of Large Language Models (LLMs) to generate OWL ontologies directly from ontological requirements. Specifically, our research question centres on the potential of LLMs in assisting human modellers, by generating OWL modelling suggestions and alternatives. We experiment with several state-of-the-art models. Our methodology incorporates diverse prompting techniques like Chain of Thoughts (CoT), Graph of Thoughts (GoT), and Decomposed Prompting, along with the Zero-shot method. Results show that currently, GPT-4 is the only model capable of providing suggestions of sufficient quality, and we also note the benefits and drawbacks of the prompting techniques. Overall, we conclude that it seems feasible to use advanced LLMs to generate OWL suggestions, which are at least comparable to the quality of human novice modellers. Our research is a pioneering contribution in this area, being the first to systematically study the ability of LLMs to assist ontology engineers.

Keywords: LLM · Ontology · Ontology Engineering

1 Introduction

Ontologies are nowadays used in many applications, spanning almost all subjects and industry domains, e.g., as schemas for Knowledge Graphs (KGs) [17] or as stand-alone models for reasoning. However, despite the long history of ontology engineering, including manual ontology engineering methodologies, modelling environments, as well as attempts to automate the process, e.g., ontology learning, ontology engineering remains a time-consuming and error-prone activity. Additionally, while being a research goal for several decades, there are still no methods or tools that allow domain experts themselves to create high-quality ontologies without the involvement of ontology engineers.

At the same time, Large Language Models (LLMs) are now exhibiting an unprecedented level of natural language understanding and generation, and due to their training data even able to produce formal output such as program code.

© The Author(s), under exclusive license to Springer Nature Switzerland AG 2024
A. Meroño Peñuela et al. (Eds.): ESWC 2024, LNCS 14664, pp. 143–161, 2024.
https://doi.org/10.1007/978-3-031-60626-7_8

There are already tools emerging for assisting programmers in formalising their requirements into program code in various languages. Analogously, we would like to investigate what contribution LLMs could make in the area of ontology engineering, i.e., in assisting humans in translating their ontological requirements into a formal ontology language, such as OWL. The objective is to pave the way for tools that can aid domain experts and ontologists, by providing precise suggestions during the ontology creation process. As concluded in [23] LLMs are unlikely to replace neither ontologies nor ontology engineers completely, while exploiting LLMs for more effective and efficient ontology engineering is essential.

In the process of ontology development, ontologists use ontology stories [6,30] and Competency Questions (CQ) [14] to specify the capabilities and content of the ontology to be developed. Ontology stories offer descriptive narratives encapsulating the requirements of a project, while specific capabilities to answer queries can be expressed using CQs. Our investigation primarily centres on the ability of LLMs to leverage these ontology stories, as well as the CQs derived from them, for ontology creation. This sets our work apart from attempts to merely translate some textual content into an ontology, since the input does not explicitly state how to model something, but rather only specifies the desired outcome. Further, due to the W3C standard OWL currently being the predominant ontology representation language, we specifically target the generation of OWL formalisations. OWL has several possible syntaxes, while we have chosen to work with Turtle in this study due to its popularity, i.e., it is likely that LLMs have encountered this syntax frequently in their training data.

Overall, this work addresses the following research questions: *To what extent can LLMs create an ontology that meets the requirements of an ontology story? Which LLMs are suitable for this task, and what prompting techniques are most effective?* The main contributions of the work are: (1) Identifying which of the current state-of-the-art LLMs are capable of producing sufficiently well-formed OWL suggestions, (2) analysing the benefits and drawbacks of a range of prompting techniques, and (3) overall assessing the feasibility of LLMs to automatically produce modelling suggestions. The remainder of the paper is organized as follows: In Sect. 2, we discuss related work. Then, in Sect. 3, we describe the methodology and experiment setup, before presenting the results in Sect. 4, and discussing implications of the results in Sect. 5. Finally, we conclude and discuss future work in Sect. 6. Supplementary material, including stories and CQs, prompts, and the results, can be found on GitHub[1].

2   Related Work

This section presents related work, in terms of ontology engineering methods, and current tool support (Sect. 2.1), automated methods for generating OWL ontologies, i.e., ontology learning, including recent approaches using LLMs (Sect. 2.2), and finally an overview of similar work in software engineering and code generation based on LLMs (Sect. 2.3).

---

[1] https://github.com/LiUSemWeb/LLMs4OntologyDev-ESWC2024.

2.1 Ontology Engineering

Ontology engineering is traditionally a manual effort. Several methodologies exist, e.g. starting from early methodologies such as METHONTOLOGY [12] and NeOn [30]. More recently, agile methods [6,25,29] have become increasingly popular, reflecting the needs of many real-world settings, and in [27] the LOT methodology is presented as a compilation of experiences of many projects, as well as methodologies and tools. While some methodologies have been proposed in combination with explicit tool support, e.g., NeOn was originally proposed together with the NeOn toolkit [30] and [27] propose a set of possible tools to use, most activities are still entirely manual, such as listing of terms and relations, formalisation into an ontology language etc. Commonly, tool support constitute an ontology engineering environment, such as Protégé[2] or TopBraid Composer[3], together with some visualisation, documentation and publishing tools. Development environments guide the user in formalising the ontology by providing common language constructs as user interface shortcuts, e.g., views, buttons, wizards, etc. However, such tools still do not provide any concrete guidance on *how* to formalise the domain, e.g., in OWL.

The notion of Ontology Design Patterns (ODP) was proposed [7,13], as a way to encode modelling best practices and thus guide the ontology engineer. Additionally, collections of best practices have been published[4]. Still, these tools do not automatically connect to the requirements, e.g., expressed ontology stories [6] or use cases [27] combined with CQs [14], but the ontology engineer must, on their own, match their modelling problem to these best practices, which is not straightforward. Thus, there is no methodology or tool currently that assists the ontology engineer in the task of matching their specific requirements to modelling best practices or suggesting potential solutions in an automated manner.

2.2 Ontology Learning and Generation

Generating ontologies automatically, e.g., from natural language text, has been a topic of research for several decades, commonly denoted *ontology learning*. Initially, the field used classic NLP techniques such as POS-tagging in combination with lexical patterns or frames, to detect occurrences of ontological constructs in text. However, with methods such as [26], treating the task as one of machine translation, deep learning has also made significant contributions to this field. Additionally, early language models have been applied for various subtasks, such as extending ontologies and mapping concepts to top-level ontologies [20].

Even more recently, LLMs have been shown to be effective in the Semantic Web area owing to their capability to capture a vast array of information during their training process. LLMs4OL [4] utilized LLMs for ontology learning, where

---

[2] https://protege.stanford.edu/.

[3] https://allegrograph.com/topbraid-composer/.

[4] See for instance the vocabularies section of https://www.w3.org/TR/ld-bp/ or the whitepaper at https://www.nist.gov/document/nist-ai-rfi-cubrcinc002pdf for OBO ontologies.

LLMs were employed to extract relations among ontology classes or instances. However, this approach was only able to extract relations among entities and did not facilitate the complete generation of an ontology. This paradigm was then used also for tasks like taxonomy discovery, and extraction of non-taxonomic relations across diverse knowledge domains. In [11] a study is made on the performance of LLMs on similarly specific tasks, e.g., NER and relation extraction, but for the biomedical domain. Other preliminary work [21] has used fine-tuned GPT models to translate restricted natural language sentences into DL axioms. However, such specific statements do not represent realistic ontology requirements or scenarios, as targeted in our work. In addition, Text2KGBench [22] and DiamondKG [1] analyzed the capability of LLMs in generating KGs from text using predefined ontologies. Similarly, [9] extends KGs and ontologies, but starting from an existing KG schema. Further, tools are appearing for supporting the practical integration of OWL with deep learning, and LLMs, e.g., [15], but such tools do not provide any guidance specifically for the ontology generation task.

Furthermore, OLaLa [16] investigated the performance of LLMs on ontology matching and showed promising results on this task. Turning, the task around, another recent work has attempted to retrofit the ontology requirements, i.e. CQs, from existing ontologies, using LLMs [2]. While this work is very interesting, as it explores the connection between CQs and OWL formalisation, and shows promising results of generating CQs even with simple prompts, it is not treating the same kind of ontology generation task as our work. Overall, we conclude that research on using LLMs for ontology learning and KG generation in the Semantic Web indicates a promising direction. These approaches indicate the potential of LLMs also for knowledge engineering, but none of the proposed methods have so far treated the specific task targeted in this paper.

2.3 LLMs for Code Generation

Taking a broader perspective, generative AI has recently greatly influenced the generation of code, including competitions like AlphaCode [19]. GitHub Copilot is an example of AI-powered coding assistance [10], that has been shown to reduce coding time for simple tasks [33]. Code Llama [28], demonstrates the growing diversity in LLM applications for coding. Additionally, models like GPT-3.5 and GPT-4, and Bard contribute to this domain, being capable of also generating code to some extent. These developments indicate a growing interest in leveraging generative AI for automating and enhancing coding, marking a paradigm shift in software development. These approaches have inspired our work, but targeting another kind of formalisation, they are not directly transferable to ontologies.

3 Methodology

Our framework for generating ontologies from ontological requirements relies on two components, an LLM and a prompting technique, which are described below. In addition, we describe the experiment setup and evaluation measures used to compare the outcomes to a baseline of manually constructed ontologies.

3.1   Large Language Models

In our research, it was crucial to thoroughly examine various LLMs to achieve generalisable results and avoid model-specific biases. Therefore, we use a representative selection of models spanning proprietary and open-source versions: GPT-3.5, GPT-4, and Bard from the closed-source spectrum, and Llama-7B, Llama-13B, Llama2-70B [32], Alpaca [31], Falcon-7B [3], Falcon-7B-Instruct [24], WizardLM [37], and Alpaca-LoRA [32][5] from the open source spectrum. We used the Microsoft Azure API to access the GPT models, for Bard we used the web interface, and open source models were run on T4 GPU and 12 GB RAM. By using this broad selection, we aimed to achieve a holistic understanding of the LLM landscape. However, it is important to note that the purpose is not to compare the effectiveness of the models but merely to determine if any of them can produce sufficiently high-quality OWL models, and the characteristics of such suggestions. Hence, we first performed initial experiments on a broad range of models but then quickly settled for a smaller set in subsequent experiments.

3.2   Prompting Methods

In the field of LLMs, a prompt is an input that guides the model's response generation. A prompting technique entails the strategic formulation of these prompts to maximize the efficacy of LLMs. It involves the deliberate structuring and phrasing of prompts to align with the model's training and capabilities.

In our research, we used multiple prompting techniques to test LLMs. Each prompt in our experiments had four sections, and constitutes a template text with variable sections where requirements are input (making them directly reusable for new tasks, also in other domains). The **header** introduces the task, including emphasising the need for a well-structured ontology that accurately captures the information in a given narrative. The **helper** section, explains foundational ontology concepts such as the correct usage of Turtle syntax for restrictions. It may explain syntax, distinctions of classes, properties, and restrictions. The subsequent **story** segment presents the ontology story, complete with its CQs, i.e., the ontological requirements, acting as the primary source of content for ontology creation. Finally, the **footer** anchors the prompt with cautionary advice, spotlighting common pitfalls and recurrent errors. These general mistakes range from overlooking classes, returning empty output, to erroneous definitions of attributes. The purpose of this layout is to present a clear roadmap for the LLM. In addition, LLMs require a memory to handle a context larger than their allowed context size. An external memory module can store information, such as previously generated code and past communications. Zero-shot is the only

---

[5] Details related to the versions and settings of these models can be found in our supplementary material.

prompting method that does not use a memory. The following details how these prompt sections are implemented for each prompting technique.

Zero-shot Prompting: This method entails a one-time interaction with the LLM, requiring no iteration or feedback loop, utilizing a prompt composed of the four components outlined in the previous section: the header, helper, story, and footer. Together the sections of the prompt provide all the information needed for the ontology construction, in a single interaction.

Sub-task Decomposed Prompting - Waterfall approach (Waterfall) : This method, adapted from [18], breaks down the ontology creation process into five stages. The first stage involves self-guidance by the LLM, followed by the extraction of classes in stage two. The third stage involves constructing a taxonomy, while the fourth stage focuses on defining object properties. Finally, in stage five, the LLM creates datatype properties. Each stage has a tailored header, helper, and footer to guide the LLM and avoid distractions. A query is dispatched to the memory at each stage to retrieve previously generated OWL code, then appending the new results, finally resulting in a complete ontology.

Sub-task Decomposed Prompting - Competency Question by Competency Question (CQbyCQ): This approach instructs the LLM to address one CQ at a time as a decomposition of the requirements. The header guides the LLM to formulate the ontology for the specific CQ, and the memory integrates the output with the previous ones. The helper, story, and footer sections remain similar to the Zero-shot approach. After modelling the final CQ the result is an integrated ontology for the entire narrative, i.e. by simply merging CQ-specific ontologies. Figure 1 illustrates a typical prompt using this prompting technique.

Header: Your task is to contribute to creating well-structured ontology information that appeared in the given story, as well as requirements and restrictions. The way you approach this is first, you pick this competency question [CQ] and read the given turtle RDF to know what the current ontology is until this stage (it can be empty at the beginning). Then, you add or change the RDF so it can answer this competency question. Your output at each stage is an independent turtle, so rewrite/edit the previous RDF and produce the new one ...
Helper: You can read these definitions to understand the concepts: Classes are the keywords/classes that are going to be node types in the knowledge graph ontology. Try to extract all classes; in addition, classes can also be defined for reification. We use Turtle Syntax for representation. Hierarchies are rdfs:subClassOf in the turtle syntax. They can be used to classify similar classes in one superclass. ...
Story: The story comes here: [Story and requirements]
Footer: Here are some possible mistakes that you might make: 1- You might forget to add prefixes at the beginning of the code. 2- Forgetting to write pivot classes at the beginning before starting to code. 3- Your output must use the previous RDF and concatenate the answer to the competency question to it. So, your output is created and merged. 4- in your output, put all of the previous RDF classes, relations, and restrictions and add yours. Your output would be passed to the next stage, so don't remove the previous code ...

Fig. 1. An excerpt of a representative example of a prompt for CQbyCQ, with the sections of the prompt marked. Grey text is constant for all stories when using the same prompting technique, while green boxes are placeholders for the input story and CQs. The first green box contains the CQ that will be modelled, and the second green box contains the ontology story and its total set of requirements. (Color figure online)

Chain of Thoughts (CoT): In the context of the CoT [36], CoT-SC [34], and GoT [5] , the concept of 'thoughts' is a series of steps that the LLM carefully crafts for itself, where the end result is an ontology. The CoT framework instructs the LLM to create a plan based on the narrative and its associated CQs. The prompt guides the execution of only the next step at a time, requiring OWL code exclusive to that step. The outcome of each step is stored in memory and added sequentially to the prompt. By following this approach, the ontology is developed progressively by following the generated plan.

Self Consistency with Chain of Thoughts (CoT-SC): CoT-SC [35] is similar to CoT but involves three distinct plans and three outputs. The LLM evaluates and picks the best plan. To ensure diversity of plans, exclusively for this prompting technique, temperature and penalty parameters are adjusted from 0 to 0.5. This allows for exploration of a broader range of potential ontologies.

Graph of Thoughts (GoT): We modified GoT [5] to enhance efficiency and cost-effectiveness. Similar to CoT-SC, the LLM generates three plans and OWL files representing unique ontologies. However, our GoT integrates ideas (instead of picking the best) and solutions from all three into a single ontology, providing a broader perspective for crafting a more effective ontology. This streamlines ontology generation and ensures a more nuanced representation of the story.

3.3   Experimental Setup

In our experiments, we analyse the ontologies produced using a specific combination of prompting techniques and LLMs. Initially we focus on small tasks, created specifically to trigger certain OWL constructs, before then moving on to more realistic tasks, represented as stories and CQs. The resulting ontologies are at each stage manually evaluated against a set of criteria. In the final stage, the same evaluation is also made for a set of student submissions from a master's course, serving as a baseline for comparison.

Initial Experiment. Studying all combinations of LLMs and prompting techniques on realistic tasks, to counteract stochastic variations, is too time-consuming as human experts are required to evaluate the LLM-generated results in detail. We, therefore, first performed a preliminary investigation to assess the performance of each LLM-prompting technique combination using a curated set of small tasks, to be able to already rule out a set of less promising combinations. The small tasks were created as concise narratives intended to produce a very small OWL solution but still test the ability of the LLM to produce models that are not necessarily straightforward and, in some cases, do not conform to common intuition. The preliminary experiment is organized into two phases.

During the first phase (Fig. 2, part 1), ontologies are constructed based on three distinct narratives, 6 prompting techniques, and using 11 different LLMs. Narrative 1 includes the challenge of modelling a restriction, where an instance of a class then violates that defined restriction, and hence a reasoner should detect the inconsistency. Narrative 2 includes the modelling of domain and range restrictions on properties, and when an individual uses this property, the individual should be classified according to the domain or range (even if this is, in common sense, counterintuitive). This design is to make sure the LLMs follow given instructions rather than basing their output on prior (common sense) knowledge. The third narrative simply requires the existence of a particular class definition. The difference is that in the third case, a new class or a set of restrictions must be added without explicit information in the narrative itself.

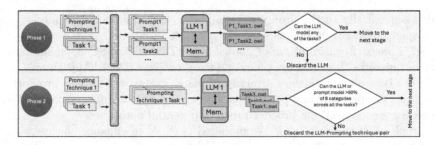

Fig. 2. The two initial experimental phases, aiming to filter out LLMs. Initially, several LLMs are assessed on small tasks. Additional criteria in phase 2 are used to filter out more LLMs, and also prompting techniques.

During this phase, minor syntax errors or other minor discrepancies will not be considered crucial since the primary goal is to pave the way for an interactive tool where such issues could be effortlessly rectified, in some cases even automatically by a syntax validation and correction process. After such syntax correction has been made, the resulting Turtle files are evaluated against one narrative-specific criteria each, similar to the ontology testing methods of error provocation and inference verification [8]. A reasoner is therefore executed over the resulting ontologies from narratives 1 and 2, and the results examined. For narrative 1, an inconsistency should be detected, and for narrative 2 the type of a specific individual is analyzed. For the third narrative, inference verification [8] is used. If the LLM was able to successfully model any one of the narratives (using any prompting techinque) it is retained for the next phase.

In the second phase, ontologies generated in phase 1, that passed the initial evaluation, are further inspected using a set of criteria derived from what content is expected based on the input narrative. We examine the OWL code generated by the models manually and assess it based on simple binary criteria to reduce the subjectivity of evaluators. The following criteria were used in this phase, where all except the last of these categories are expected in the result of each narrative: (1) Presence of an 'EquivalentClass' restriction. (2) Presence of a reification class, i.e. a class not explicitly mentioned in the narrative but needed to model an n-ary relation. (3) Correctness of the Turtle syntax. (4) Presence of domain and range for properties (at least one restriction). (5) Presence of a class hierarchy axiom (rdfs:subClassOf). (6) Semantic coherence of all class hierarchy axioms. (7) Presence of a datatype property. (8) Presence of instances (if specified by the story, one narrative does not require it). Then, by averaging the scores over these binary criteria across three tasks, we derive a score for each LLM-Prompting technique combination (c.f. Fig. 2, phase 2), and LLM-prompting technique combinations with scores over 0.9 are retained.

Main Experiment. The primary objective of our main experiment is to evaluate the quality of OWL files generated by LLMs based on realistic ontological requirements using different prompting techniques. Based on our initial experiment, we have pre-selected a set of LLM-prompting combinations that seem to give reasonable results, and now they are investigated further, i.e. using the experimental setup in Fig. 3. Next, we compare the scores of the LLM-generated ontologies to the same scoring of the first and last student submissions on a course task, using exactly the same ontology stories.

Our evaluation criteria are centred on the extent to which the ontology can address the CQs. We assess this by determining whether a SPARQL query can be written, representing each CQ, for retrieving relevant data as expressed using the ontology, which has been previously suggested as a suitable ontology testing method (c.f. CQ Verification [8]). If the ontology enables successful data retrieval, it passes the test for that CQ. This is then aggregated over the CQs of each story, calculating the proportion of successfully modelled CQs. Two variants of the assessment are used, i.e. either with or without considering minor issues. Minor issues are defined as situations where (1) a single syntactic error is detected, or (2) a single object property or datatype property is missing, preventing the formulation of a correct SPARQL query by one single triple pattern. The rationale behind these being minor issues is that if all other components, e.g., classes, restrictions, and other properties, are present, then adding one single property is an easy task even for a novice modeller.

For scoring the solutions we calculate the following: Let O be the generated model, CQ_i represent i-th CQ, n be the total number of CQs, and $f(O, CQ_i)$ be a function that evaluates the model O with respect to CQ_i and returns a value from $\{0,1\}$. The total score is then given by:

$$\text{score} = \frac{\sum_{i=1}^{n} f(O, CQ_i)}{n}$$

Baseline Dataset. The three ontology stories, with associated CQs, used in the main experiment originate from a series of courses, at master's and PhD level as well as conference tutorials, (about 15–20 course instances in total). The tasks were initially developed to allow for experimenting with tool and methodology support for ontology engineering, hence, they are comparable in size (all three stories are accompanied with 14 CQs and one additional statement) and level of difficulty and cover a set of similar modelling problems (set in music, theater, and hospital domains respectively). In particular, all three stories cover a fixed set of modelling challenges, such as creating a coherent taxonomy, reification of relations, use of domain and range, time-dependent relations, and restrictions enabling instance classification, etc. For example, one story set in the theatre domain implies the reification and time indexing of the participation relation when actors are engaged in an ensemble, while in the hospital story, the participation is instead present through employment and membership in a union during a certain time. This makes the tasks also ideal for experimenting with LLMs, since they on one hand cover a broad set of modelling challenges, but also represent three domain-specific variants of the challenges.

In the experiment setup of this paper we specifically used the solutions created by student groups in a master's course in 2009 (10 groups of students in

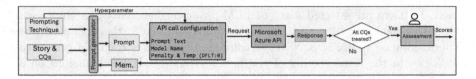

Fig. 3. Architecture of the pipeline for generating and assessing the generated ontologies in the main experiment. The prompting technique determines the hyperparameters of the LLM and creates a prompt text using the memory and a story, to send a request to the MS Azure API. At the end of the pipeline, the human evaluator assesses the ontology and a score for the prompting technique-LLM pair is generated.

total). The students were part of a master's program on information engineering and management, where all had some logic, modelling and programming background but no ontology experience. Before encountering the tasks, the students had a series of lectures on OWL and ontology engineering, and a basic introduction to OWL and the Protégé tool, with some hands-on sessions.

For each assignment (ontology story) students submitted their first attempt, received feedback, and in most cases had to improve their solutions before passing the assignment. Most groups submitted their solutions at least twice, in some cases up to 5 times, to pass. Students worked primarily in pairs (but sometimes 3, or alone). The three stories were each modelled by 10 groups, and picking the first students' submissions and the last ones gives us a dataset of 60 OWL files (i.e. 3 stories, modelled by 10 groups, and 2 solutions per group). The final submissions contained approximately 23 ± 3 classes, 24 ± 5 object properties, and 8 ± 3 datatype properties. We argue that the first submissions would be comparable to what an average junior programmer would come up with after perhaps studying some online tutorials, without task-specific guidance or feedback. While the last student submission represents an acceptable solution to the modelling problem, i.e., sufficient quality with only a few minor issues. This gives us two baselines for comparison with the LLM-generated models, in terms of quality.

4   Experimental Results

For our experiments, we first make a basic assessment of the output quality of OWL ontologies generated based on the three small narratives, allowing us to make an initial selection of LLMs and prompting techniques. We then assess the quality of the LLM output, on the three larger tasks, and compare these results against the ontologies produced by students, using these as a baseline.

4.1   Initial Experiment Results

Phase One: The initial phase of our experiment provided insightful observations regarding the performance of various LLMs in conjunction with different prompting techniques. The results indicate that the open-source and smaller models (<20B parameters) tested, i.e., LlaMA7B, LlaMA13B, WizardLM, Falcon7B, Alpaca-LoRA, and CodeLlama-13b, lack the capability to provide accurate suggestions. These models consistently failed to produce relevant output to any of the small tasks in this phase. The generated responses were either too brief, ambiguous, only code fragments, or irrelevant text, such as trivial conversations, altogether indicating a lack of ability to perform the specified tasks.

When considering larger models, LlaMA2-70B and Bard exhibited a notable inability to generate outputs in the Turtle syntax. During the code review, we identified numerous issues such as incomplete code, incorrect usage of prefixes, irrelevant modelling for the given story, and numerous syntax errors. In stark contrast, GPT-3.5 and GPT-4 emerged as the only models capable of successfully completing the tasks. They consistently generated outputs that aligned with the expected prompt, demonstrating a considerably higher level of proficiency in ontology generation compared to the other models.

Furthermore, an analysis of the prompting techniques revealed the Sub-task Decomposed Prompting approach, the waterfall method, failed to produce satisfactory results across the three narratives, even when applied with GPT-3.5, and failed in two out of three narratives with GPT-4. This suggests that, despite refinements of the prompts after each failure of the LLM, this technique is not effective for the task at hand. Additionally, the CQbyCQ technique proved too complex for GPT-3.5, resulting in the LLM producing irrelevant output, or too brief code fragments, leading to failures across all small tasks.

In summary, the results from phase one of the preliminary experiment underscore the significant variation in performance across different LLMs and prompting techniques. Only GPT-3.5 and GPT-4 provided sufficiently high-quality results to proceed to the next phase.

Phase Two: For the retained models, the outputs were then evaluated based on the established binary criteria (see Sect. 3.3, phase two). In aspects such as the implementation of an 'owl:EquivalentClass'-restriction, presence of a reification class, and the correctness of the Turtle syntax, GPT-3.5 exhibited considerable shortcomings in comparison to GPT-4. For example, across all prompting techniques, GPT-3.5 missed almost half of the necessary reifications while GPT-4 correctly modelled 95% of them. Overall, during this phase, GPT-4 outperformed GPT-3.5 by approximately 15% on average for all criteria and prompting techniques. Conversely, in criteria involving the expression of domain and range for properties, the creation of a taxonomy using 'rdfs:subClassOf', semantic coherence in class hierarchies, the definition of datatype properties, and the provision of instances, GPT-4 and GPT-3.5 showed similar levels of competence. This suggests that while GPT-3.5 is capable in certain aspects of ontology modelling, it falls short in more complex and nuanced tasks. Regarding, prompting techniques, the waterfall and Zero-shot techniques performed poorly on GPT3.5 and GPT4, with scores below the threshold (0.9), and were therefore excluded.

Considering the overall performance in phase two, it was decided to proceed exclusively with GPT-4 as LLM and CQbyCQ, CoT, CoT-SC, and GoT prompts in the subsequent phases of our research. Again, it is worth noting that the purpose of the study is to explore the feasibility of LLMs to assist in OWL modelling, not to compare and quantify the capabilities of different models.

4.2   Main Experiment Results

Our experiment aimed to evaluate GPT-4 with remaining prompting techniques. We ran the model three times for each technique and story, analysed all three models with respects to addressed CQs (c.f. criteria in Sect. 3.3). The analysis involved manually writing SPARQL queries that answer each CQ[6]. An example SPARQL query for a reified relation in the hospital story is shown in Listing 1.1. Scores were averaged over the results of the runs, thus reducing the effects of random variations. Then, we compared these averaged outputs from our framework against the average of the 10 student groups' submissions.

Listing 1.1. Example of typical test case in the form of a SPARQL query.

```
# CQ: What role does a certain person have within a certain union
# group at a certain point in time?
SELECT ?role ?person ?union
WHERE {
  ?membership rdf:type :UnionMembership . #Reified relation
  ?membership :memberOf ?union .
  ?membership :member ?person .
  ?membership :role ?role .
  ?membership :startTime ?start .
  ?membership :endTime ?end .
  FILTER (?start > "..."^^xsd:dateTime && ?end < "..."^^xsd:dateTime) }
```

In Fig. 4, each section depicts one ontology story and compares the performance of students with the LLMs outputs. This comparison focuses on to what extent the CQs have been addressed and presents scores as the proportion of the CQs that passed the same test. However, we also provide a modified view that ignores minor errors (annotated with 'IG'), as defined in Sect. 3.3. The latter illustrates how overlooking minor mistakes, e.g., those which even a novice ontology engineer can spot and fix, affects the overall effectiveness of the LLMs.

An observation from Fig. 4 is that by ignoring minor errors, CQbyCQ outperforms all other prompting methods and students' first submissions, while the students' last submissions typically yielded the best output. When comparing students' first and last submissions across all CQs, there was an average improvement of approximately 20%. This improvement showcases the learning process of students during multiple resubmissions and feedback. However, the CQbyCQ scores were closer in performance to the students' last submissions than their first submissions. This implies that this technique is particularly effective, aligning closely with the understanding students develop over time.

By disregarding minor issues, as shown in Fig. 4, there is a considerable increase in the quality of the LLM-generated OWL files. This improvement sug-

---

[6] The test is passed if a query can be formulated, i.e., no test data is used, and the complexity of the queries has not been analysed so far.

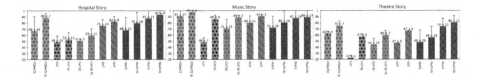

Fig. 4. Comparison of average scores (STD and AVG) of GPT-4, combined with the selected prompting techniques, against students' submissions, expressed as proportion of CQs they sufficiently modelled. 'StuFS' represents students' first submissions, and 'StuLS' students' last submissions. 'IG' indicates results when minor issues are ignored.

gests that many errors in the LLM's outputs are minor. The fact that overlooking these minor issues leads to a substantial boost in performance highlights the potential of LLMs in ontology generation. It underscores that with minimal human intervention for error correction, regarding minor issues in the suggested model, LLMs can be highly effective in assisting the modelling process.

To gain a deeper understanding of the types of CQs that posed challenges for both GPT-4 and students in their modelling efforts, we categorized the CQs into four distinct types, based on the intended modelling solution:

1. Simple Datatype Property: Includes CQs that can be addressed by adding one or more datatype properties, e.g., CQs related to the date of an event.
2. Simple Object Property: Involves defining a connection between two classes by creating an object property, e.g., providing the author of a book.
3. Reification: CQs that require the creation of an abstract class to connect other classes, often in complex situations, e.g., the role of a person in an event at a specific time is a question of reifying the 'role-playing situation'.
4. Restrictions: This involves imposing restrictions on classes or properties, e.g., at least one article is always presented at each seminar.

In Table 1 we present the performance of GPT-4 using the previous prompting techniques, as well as student submission, categorized by the four types of CQs (columns) presented above, where minor issues are again disregarded. Each row corresponds to an average score of different OWL outputs by models or students. The key findings suggest the performance of the CQbyCQ technique stands out in its ability to model simple object properties, datatype properties and apply reification. It consistently performs at a level that is better or comparable to other techniques. However, its effectiveness in creating restrictions is less reliable.

The techniques CoT, GoT, and CoT-SC established restrictions more effectively than other techniques, where CoT consistently succeeded in modelling all restrictions. However, considering their overall performance, the picture is more unclear, and additional experiments would be needed to distinguish their benefits and drawbacks in detail.

Table 1. Scores of the solutions on four different categories of CQs: datatype property (DP), object property (OP), reification (Reif) and restrictions (Rest). Minor issues ignored. Rows represent solutions by prompting techniques/students' scores.

	Theatre Story				Music Story				Hospital Story			
	DP	OP	Reif	Rest	DP	OP	Reif	Rest	DP	OP	Reif	Rest
CQbyCQ	**.93**	**.78**	.55	.33	**1.0**	**.94**	**1.0**	**1.0**	**1.0**	.94	**1.0**	0
CoT	.80	.50	.22	**1.0**	.80	.78	**1.0**	**1.0**	.53	.50	.44	**1.0**
CoT-SC	**.93**	.55	0	**1.0**	.80	.89	**1.0**	**1.0**	.53	.66	.66	.33
GoT	.86	.61	.44	**1.0**	.80	**.94**	**1.0**	**1.0**	.73	**1.0**	.66	.66
StuFS	.64	.63	.43	.10	.84	.83	.87	.30	.78	.85	.73	.60
StuLS	.92	**.78**	**.73**	.50	.94	.92	.90	.50	.94	**1.0**	.90	.60

5 Discussion

According to the findings, the utilisation of LLMs yields promising results in the development of ontology engineering support. However, the generalisability and verifiability of our results can be questioned due to the closed-source nature of the LLMs. For instance, not all hyperparameters of the model are accessible, and model updates may not necessarily improve performance. Even though the previous GPT models are still accessible through MS Azure, hence at the moment the experiments can be repeated, there is no future guarantee. Therefore, newer versions need to be evaluated before being used, especially when it comes to the performance of prompting techniques. In addition, future versions of open-source models will hopefully reach similar performance as seen by GPT-4 in this study. Another potential future direction would be to fine-tune the models to produce OWL output and treat ontology stories and CQs. However, generating broad training data for such fine-tuning might be a challenge.

Regarding validity, there is another potential concern, namely the leakage problem, i.e. the ontology stories used in the experiments already being seen in the training data of LLMs. The narratives for the initial experiment were written for sake of the experiment, but are short enough that it cannot be guaranteed they are not similar to modelling problems already available online. Ontology stories of the main experiment were publicly available, e.g., on course websites, but to the best of our knowledge, no modelling solutions have been publicly available, which significantly reduces the risk of potential leakage. However, further experimentation is necessary to confirm the main experiment results to mitigate this risk, e.g., by applying the method to novel use cases in new domains.

A potential drawback to this effort concerns expenses. Our experiments used realistic, but still small-scale input, and should the length of the stories increase GPT-4 would become a pricey tool to use. Moreover, fine-tuning GPT-4 could be even more expensive. Despite its extensive context size, GPT-4 is also still prone to forgetting certain aspects of the context. This poses the risk of inadequate modelling for large stories at a high cost.

Nevertheless, we still believe that ontology engineering will soon undergo a similar transformation as we have seen for programming. Where on the one hand, ontology engineering will be made accessible to much broader groups, and on the other hand, modelling will be done at an unprecedented speed, also by experienced ontology engineers, due to being able to automate simpler tasks in the modelling process. However, the risks involved are similar to those of using generative black-box methods in general, since the output may comprise an unknown amount of bias, creating unexpected and unwanted side effects when applying the ontologies in downstream tasks. Also, the performance may vary significantly depending on the domain. The tasks used for our experiments were set in common domains, such as music and theatre, while more specialised domains might pose more complex challenges to an LLM modelling assistant. Additionally, the generalisability of prompt structures needs further research. While our prompts can certainly be reused for new tasks and domains, this study is too small to ensure generalisability of the conclusions on prompt structure.

6   Conclusions and Future Work

In our study, we explored a set of prompting techniques, and assessed their effectiveness for generating OWL ontologies, across a range of LLMs. By comparing the ontologies generated by LLMs with student solutions, using consistent evaluation measures, we established a baseline for performance assessment. We conclude that, at the moment, only GPT-3.5 and GPT 4 produce reasonable OWL output in the first place. Further experimentation with GTP-4 revealed that, when used with the CQbyCQ prompting technique, GPT-4 outperforms the average quality of the initial submissions of students (novice ontology engineers) and had a performance similar to their final submissions after multiple feedback rounds, when ignoring minor errors. While for creating more complex OWL constructs, e.g., restrictions, other prompting techniques yield better results. This allows us to conclude that GPT-4, and a combination of prompting techniques, is currently most likely the best avenue ahead for creating an OWL modelling assistant using out-of-the-box models.

Looking ahead, our next objective is therefore to develop a Protégé plugin using such combinations and evaluate its efficacy in reducing the time taken for ontology development by expert ontologists, as well as the level of support that can be provided for novice modellers. Another avenue of future work would be to investigate fine-tuning of models for ontology engineering tasks, and to assess the performance of upcoming releases of open-source models.

Acknowledgement. This project has received funding from the European Union's Horizon Europe research and innovation programme under grant agreement no. 101058682 (Onto-DESIDE), and is supported by the strategic research area Security Link. The student solutions used in the research were collected as part of a master's course taught by Assoc. Prof. Blomqvist while employed at Jönköping University.

ChatGPT was used to enhance the readability of some of the text and improve the language of this paper, after the content was first added manually. All material was then checked manually before submission.

Disclosure of Interests. The authors have no competing interests to declare that are relevant to the content of this article.

Appendix

Motivations, Limitations and Negative Results

Prompt Components: As mentioned in the methodology section 3, there are four sections in each prompt. At a quick glance, the header and story sections appear to be necessary since we provide a brief prompt and the story requirements. The helper and footer sections may be considered optional. However, removing the helper section causes the LLM to completely avoid modelling reifications and misplacing properties, such as putting a datatype property as a range for an object property. The helper begins by outlining strategies to establish a taxonomy, which is otherwise often ignored by LLMs.

The footer, or pitfall section, also enhances the output significantly. It offers the LLMs with common mistakes that they produce. Common errors that are mentioned as pitfalls to avoid are: (1) Providing an empty output of the given prompt. (2) Avoid the use of the Turtle syntax and instead provide a list of items in Python syntax. (3) Avoiding to provide an OWL output without establishing any taxonomy of classes. (4) In the thoughts prompting techniques, avoiding to run the complete plan (several steps) at a current step, since LLMs can ignore instructions and give the complete answer at the first step. (5) Providing explanations instead of providing the code.

Ontology Design Patterns (ODPs) serve as guides for ontology engineer to model an ontology. However, adding examples to prompts seems to degrade output performance. Despite fitting the prompt and story, 32K context LLMs tend to forget the ontology story (we tried with the 128K context GPT4-turbo model and it failed). This could be because the large context is distracting the current LLMs (this could be caused by the low performance of attention layers in LLMs). We used the term "distraction" since the model starts modelling the ODPs in the output instead of the given task.

Limitations: This study, while insightful, has several limitations. Our choice of evaluation method was additionally influenced by the time constraints faced by human experts in manually evaluating the outputs. While this approach was necessary given the available resources, it may not capture the full depth and nuances of LLM-generated ontologies compared to a more thorough, even though time-consuming, manual evaluation.

Due to their extensive branching, the tree of thoughts and the full version of the graph of thoughts techniques proved expensive. This complexity led to slower processing times and increased costs, limiting their practicality for larger-scale or time-sensitive applications.

We used the Microsoft Azure API to access GPT-3.5 and GPT-4, versions 613 trained until 2021. Consequently, our analysis did not consider any advancements or updates in these models post-2021, including the introduction of seed features in newer updates. This might limit the relevance of our findings in the context of the latest LLM capabilities. The accessibility of hyperparameters in GPT-4 and GPT-3.5 is limited, which presented challenges in our experiment. Despite setting the temperature and penalty parameters to zero (except in plan generation for GoT and CoT-SC, where they were set to 0.5), we observed inconsistencies in the outcomes when using identical prompts. This variability underscores the significance of utilizing open-source LLMs for achieving more consistent and reliable LLM performance rather than depending on unpredictable factors.

We faced another setback in our attempt to produce a more efficient OWL code to reduce context size or general improvement of modelling. For example, in CQbyCQ, when a CQ is addressed, we simply merge it with the previous CQs instead of asking LLM to merge if this CQ has not been addressed. This choice was made since LLMs often forgot to merge classes (or properties) from the previous section, which resulted in incomplete modelling.

Lastly, we encountered another challenge by experimenting with few-shot prompting techniques. In few-shot prompting, a few examples are provided to LLMs as an example. We faced difficulty finding examples of ontology modelling that were not too similar to the ontology story, as this could potentially provide an answer to the LLM. However, this challenge may lead to a similar experiment as the one we mentioned earlier in the usage of ODPs (LLM distraction due to large context size).

Initial Experiment Result Details

Due to space limitations we were not able to present all details of the initial experiment in the main paper body, merely a conclusion summary. The detailed results of the initial experiment, phase 2, are instead reflected here. In Table 2, the LLM-Prompting scores are presented, averaged over the three tasks and 8 criteria, and a threshold of 0.9 is chosen to pass.

Table 2. After conducting the initial experiment phase two, it was decided that CoT, CoT-SC, CQbyCQ, and GoT would move to the next stage (score > 0.9). GPT-3.5 was excluded as its performance was found to be equal to or less than GPT-4.

Prompting Technique:	Zero-shot	Waterfall	CoT	CoT-SC	CQbyCQ	GoT
Score using GPT-3.5	0.77	0.86	**0.91**	0.6	0.77	0.73
Score using GPT-4	0.86	0.86	**0.91**	**0.96**	1	**0.92**

References

1. Alharbi, R., et al.: Exploring the role of generative AI in constructing knowledge graphs for drug indications with medical context. In: 15th International Semantic Web Applications and Tools for Healthcare and Life Sciences (SWAT4HCLS 2024) (2024). (to appear)
2. Alharbi, R., Tamma, V., Grasso, F., Payne, T.: An experiment in retrofitting competency questions for existing ontologies. arXiv preprint arXiv:2311.05662 (2023)

3. Almazrouei, E., et al.: Falcon-40B: an open large language model with state-of-the-art performance (2023). https://huggingface.co/tiiuae/falcon-40b
4. Babaei Giglou, H., D'Souza, J., Auer, S.: Llms4ol: large language models for ontology learning. In: Payne, T.R., et al. (eds.) ISWC 2023. LNCS, pp. 408–427. Springer, Cham (2023). https://doi.org/10.1007/978-3-031-47240-4_22
5. Besta, M.: Graph of thoughts: solving elaborate problems with large language models. In: Proceedings of the AAAI Conference on Artificial Intelligence, vol. 38, no. 16, pp. 17682–17690 (2024)
6. Blomqvist, E., Hammar, K., Presutti, V.: Engineering ontologies with patterns-the extreme design methodology. In: Ontology Engineering with Ontology Design Patterns. IOS Press (2016)
7. Blomqvist, E., Sandkuhl, K.: Patterns in ontology engineering: classification of ontology patterns. In: ICEIS, vol. 3, pp. 413–416. SciTePress (2005). https://doi.org/10.5220/0002518804130416. ISBN: 972-8865-19-8. INSTICC
8. Blomqvist, E., Seil Sepour, A., Presutti, V.: Ontology testing-methodology and tool. In: ten Teije, A., et al. (eds.) Knowledge Engineering and Knowledge Management. EKAW 2012. LNCS, vol. 7603, pp. 216–226. Springer, Heidelberg (2012). https://doi.org/10.1007/978-3-642-33876-2_20
9. Caufield, J.H., et al.: Structured prompt interrogation and recursive extraction of semantics (SPIRES): a method for populating knowledge bases using zero-shot learning. Bioinformatics **40**(3), btae104 (2024). https://doi.org/10.1093/bioinformatics/btae104
10. Chen, M., et al.: Evaluating large language models trained on code. arXiv preprint arXiv:2107.03374 (2021)
11. Chen, Q., et al.: Large language models in biomedical natural language processing: benchmarks, baselines, and recommendations (2024). https://arxiv.org/abs/2305.16326
12. Fernández, M., Gómez-Pérez, A., Juristo, N.: Methontology: from ontological art towards ontological engineering. In: Proceedings of the AAAI97 Spring Symposium Series on Ontological Engineering (1997)
13. Gangemi, A.: Ontology design patterns for semantic web content. In: Gil, Y., Motta, E., Benjamins, V.R., Musen, M.A. (eds.) ISWC 2005. LNCS, vol. 3729, pp. 262–276. Springer, Heidelberg (2005). https://doi.org/10.1007/11574620_21
14. Grüninger, M., Fox, M.S.: The role of competency questions in enterprise engineering. In: Rolstadås, A. (eds.) Benchmarking — Theory and Practice. IFIP Advances in Information and Communication Technology, pp. 22–31. Springer, MA (1995). https://doi.org/10.1007/978-0-387-34847-6_3
15. He, Y., Chen, J., Dong, H., Horrocks, I., Allocca, C., Kim, T., Sapkota, B.: Deeponto: A python package for ontology engineering with deep learning (2024). (To appear in the Semantic Web Journal)
16. Hertling, S., Paulheim, H.: OLaLa: ontology matching with large language models. In: Proceedings of the 12th Knowledge Capture Conference 2023. K-CAP '23, pp. 131–139. Association for Computing Machinery, New York, NY (2023). https://doi.org/10.1145/3587259.3627571
17. Hogan, A., et al.: Knowledge Graphs. Morgan & Claypool Publishers, San Rafael (2021)
18. Khot, T., et al.: Decomposed prompting: a modular approach for solving complex tasks. arXiv preprint arXiv:2210.02406 (2022)
19. Li, Y., et al.: Competition-level code generation with alphacode. Science **378**(6624), 1092–1097 (2022)

20. Lopes, A., Carbonera, J., Schmidt, D., Garcia, L., Rodrigues, F., Abel, M.: Using terms and informal definitions to classify domain entities into top-level ontology concepts: an approach based on language models. Knowl. Based Syst. **265**, 110385 (2023). https://doi.org/10.1016/j.knosys.2023.110385, https://www.sciencedirect.com/science/article/pii/S0950705123001351

21. Mateiu, P., Groza, A.: Ontology engineering with large language models (2023). https://arxiv.org/abs/2307.16699

22. Mihindukulasooriya, N., Tiwari, S., Enguix, C.F., Lata, K.: Text2kgbench: a benchmark for ontology-driven knowledge graph generation from text. In: Payne, T.R., et al. (eds.) ISWC 2023. LNCS, vol. 14266, pp. 247–265. Springer, Cham (2023). https://doi.org/10.1007/978-3-031-47243-5_14

23. Neuhaus, F.: Ontologies in the era of large language models-a perspective. Appl. Ontol. **18**(4), 399–407 (2023)

24. Penedo, G., et al.: The refinedweb dataset for falcon LLM: outperforming curated corpora with web data, and web data only. arXiv preprint arXiv:2306.01116 (2023)

25. Peroni, S.: A simplified agile methodology for ontology development. In: Dragoni, M., Poveda-Villalón, M., Jimenez-Ruiz, E. (eds.) OWLED ORE 2016 2016. LNCS, vol. 10161, pp. 55–69. Springer, Cham (2016). https://doi.org/10.1007/978-3-319-54627-8_5

26. Petrucci, G., Rospocher, M., Ghidini, C.: Expressive ontology learning as neural machine translation. J. Web Seman. **52**, 66–82 (2018)

27. Poveda-Villalón, M., Fernández-Izquierdo, A., Fernández-López, M., García-Castro, R.: Lot: an industrial oriented ontology engineering framework. Eng. Appl. Artif. Intell. **111**, 104755 (2022). https://doi.org/10.1016/j.engappai.2022.104755, https://www.sciencedirect.com/science/article/pii/S0952197622000525

28. Roziere, B., et al.: Code llama: open foundation models for code. arXiv preprint arXiv:2308.12950 (2023)

29. Shimizu, C., Hammar, K., Hitzler, P.: Modular ontology modeling. Semant. Web **14**(3), 459–489 (2023)

30. Suárez-Figueroa, M., Gómez-Pérez, A., Motta, E., Gangemi, A. (eds.): Ontology Engineering in a Networked World. Springer, Cham (2012)

31. Taori, R., et al.: Stanford alpaca: an instruction-following llama model. https://github.com/tatsu-lab/stanford_alpaca (2023)

32. Touvron, H., et al.: Llama: open and efficient foundation language models. arXiv preprint arXiv:2302.13971 (2023)

33. Vaithilingam, P., Zhang, T., Glassman, E.L.: Expectation vs. experience: evaluating the usability of code generation tools powered by large language models. In: Chi Conference on Human Factors in Computing Systems Extended Abstracts, pp. 1–7 (2022)

34. Wang, L., et al.: Plan-and-solve prompting: improving xero-shot chain-of-thought reasoning by large language models. In: Rogers, A., Boyd-Graber, J., Okazaki, N. (eds.): Proceedings of the 61st Annual Meeting of the Association for Computational Linguistics, vol. 1: Long Papers, pp. 2609–2634. Association for Computational Linguistics, Toronto (2023). https://doi.org/10.18653/v1/2023.acl-long.147

35. Wang, X., et al.: Self-consistency improves chain of thought reasoning in language models. arXiv preprint arXiv:2203.11171 (2022)

36. Wei, J., et al.: Chain-of-thought prompting elicits reasoning in large language models. Adv. Neural. Inf. Process. Syst. **35**, 24824–24837 (2022)

37. Xu, C., et al.: Wizardlm: empowering large language models to follow complex instructions. arXiv preprint arXiv:2304.12244 (2023)

ESLM: Improving Entity Summarization by Leveraging Language Models

Asep Fajar Firmansyah[1,2]([✉]) [iD], Diego Moussallem[1,3] [iD],
and Axel-Cyrille Ngonga Ngomo[1] [iD]

[1] Paderborn University, Warburger Str. 100, 33098 Paderborn, Germany
{diego.moussallem,axel.ngonga}@upb.de
[2] The State Islamic University Syarif Hidayatullah Jakarta, Jakarta, Indonesia
asep.fajar.firmansyah@upb.de, asep.airlangga@uinjkt.ac.id
[3] Jusbrasil, Salvador, Brazil

Abstract. Entity summarizers for knowledge graphs are crucial in various applications. Achieving high performance on the task of entity summarization is hence critical for many applications based on knowledge graphs. The currently best performing approaches integrate knowledge graphs with text embeddings to encode entity-related triples. However, these approaches still rely on static word embeddings that cannot cover multiple contexts. We hypothesize that incorporating contextual language models into entity summarizers can further improve their performance. We hence propose ESLM (Entity Summarization using Language Models), an approach for enhancing the performance of entity summarization that integrates contextual language models along with knowledge graph embeddings. We evaluate our models on the datasets DBpedia and LinkedMDB from ESBM version 1.2, and on the FACES dataset. In our experiments, ESLM achieves an F-measure of up to 0.591 and outperforms state-of-the-art approaches in four out of six experimental settings with respect to the F-measure. In addition, ESLM outperforms state-of-the-art models in all experimental settings when evaluated using the NDCG metric. Moreover, contextual language models notably enhance the performance of our entity summarization model, especially when combined with knowledge graph embeddings. We observed a notable boost in our model's efficiency on DBpedia and FACES. Our approach and the code to rerun our experiments are available at https://github.com/dice-group/ESLM.

Keywords: Entity Summarization · Language Models · Knowledge Graph Embeddings

1 Introduction

Entity summarizers are extensively utilized across user-facing applications driven by knowledge graphs (e.g., Web search [10], RDF browsers [7], and recommender systems [22]) to provide succinct summaries of entities, and hence facilitate user comprehension. Recent methods (e.g., ESA [28], DeepLENS [15], GATES [8], and ESCS [4]) have achieved improved effectiveness by employing deep learning algorithms to encode triples containing entity descriptions and generate accurate summaries. For instance, ESA combines word embeddings [2] with Knowledge Graph Embeddings (KGEs) (e.g., computed using TransE [3]) to transform the predicate and object of each triple into vectors. By combining word embeddings and KGEs, ESA significantly outperforms previous methods based on unsupervised learning [5,9,25,27].

While approaches based on embeddings are effective, they exhibit an important limitation: They all rely on static word embeddings such as Word2Vec [2] and GloVe [20]. These static models are unable to account for the various contexts of a word, particularly in cases where homonyms appear in different entity descriptions. The motivation behind our work was hence to validate the following hypothesis: *Integrating contextual language models (LMs) into entity summarization methods can enhance their performance.* We hence present ESLM (Entity Summarization using Language Models), an entity summarization that leverages contextual LMs alongside the topological information of KGEs. In our implementation of ESLM, we use the contextual LMs BERT [6] and ERNIE [23] because of the differing representations they compute: BERT acquires dynamic representations based on the transformer architecture using purely textual data, while ERNIE enhances these representations with external knowledge from KGs [32]. Additionally, ESLM incorporates a transformed-based large LM based on the T5 model [21]. We conduct a comprehensive evaluation of ESLM using the ESBM (Entity Summarization BenchMark, version 1.2) dataset [13], which comprises datasets based on DBpedia and LinkedMDB. We also evaluate our approach on the FACES [9] dataset. ESLM consistently outperforms the current state-of-the-art (SOTA) methods on these datasets w.r.t. the normalized discounted cumulative gain (NDCG) measure. Additionally, ESLM achieves an F-measure of up to 0.591 on the DBpedia dataset.

Our contributions to entity summarization are as follows:

- We introduce a new approach for entity summarization dubbed ESLM, which leverages contextual LMs combined with KGE to enhance the performance of entity summarization.
- We conduct a detailed ablation study to analyze the impact of contextual LMs and their integration with KGE within our model. This study aids in understanding the contribution of key components to the overall performance of ESLM.
- We conclude from our findings that the utilization of contextual LMs and their integration with KGE significantly outperforms SOTA, particularly on the DBpedia and FACES datasets.

The rest of this paper is organized as follows: Sect. 2 provides a summary of related works. Section 3 defines the entity summarization problem formally and introduces our ESLM model. Section 4 details the implementation of our approach. Our evaluation is described in Sect. 5. Finally, Sect. 6 concludes the paper and suggests directions for future work.

2   Related Work

2.1   Entity Summarization

Unsupervised learning has previously been used in entity summarization tasks on single and combined features [14] such as frequency, centrality, informativeness, and similarity. For example, RELIN [5] computes the relatedness between RDF triples and uses the informativeness measurement of each triple in a random surfer model. DIVERSUM [24] employs the concept of diversification to address the entity summarization problem, incorporating it into its summarizing algorithm. FACES [9] utilizes all the aforementioned dimensional features to generate entity summaries.

Recently, deep learning techniques are being used for entity summarization. ESA [28], the first model to use deep learning for this purpose, employs bidirectional long short-term memory (BiLSTM) networks. It combines a word embedding technique [2] with TransE [3] to encode the predicate and object of a triple, identifying these components as crucial for summarizing an entity's triples [27]. ESA applies BiLSTM with an attention mechanism, selecting the top-k triples for the entity summary. In contrast, DeepLENS [15] relies solely on word embeddings, specifically fastText [11] to encode triples containing text-based entity descriptions. The authors argue that word embeddings provide richer textual semantics than KGEs for this task. They proceed to show that DeepLENS outperforms ESA on benchmark datasets like ESBM (version 1.2) [13]. GATES [8] combines GloVe word embeddings [20] with KGE (such as ComplEx [26]) using graph neural networks (GNNs). This method aims to enhance the quality of entity summaries by encoding topological information through KGE. GATES outperforms both DeepLENS and ESA on the ESBM (version 1.2) and FACES datasets. Most recently, ESCS [4] was introduced. It employs an approach similar to DeepLENS by using Word2Vec [18], and introduces a novel method for the computation of triple scores and the construction of summary sets for any target entity. This method is based on the idea of salience, which is computed by evaluating the similarity between an entity's semantic embeddings and a particular property (predicate). Additionally, it computes and exploits the complementarity of predicates and aims to optimise the complementary of the relationships it returns in entity summaries.

2.2   Contextual Language Models

In recent years, contextual LMs have significantly impacted various downstream tasks, such as question answering (QA) [31], text summarization [17,19], and relation extraction [1]. These models achieve the current SOTA performance in numerous Natural Language Processing (NLP) tasks. Contextual LMs, as opposed to static word embeddings, provide each token with a representation derived from the entire input sequence. This approach allows them to capture more contextual information than static embeddings. For example, BERT [6] is a pre-trained contextual LM that is built upon a multi-layer bidirectional transformer encoder. The model is trained on a corpus that includes the English Wikipedia and the BooksCorpus. During training, BERT employs a masked language modeling technique, wherein certain tokens within an input sequence are randomly replaced with a [MASK] token. BERT consequently learns to predict these masked tokens based on the context provided by the unmasked tokens in the sequence. To further enhance LMs, ERNIE [32] incorporates knowledge graphs into the computation of word embeddings. ERNIE constructs entity representations from words, encoding them using KGE models such as TransE. Similarly to BERT, it then utilizes masked LMs and next-sentence prediction for pre-training, and for the extraction of lexical and syntactic information from text tokens. Like BERT, ERNIE uses English Wikipedia for pre-training and aligns text with Wikidata through KGE.

Unlike BERT and ERNIE, T5 models every NLP problem as a text-to-text problem [21]. The model is trained using a denoising autoencoder objective, where it learns to reconstruct the original text from a corrupted version. The T5 model does not require task-specific heads. It distinguishes tasks in the input text through prefixes that guide the model during the output computation and generation. For example, the user may provide an input in the form of `classify: text` to prompt the model to output a class label for the input text.

3 Approach

3.1 Problem Statement

Entity Description. Let E be a set of entities, R be a set of relations, C be a set of classes and L denote a set of literals. A knowledge graph $T \subseteq E \times R \times (C \cup L \cup E)$ is a set of triples (s, p, o). s is called the subject, p the predicate, and o the object of the triple. We define an entity description, $Desc(e, T)$, as follows:

$$Desc(e, T) = \{(e, r, o) \in T \vee (s, r, e) \in T\} \tag{1}$$

where $s, e \in E$, $o \in (C \cup L \cup E)$, and r is a predicate. An example of such as description is provided in Fig. 1. Note that the triples (3WAY FM, Type, Radio Station) and (Warrnambool, Broadcast Area of, 3WAY FM) belong to the description of 3WAY FM, i.e., as per Eq. 1, the target entity e can function as either a subject or an object in the elements of its description $Desc(e, T)$.

Entity Summarization. Let e be an entity e, $Desc(e, T)$ be its entity description, and $k \in \mathbb{N}$ be a size constraint. We define an entity summary $ES(e)$ as a subset of $Desc(e, T)$ with $|ES(e)| \leq k$, where $k = 5$ or $k = 10$ is often used in practice. The purpose of entity summarization techniques is to compute $ES(e)$ by selecting the k best suited triples from $Desc(e, T)$.

3.2 ESLM Model

The ESLM relies on a Transformer-based language model [16], allowing for context-aware processing and prediction. The attention mechanism allows the model to focus on the most relevant aspects of data sequences. Additionally, ESLM employs a multi-layer perceptron (MLP) for accurate triple scoring, which aids in selecting the most relevant triples for each entity. Moreover, we enrich the model with KGEs to augment the model's effectiveness, leveraging the rich semantic information from the knowledge graphs. The architecture of ESLM is detailed in Fig. 1. In the following subsections, we discuss each of our model's components.

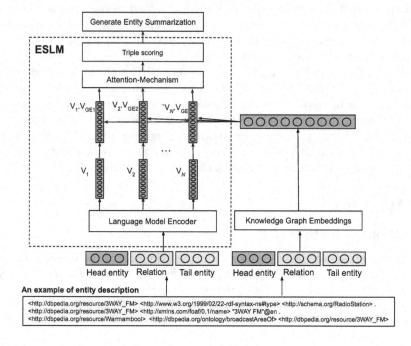

Fig. 1. The ESLM model architecture

Language Model Encoder. This encoder relies on a pre-trained model based on the Transformer architecture (e.g., BERT, ERNIE, or T5), which is further fine-tuned using labeled data from the input representations. The encoder's configuration includes the number of layers (i.e., transformer blocks) denoted by N_L set to 12, the number of hidden layers marked by H set to 768, and the number of self-attention heads denoted by AH set to 12.

Each component (subject (s), predicate (p), object (o)) from the triple t is represented in their textual form. When RDF resources are identified as IRIs (Internationalized Resource Identifiers), we utilize their `rdfs:label` for textual representation; otherwise, the local name of the IRI is used as the textual representation. The local name, extracted from the IRI segment after the last slash ('/'), acts as a unique, concise identifier within the IRI's namespace, serving as a human-readable term when a label is absent. Using the text form allows the LM to process and vectorize IRIs effectively, ensuring that the lexical and semantic nuances of the triples are captured during encoding. As proposed in [30], each triple t within the entity description $Desc(e, T)$ is structured as a single sequence-packed sentence (s, p, o). Furthermore, all these single sequences are formatted as input representations for the ESLM model. Every sequence begins with the special classification token ([CLS]). Additionally, we use a special separator token ([SEP]) to distinctly separate each component of the triple. For example, a triple t such as (3WAY FM, Type, Radio Station) is transformed into textual input such as (3WAY FM [SEP] Type [SEP] Radio Station) and then translated into a series of tokens denoted as $\{Tok_1, Tok_2, \ldots, Tok_N\}$, where N is the count of tokens. Furthermore, the input tokens of t are converted into embeddings by implementing a transformer-based encoder. Moreover, the LM encoder generates two kinds of outputs. The first is the final hidden state of the special classification token, denoted as ([CLS]) and represented by $C \in \mathbb{R}^H$, which is an aggregate representation of the entire input sequence often used in classification tasks. The second output consists of the last hidden states for each input token in the sequence. These are denoted as V_i, leading to set a $V = \{V_1, V_2, \ldots, V_N\}$, where each $V_i \in \mathbb{R}^H$ corresponds to the ith input token. Here, N is the sequence length, and H is the hidden state dimension.

Attention Mechanism. In the ESLM model, the attention mechanism is applied through a three-step process to assess the importance of each token in the LM encoder output. Initially, attention scores are assigned to each token using a linear transformation represented by $A_{weights} = VW_{attn} + b_{attn}$. Here, V is the LM encoder output. Meanwhile, W_{attn} and b_{attn} are the weights and bias of the linear transformation, which are hence normalized to form a valid probability distribution. The Softmax function is then applied to transform these scores into probabilities that sum up to 1, reflecting the relative significance of each token in the sequence. This mechanism enables the model to focus selectively on the most relevant parts of the input, enhancing its language understanding and generation capabilities. The complete equation is shown in Eq. 2:

$$A_{weights} = Softmax(VW_{attn} + b_{attn}). \tag{2}$$

Subsequently, these normalized scores are applied to the output V of the LM encoder using element-wise multiplication, which is shown by $\odot$, resulting in the attention-weighted output A, which can be seen in Eq. 3:

$$A = A_{weights} \odot V. \tag{3}$$

This process allows the model to focus selectively on the most relevant parts of the input sequence, enhancing its ability to interpret and generate language by considering both the context and significance of each element in the sequence.

Triple Scoring. This step relies on the MLP, which is applied to learn the output from the attention mechanism process A into a form that is more amenable for the subsequent operations—calculating the mean and applying the softmax function. The mean operation suggests an aggregation or summarization of the features learned by the MLP, and the softmax function indicates that the final goal might be to interpret these aggregated features probabilistically, possibly for a task like regression or probabilistic classification. The scoring function for a triple is denoted S_t, where $S_t \in \mathbb{R}$, is calculated as shown in Eq. 4:

$$S_t = softmax(mean(MLP(A))). \tag{4}$$

3.3   ESLM Model Enrichment

Figure 1 demonstrates that the triple-scoring computation is based on semantic information generated by a pre-trained LM (PLM) that does not consider the structural information of KGs. We now enhance our model by incorporating KGEs based on the ComplEx method to obtain information from the structure information of the KG.

KGE Using ComplEx. Let $GE : E \cup R \cup L \cup C \to \mathbb{C}^d$ be an embedding function that computes vectors for the elements of a knowledge graph. In this work, we use ComplEx with $d = 300$, i.e., every component of a triple (s, p, o) is represented by a 300-dimensional complex-valued vector. The vector for a given triple, $V_{GE_{ti}}$, is constructed by concatenating the embeddings for each component of the triple (s_{ti}, p_{ti}, o_{ti}), thus creating a single vector that holds all the structural information for that triple, as indicated in Eq. 5:

$$V_{GE_{ti}} = GE(s_{ti})\|GE(p_{ti})\|GE(o_{ti}). \tag{5}$$

Integration of LM with KGE. To leverage both structured and unstructured information, the model concatenates the last hidden vectors V_{ti} from the LM encoder with the KGE vectors $V_{GE_{ti}}$, resulting in a new vector V'_{ti} as shown in Eq. 6. This concatenated vector V'_{ti} holds both the contextualized semantic information from the LM and the structural relationships from the KG, yielding a more comprehensive representation for each triple:

$$V'_{ti} = V_{ti}\|V_{GE_{ti}}. \tag{6}$$

By concatenating these vectors, we literally fuse textual information (V_{ti}) with knowledge graph information ($V_{GE_{ti}}$). An attention mechanism is applied to the concatenated vectors V'_{ti} to focus on the most relevant parts of the combined embedding while generating the summary. Therewith, the model learns the relation between the embedding dimensions in LM and KGE representations. As shown in our experiments, this enables our model to better discern the salience of triples for entities to summarise. The exact computation of this operation is given by the formulas outlined in Eqs. 2 and 3, and involve calculating attention weights and context vectors. With the attention-enriched vectors V', the model computes the triple scoring as specified in Eq. 4. The scoring process evaluates the relevance of each triple for the summary using the enriched embeddings that now incorporate both semantic and structural insights.

3.4   Entity Summarization

Finally, the entity summary $ES(e)$ is generated for an entity e by selecting the top-k triples from $Desc(e, T)$, which is a set of entity descriptions for entity e in a KG. The parameter k represents a size constraint, indicating the number of triples to be included in the summary. The selection is based on Eq. 7, as defined by [8].

$$\forall t' \in ES(e) \backslash Desc(e, T) : S_{t_i} \leq \min_{t_i \in Desc(e,T)} S_{t_i}. \tag{7}$$

4   Experimental Setup

4.1   Baselines

Since supervised learning with a deep learning approach in entity summarization tasks substantially outperforms unsupervised learning-based entity summarization tasks, we only consider the following methods as the baselines:

1. **ESA** [28] exploits graph embedding to encode triples of entity descriptions. Triple scores are calculated by leveraging normalized attention weights based on output vectors of BiLSTM computation.
2. **AutoSUM** [29] improves the ESA model by incorporating multi-user preference simulations such as entity and user phase attention.
3. **NEST** [12] leverages a KG encoder that represents structural and textual representations from KGs, employing joint learning from salience and diversified summary learning to produce the entity summary.
4. **DeepLENS** [15] uses textual semantics for triple encoding, and employs a deep learning model (such as BiLSTM) and MLP to generate scores for triples of entity descriptions.
5. **GATES** [8] computes triple scores using a combination of information from textual and structural representations generated by GNNs. Additionally, ensemble learning is used to improve triple-scoring performance.
6. **ESCS** [4] exploits description complementary and salience learning to score components of the input knowledge graphs. It then employs joint learning to calculate triple scores.

We used the DeepLENS, AutoSUM, and ESA experimental results presented in [8] as the code for these evaluation is open and the evaluation can be replicated. However, we could not use the NDCG approach to compare our model to NEST and ESCS due to unavailable codes.

4.2 Datasets

We used two types of benchmark datasets, including the ESBM (version 1.2)[1] and FACES. The ESBM comprises 125 and 50 entities which are from the DBpedia and LinkedMDB datasets, respectively. From the FACES dataset, we utilize 50 entities that are from the DBpedia dataset. The DBpedia dataset in ESBM separates the entities into five classes: agent, event, location, species, and work. The LinkedMDB dataset contains two distinct categories: films and persons. Each entity in every dataset is described by at least 20 triples. Additionally, FACES contains at least four manually generated entity summaries, while the ESBM provides six manually constructed ground truth summaries for each entity.

4.3 Experimental Settings

The ESLM is initialized by incorporating pre-trained LMs to leverage the knowledge already acquired. Specifically, we utilized *bert-base-uncased*[2] for integrating the BERT model and *ernie-2.0-en*[3] for ERNIE. Both models comprise 12 layers, 12 self-attention heads, and 768 hidden layers H. For the T5 model integration within ESLM, we employed the *t5-base'* LM[4]. The fine-tuning process was implemented on ESLM with varied learning rates $\in \{1 - 5 \times 10^{-5}\}$. Furthermore, we employed the binary cross-entropy (BCE) loss function as the primary criterion for training. The BCE loss is particularly well-suited for binary classification tasks, which aligns with the nature of our entity summarization problem where the model needs to predict the relevance of each entity triple within a given context. During training, the AdamW optimizer was employed alongside BCE. Throughout our evaluation, we used a five-fold cross-validation. In subsequent experiments, we used the ComplEx [26] method generated by the DGLKE framework[5] to enrich a PLM with KGE.

We conducted a statistical significance test to check whether the findings obtained by our models were significantly different from those of SOTA methods. In particular, we utilized the Wilcoxon signed ranked test with a 95% significance level. All experiments were conducted on a 64-core AMD EPYC 7713 CPU (2.0 GHz) with 1024 GB of RAM, running on Debian with CUDA using 2 NVIDIA GeForce RTX A5000 24 GB GPUs.

5 Results and Analysis

5.1 Comparison with State-of-the-Art Approaches

The baselines ESCS, GATES, DeepLENS, AutoSUM, NEST, and ESA are together referred to as SOTA. Table 1 shows average F-measure scores for differ-

---

[1] https://github.com/nju-websoft/ESBM/tree/master/v1.2.
[2] https://huggingface.co/bert-base-uncased.
[3] https://huggingface.co/nghuyong/ernie-2.0-base-en.
[4] https://huggingface.co/t5-base.
[5] https://github.com/awslabs/dgl-ke.

ent entity summarization models across three datasets: DBpedia, LinkedMDB, and FACES. These models include baseline models and our approach (ESLM) model, evaluated at the cut-offs $k = 5$ and $k = 10$, representing the top-5 and top-10 retrieved sets of triples, respectively. A clear pattern emerges: models generally improve their scores as the k value increases. This suggests a greater capacity to capture relevant triples of entities within a larger retrieval window. Notably, the ESLM model, which employs a combination of BERT, ERNIE, and T5 LMs determined to be suitable for entity summarization through an ablation study (see Sect. 5.3), outperforms others in DBpedia and FACES at both $k = 5$ and $k = 10$, indicating its superior summarization performance on these datasets.

The lower F-measure scores of ESLM model on the LinkedMDB dataset, specifically at $k = 5$ and $k = 10$, indicate that it may not be as well calibrated to the specific domain of the dataset as the ESCS and DeepLENS models. Although ESLM's performance shows improvement when considering a broader range of top predictions, it still falls behind models such as AutoSUM, GATES, and ESCS. The results underscore the need for ESLM to optimize its feature extraction and integrate KGE more effectively to enhance its summarization quality, indicating room for further refinement of the model's approach.

Table 1. Average F-measure score based on our model testing via five-fold cross-validation processes to all entities of the benchmark.

	DBpedia		LinkedMDB		FACES	
	k = 5	k = 10	k = 5	k = 10	k = 5	k = 10
ESA	0.332	0.532	0.353	0.435	0.153	0.261
NEST	0.354	0.540	0.332	0.465	0.272	0.346
AutoSUM	0.372	0.555	0.430	0.520	0.241	0.316
DeepLENS	0.404	0.575	0.469	0.489	0.130	0.248
GATES	0.423	0.574	0.437	**0.535**	0.254	0.324
ESCS	0.415	0.582	**0.494**	0.512	-	-
ESLM	**0.427**	**0.591**	0.467	0.498	**0.301**	**0.369**

NDCG scores for ESLM and the baselines across DBpedia, LinkedMDB, and FACES datasets are presented in Table 2. NDCG is a performance metric for the quality of ranked order outputs, with higher scores indicating that a model is effectively ranking highly relevant triples of entities at the top of the list. Our model ESLM clearly outperforms other methods in all datasets at both $k = 5$ and $k = 10$, with its peak score on DBpedia at $k = 10$ being 0.913. In comparison, while DeepLENS, GATES, and AutoSUM demonstrate strong performance with scores over 0.800 in several instances, they fall short of ESLM's consistency across datasets and cutoff values. The improvement from $k = 5$ to $k = 10$ for all models suggests that they are more adept at providing quality entity summarizations when more results are included. In particular, ESLM's superior NDCG scores at both levels indicate its robust capacity to synthesize and rank entity information effectively using LM-driven approaches. This performance implies that ESLM has a significant advantage in tasks requiring nuanced discernment of entity triple relevance, especially in larger result sets.

Table 2. Average NDCG score based on our model testing via five-fold cross-validation processes to all entities of the benchmark.

	DBpedia		LinkedMDB		FACES	
	k = 5	k = 10	k = 5	k = 10	k = 5	k = 10
ESA	0.755	0.846	0.737	0.799	0.601	0.707
AutoSUM	0.797	0.882	0.809	0.856	0.693	0.768
DeepLENS	0.825	0.905	0.855	0.888	0.585	0.715
GATES	0.798	0.893	0.804	0.881	0.697	0.759
ESLM	**0.850**	**0.913**	**0.868**	**0.893**	**0.735**	**0.793**

We also ran a statistical test to see whether our method significantly outperforms the current SOTA benchmarks. We regard a p-value less or equal to 0.05 as significant. As shown in Table 3, ESLM performs significantly better than the compared models in most cases. In particular, the results show that ESLM outperforms the existing SOTA significantly in at least one experimental setting on the DBpedia dataset from ESBM and the FACES datasets. Additionally, ESLM significantly outperforms ESA, AutoSUM, and GATES across all evaluated $k = 5$ in the LinkedMDB dataset. The provided p-values confirm the statistical significance of our findings, with ESLM demonstrating a clear advantage over ESA, AutoSUM, DeepLENS, and GATES in almost all settings.

Table 3. Comparison of ESLM and SOTA using a Wilcoxon-rank test on the F-measure scores. The leftmost column shows the approaches ESLM was compared with. The values in the table are p-values.

	DBpedia		LinkedMDB		FACES	
	k = 5	k = 10	k = 5	k = 10	k = 5	k = 10
ESA	0.000	0.000	0.000	0.004	0.000	0.000
AutoSUM	0.000	0.000	0.028	-	0.003	0.004
DeepLENS	0.013	0.015	-	0.672	0.000	0.000
GATES	0.505	0.002	0.000	-	0.016	0.001

5.2   Example Findings

This section describes two examples found in our on test results. Due to space constraints, we chose the strongest baseline, GATES, and our model, ESLM, as well as a form of ground truth. In Fig. 2, we see a direct comparison between entity summaries produced by the ESLM and GATES models for a given entity, Ludwigsburg University. The output of ESLM is well aligned with the ground truth summary, capturing a subset of triples that directly correspond to the essential information about the entity. This is reflected in its higher F-measure score of 0.666, surpassing GATES' score of 0.600. A key observation is that GATES omits crucial label and type information in its summary, which indicates a gap in capturing and presenting significant details. ESLM's ability to include these important triples results in a more accurate and comprehensive summary enhances the model's utility in applications that depend on detailed entity representations.

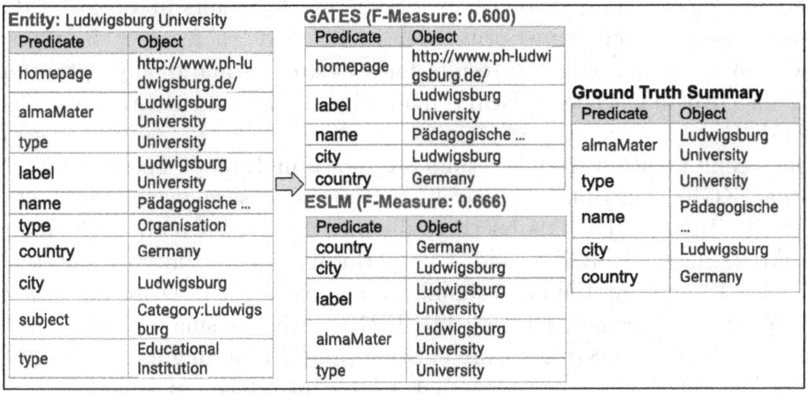

Entity: Ludwigsburg University		GATES (F-Measure: 0.600)			
Predicate	Object	Predicate	Object		
homepage	http://www.ph-ludwigsburg.de/	homepage	http://www.ph-ludwigsburg.de/	**Ground Truth Summary**	
almaMater	Ludwigsburg University	label	Ludwigsburg University	Predicate	Object
type	University	name	Pädagogische ...	almaMater	Ludwigsburg University
label	Ludwigsburg University	city	Ludwigsburg	type	University
name	Pädagogische ...	country	Germany	name	Pädagogische ...
type	Organisation	**ESLM (F-Measure: 0.666)**			
country	Germany	Predicate	Object	city	Ludwigsburg
city	Ludwigsburg	country	Germany	country	Germany
subject	Category:Ludwigsburg	city	Ludwigsburg		
		label	Ludwigsburg University		
type	Educational Institution	almaMater	Ludwigsburg University		
		type	University		

Fig. 2. Effectiveness of ESLM compared to GATES in use case one.

Figure 3 shows another example where ESLM outperforms GATES concerning entity summary results. Here, ESLM's F-measure improves from 0.533 to 0.733 by providing a wide range of information on the target entity, whereas the GATES lacks diversity in the information it provides, as it tends to repeat information about places.

According to both use cases, ESLM demonstrates that the use of LMs for entity summarization tasks relatively improves the quality of the summaries.

5.3   Ablation Study

In a preliminary assessment of the ESLM model, we compared the performance of pre-trained LMs such as BERT, ERNIE, and T5 on the DBpedia, LinkedMDB, and FACES datasets without the enhancement of KGE. The F-measure served

Entity: Battle of Rottofreddo

Predicate	Object
Type	MilitaryConflict
isPartOfMilitaryConflict	War of the Austrian Succession
Result	French victory
Label	Battle of Rottofreddo
Date	1746-08-12
Place	Province of Piacenza
Place	Italy
Place	Rottofreno
Subject	Category:1746 in France
Subject	Category:1746 in Italy
Subject	Category:Conflicts in 1746

GATES (F-Measure: 0.533)

Predicate	Object
Type	MilitaryConflict
Date	1746-08-12
Place	Province of Piacenza
Place	Italy
Place	Rottofreno

ESLM (F-Measure: 0.733)

Predicate	Object
Date	1746-08-12
Type	MilitaryConflict
Result	French victory
isPartOfMilitaryConflict	War of the Austrian Succession
Place	Italy

Ground Truth Summary

Predicate	Object
Type	MilitaryConflict
Result	French victory
Label	Battle of Rottofreddo
Date	1746-08-12
Place	Italy

Fig. 3. Effectiveness of ESLM compared to GATES in use case two.

as the basis for this comparison. As evidenced by the results in Table 4, ERNIE demonstrates consistent superiority over BERT and T5 for $k = 5$ across all datasets. When examining the results for $k = 10$, T5 marginally outperforms BERT on DBpedia, whereas BERT maintains a slight advantage over T5 on LinkedMDB.

The second ablation study's results are shown in Table 4. These results indicate that KGEs contribute positively to the F-measure performance of the BERT, ERNIE, and T5 LMs on three distinct datasets. On the DBpedia and LinkedMDB datasets, BERT and T5's performances are notably boosted by KGE, with T5 showing the most significant enhancement in DBpedia for $k = 5$, increasing its score from 0.410 to 0.427. ERNIE exhibits smaller gains with the addition of KGE, suggesting its pre-existing pre-training might already include some of the relational knowledge that KGEs introduce. In contrast, on the FACES dataset, T5 especially gains at $k = 10$. These results suggest that the degree of alignment between a model's pre-training and a dataset can be a crucial determinant of the effectiveness of integrating KGEs. The consistent increase in F-measure scores from $k = 5$ to $k = 10$ suggests that KGEs provide supplementary information that can improve the models' capability to effectively rank relevant information.

We also used NDCG scores to examine the BERT, ERNIE, and T5 models on the ESLM model, evaluating their ranking effectiveness both with and without KGE, as shown in Table 5. BERT exhibits marginal improvements with KGE integration, with the NDCG score slightly increasing across the DBpedia and LinkedMDB datasets for both the top-5 and top-10 results but showing minimal change on the FACES dataset. ERNIE's performance with KGE is mixed. It experiences a slight decrease in the DBpedia dataset but improvements in LinkedMDB and FACES, particularly at $k = 10$ in LinkedMDB. T5 demonstrates the most significant gains from KGEs across all datasets, especially at $k = 10$ in LinkedMDB, suggesting that T5's architecture effectively utilizes the

Table 4. Highest F-measure performance of BERT, ERNIE, and T5 on ESLM

Models	DBpedia		LinkedMDB		FACES	
	k = 5	k = 10	k = 5	k = 10	k = 5	k = 10
BERT	0.411	0.574	0.444	0.494	0.286	0.355
BERT + KGE	0.417	0.586	0.445	0.482	**0.301**	0.347
ERNIE	0.421	0.583	0.448	0.482	0.292	0.348
ERNIE + KGE	0.423	0.586	**0.467**	0.494	0.295	**0.369**
T5	0.410	0.584	0.442	0.486	0.287	0.352
T5 + KGE	**0.427**	**0.591**	0.455	**0.498**	0.300	0.361

additional relational knowledge provided by KGEs to enhance ranking accuracy. The consistently higher scores at $k = 10$ across all models imply that the models perform better when evaluating a larger set of predictions, which is crucial for tasks involving entity summarization where multiple correct answers are possible.

Table 5. Highest NDCG scores of BERT, ERNIE, and T5 on ESLM

Models	DBpedia		LinkedMDB		FACES	
	k = 5	k = 10	k = 5	k = 10	k = 5	k = 10
BERT	0.841	0.904	0.836	0.848	0.736	**0.797**
BERT + KGE	0.842	0.908	0.835	0.840	0.733	0.796
ERNIE	0.841	0.910	0.831	0.837	0.740	0.788
ERNIE + KGE	0.838	0.909	0.867	0.858	**0.744**	0.795
T5	**0.852**	0.908	0.862	0.869	0.742	0.785
T5 + KGE	0.850	**0.913**	**0.868**	**0.893**	0.736	0.790

5.4   Computational Requirements and Efficiency of ESLM

Table 6 outlines the training times and processing efficiencies of among ESLM models with and without KGE, trained on the DBpedia, LinkedMDB, and FACES datasets over 50 epochs using 1 GPU. The integration of contextual LMs and KGEs slightly increases training times across all models, suggesting a modest rise in computational requirements. The T5 model is the most time-consuming. All configurations process a comparable number of entities per second, with a negligible increase when KGE is included. This observation is important as it implies that the addition of KGEs enhances the performance of supervised entity summarization approaches without increasing their computational needs.

Table 6. Comparative Analysis of Training Times and Entity Processing Efficiency among ESLM models. All times are seconds, and the total number of epochs is 50.

Models	Topk	Input Triples	Output Triples	Training Time Total	Mean	Prediction Time Single Triples
BERT	5	4436	750	329.56	6.59	0.060
BERT + KGE	5	4436	750	404.50	8.09	0.060
ERNIE	5	4436	750	327.39	6.55	0.067
ERNIE + KGE	5	4436	750	328.88	6.58	0.070
T5	5	4436	750	402.96	8.06	0.071
T5 + KGE	5	4436	750	411.86	8.24	0.072
BERT	10	4436	1500	333.62	6.67	0.060
BERT + KGE	10	4436	1500	329.91	6.60	0.059
ERNIE	10	4436	1500	333.32	6.67	0.069
ERNIE + KGE	10	4436	1500	329.24	6.58	0.069
T5	10	4436	1500	403.83	8.07	0.070
T5 + KGE	10	4436	1500	413.38	8.27	0.073
BERT	5	2148	125	184.06	3.68	0.123
BERT + KGE	5	2148	125	185.14	3.70	0.125
ERNIE	5	2148	125	184.85	3.70	0.144
ERNIE + KGE	5	2148	125	185.82	3.72	0.144
T5	5	2148	125	188.05	3.76	0.151
T5 + KGE	5	2148	125	189.40	3.79	0.154
BERT	10	2148	250	185.65	3.71	0.173
BERT + KGE	10	2148	250	185.71	3.71	0.123
ERNIE	10	2148	250	185.92	3.72	0.157
ERNIE + KGE	10	2148	250	185.82	3.72	0.145
T5	10	2148	250	189.70	3.79	0.153
T5 + KGE	10	2148	250	188.20	3.76	0.155
BERT	5	2152	125	186.47	3.73	0.122
BERT + KGE	5	2152	125	186.22	3.72	0.126
ERNIE	5	2152	125	186.32	3.73	0.142
ERNIE + KGE	5	2152	125	186.37	3.73	0.154
T5	5	2152	125	188.09	3.76	0.171
T5 + KGE	5	2152	125	189.31	3.79	0.154
BERT	10	2152	250	188.55	3.77	0.124
BERT + KGE	10	2152	250	186.85	3.74	0.126
ERNIE	10	2152	250	187.27	3.75	0.147
ERNIE + KGE	10	2152	250	187.02	3.74	0.146
T5	10	2152	250	191.41	3.83	0.156
T5 + KGE	10	2152	250	190.75	3.82	0.157

6   Conclusion and Future Work

In this study, we introduced ESLM, an entity summarization method leveraging LMs enhanced with KGEs. Our analysis showed that ERNIE-based implementations of ESLM outperform BERT-based approaches, with further improvements when these models are enriched with KGEs, particularly in the T5 model. Our results also suggest that ESLM achieves significantly better results than the current SOTA methods, as evidenced by ESLM superior performance on benchmark datasets such as DBpedia and FACES. Despite these advancements, we recognize a limitation in the scale of current gold standard datasets, such as ESBM and FACES. This highlights a broader issue in the field's lack of comprehensive benchmarking datasets for entity summarization. To address this challenge, we plan to develop an extensive silver dataset to support the creation of robust and reliable entity summarization models. Additionally, we aim to explore the integration of ESLM with graph neural networks, potentially enhancing our model's capabilities further.

Acknowledgements. This work has been supported by the Ministry for Economic Affairs, Innovation, Digitalisation and Energy of North Rhine-Westphalia (MWIDE NRW) within the project Climate bOWL (grant no. 005-2111-0020), the German Federal Ministry of Education and Research (BMBF) within the projects KIAM (grant no. 02L19C115), COLIDE (grant no. 01I521005D), the European Union's Horizon Europe research and innovation programme (grant no. 101070305), the Deutsche Forschungsgemeinschaft (DFG, German Research Foundation): TRR 318/1 2021 – 438445824, and Mora Scholarship from the Ministry of Religious Affairs, Republic of Indonesia.

References

1. Ali, M., Saleem, M., Ngomo, A.C.N.: Unsupervised relation extraction using sentence encoding. In: Verborgh, R., et al. (eds.) ESWC 2021. LNCS, pp. 136–140. Springer, Cham (2021). https://doi.org/10.1007/978-3-030-80418-3_25
2. Bengio, Y., Ducharme, R., Vincent, P., Janvin, C.: A neural probabilistic language model. J. Mach. Learn. Res. **3**, 1137–1155 (2003). http://jmlr.org/papers/v3/bengio03a.html
3. Bordes, A., Usunier, N., Garcia-Duran, A., Weston, J., Yakhnenko, O.: Translating embeddings for modeling multi-relational data. In: Neural Information Processing Systems (NIPS), pp. 1–9 (2013)
4. Chen, L., et al.: Entity summarization via exploiting description complementarity and salience. IEEE Trans. Neural Netw. Learn. Syst. (2022)
5. Cheng, G., Tran, T., Qu, Y.: RELIN: relatedness and informativeness-based centrality for entity summarization. In: Aroyo, L., et al. (eds.) ISWC 2011. LNCS, vol. 7031, pp. 114–129. Springer, Heidelberg (2011). https://doi.org/10.1007/978-3-642-25073-6_8
6. Devlin, J., Chang, M., Lee, K., Toutanova, K.: BERT: pre-training of deep bidirectional transformers for language understanding. CoRR **abs/1810.04805** (2018), http://arxiv.org/abs/1810.04805
7. Ermilov, T., Moussallem, D., Usbeck, R., Ngomo, A.C.N.: Genesis: a generic RDF data access interface. In: Proceedings of the International Conference on Web Intelligence, pp. 125–131. WI 2017, Association for Computing Machinery, New York, NY, USA (2017). https://doi.org/10.1145/3106426.3106514
8. Firmansyah, A.F., Moussallem, D., Ngomo, A.N.: GATES: using graph attention networks for entity summarization. In: Gentile, A.L., Gonçalves, R. (eds.) K-CAP 2021: Knowledge Capture Conference, Virtual Event, USA, December 2-3, 2021, pp. 73–80. ACM (2021). https://doi.org/10.1145/3460210.3493574

9. Gunaratna, K., Thirunarayan, K., Sheth, A.: Faces: diversity-aware entity summarization using incremental hierarchical conceptual clustering. In: Proceedings of the Twenty-Ninth AAAI Conference on Artificial Intelligence, pp. 116–122. AAAI Press (2015)

10. Hasibi, F., Balog, K., Bratsberg, S.E.: Dynamic factual summaries for entity cards. In: Proceedings of the 40th International ACM SIGIR Conference on Research and Development in Information Retrieval, pp. 773–782. SIGIR 2017, Association for Computing Machinery, New York, NY, USA (2017). https://doi.org/10.1145/3077136.3080810

11. Joulin, A., Grave, E., Bojanowski, P., Douze, M., Jégou, H., Mikolov, T.: Fasttext.zip: compressing text classification models. CoRR **abs/1612.03651** (2016), http://arxiv.org/abs/1612.03651

12. Li, J., et al.: Neural entity summarization with joint encoding and weak supervision. In: Bessiere, C. (ed.) Proceedings of the Twenty-Ninth International Joint Conference on Artificial Intelligence, IJCAI 2020, pp. 1644–1650. ijcai.org (2020). https://doi.org/10.24963/ijcai.2020/228

13. Liu, Q., Cheng, G., Gunaratna, K., Qu, Y.: ESBM: an entity summarization Bench-Mark. In: Harth, A., et al. (eds.) ESWC 2020. LNCS, vol. 12123, pp. 548–564. Springer, Cham (2020). https://doi.org/10.1007/978-3-030-49461-2_32

14. Liu, Q., Cheng, G., Gunaratna, K., Qu, Y.: Entity summarization: state of the art and future challenges. J. Web Semant. **69**, 100647 (2021). https://doi.org/10.1016/j.websem.2021.100647

15. Liu, Q., Cheng, G., Qu, Y.: Deeplens: deep learning for entity summarization. CoRR **abs/2003.03736** (2020). https://arxiv.org/abs/2003.03736

16. Liu, S., Chen, Y., Liu, K., Zhao, J.: Exploiting argument information to improve event detection via supervised attention mechanisms. In: Barzilay, R., Kan, M.Y. (eds.) Proceedings of the 55th Annual Meeting of the Association for Computational Linguistics (Volume 1: Long Papers), pp. 1789–1798. Association for Computational Linguistics, Vancouver, Canada (2017). https://doi.org/10.18653/v1/P17-1164, https://aclanthology.org/P17-1164

17. Liu, Y.: Fine-tune BERT for extractive summarization. CoRR **abs/1903.10318** (2019). http://arxiv.org/abs/1903.10318

18. Mikolov, T., Chen, K., Corrado, G., Dean, J.: Efficient estimation of word representations in vector space. In: Bengio, Y., LeCun, Y. (eds.) 1st International Conference on Learning Representations, ICLR 2013, Scottsdale, Arizona, USA, May 2-4, 2013, Workshop Track Proceedings (2013). http://arxiv.org/abs/1301.3781

19. Patil, P., Rao, C., Reddy, G., Ram, R., Meena, S.M.: Extractive text summarization using BERT. In: Gunjan, V.K., Zurada, J.M. (eds.) Proceedings of the 2nd International Conference on Recent Trends in Machine Learning, IoT, Smart Cities and Applications. LNNS, vol. 237, pp. 741–747. Springer, Singapore (2022). https://doi.org/10.1007/978-981-16-6407-6_63

20. Pennington, J., Socher, R., Manning, C.: GloVe: global vectors for word representation. In: Proceedings of the 2014 Conference on Empirical Methods in Natural Language Processing (EMNLP), pp. 1532–1543. Association for Computational Linguistics, Doha, Qatar (2014). https://doi.org/10.3115/v1/D14-1162

21. Raffel, C., Shazeer, N., Roberts, A., Lee, K., Narang, S., Matena, M., Zhou, Y., Li, W., Liu, P.J.: Exploring the limits of transfer learning with a unified text-to-text transformer. J. Mach. Learn. Res. **21**(1) (2020)

22. Sacenti, J.A., Fileto, R., Willrich, R.: Knowledge graph summarization impacts on movie recommendations. J. Intell. Inf. Syst. **58**(1), 43–66 (2022)

23. Sun, Y., et al.: ERNIE 2.0: a continual pre-training framework for language understanding. In: The Thirty-Fourth AAAI Conference on Artificial Intelligence, AAAI 2020, The Thirty-Second Innovative Applications of Artificial Intelligence Conference, IAAI 2020, The Tenth AAAI Symposium on Educational Advances in Artificial Intelligence, EAAI 2020, New York, NY, USA, February 7-12, 2020, pp. 8968–8975. AAAI Press (2020). https://ojs.aaai.org/index.php/AAAI/article/view/6428

24. Sydow, M., Pikula, M., Schenkel, R.: DIVERSUM: towards diversified summarisation of entities in knowledge graphs. In: Workshops Proceedings of the 26th International Conference on Data Engineering, ICDE 2010, March 1-6, 2010, Long Beach, California, USA, pp. 221–226. IEEE Computer Society (2010). https://doi.org/10.1109/ICDEW.2010.5452707

25. Thalhammer, A., Lasierra, N., Rettinger, A.: LinkSUM: using link analysis to summarize entity data. In: Bozzon, A., Cudre-Maroux, P., Pautasso, C. (eds.) ICWE 2016. LNCS, vol. 9671, pp. 244–261. Springer, Cham (2016). https://doi.org/10.1007/978-3-319-38791-8_14

26. Trouillon, T., Welbl, J., Riedel, S., Gaussier, É., Bouchard, G.: Complex embeddings for simple link prediction. In: International Conference on Machine Learning, pp. 2071–2080. PMLR (2016)

27. Wei, D., Gao, S., Liu, Y., Liu, Z., Hang, L.: MPSUM: entity summarization with predicate-based matching. CoRR **abs/2005.11992** (2020). https://arxiv.org/abs/2005.11992

28. Wei, D., Liu, Y.: ESA: entity summarization with attention. CoRR **abs/1905.10625** (2019). http://arxiv.org/abs/1905.10625

29. Wei, D., et al.: AutoSUM: automating feature extraction and multi-user preference simulation for entity summarization. In: Lauw, H.W., Wong, R.C.-W., Ntoulas, A., Lim, E.-P., Ng, S.-K., Pan, S.J. (eds.) PAKDD 2020. LNCS (LNAI), vol. 12085, pp. 580–592. Springer, Cham (2020). https://doi.org/10.1007/978-3-030-47436-2_44

30. Yao, L., Mao, C., Luo, Y.: KG-BERT: BERT for knowledge graph completion. CoRR **abs/1909.03193** (2019). http://arxiv.org/abs/1909.03193

31. Zaib, M., Tran, D.H., Sagar, S., Mahmood, A., Zhang, W.E., Sheng, Q.Z.: BERT-CoQAC: BERT-based conversational question answering in context. In: Ning, L., Chau, V., Lau, F. (eds.) PAAP 2020. CCIS, vol. 1362, pp. 47–57. Springer, Singapore (2021). https://doi.org/10.1007/978-981-16-0010-4_5

32. Zhang, Z., Han, X., Liu, Z., Jiang, X., Sun, M., Liu, Q.: ERNIE: enhanced language representation with informative entities. In: Korhonen, A., Traum, D.R., Màrquez, L. (eds.) Proceedings of the 57th Conference of the Association for Computational Linguistics, ACL 2019, Florence, Italy, July 28- August 2, 2019, Volume 1: Long Papers, pp. 1441–1451. Association for Computational Linguistics (2019). https://doi.org/10.18653/v1/p19-1139

Explanation of Link Predictions on Knowledge Graphs via Levelwise Filtering and Graph Summarization

Roberto Barile[1]([✉])(iD), Claudia d'Amato[1,2](iD), and Nicola Fanizzi[1,2](iD)

[1] Dipartimento di Informatica – University of Bari Aldo Moro, Bari, Italy
{roberto.barile,claudia.damato,nicola.fanizzi}@uniba.it
[2] CILA – University of Bari Aldo Moro, Bari, Italy

Abstract. Link Prediction methods aim at predicting missing facts in Knowledge Graphs (KGs) as they are inherently incomplete. Several methods rely on Knowledge Graph Embeddings, which are numerical representations of elements in the Knowledge Graph. Embeddings are effective and scalable for large KGs; however, they lack explainability. KELPIE is a recent and versatile framework that provides post-hoc explanations for predictions based on embeddings by revealing the facts that enabled them. Problems have been recognized, however, with filtering potential explanations and dealing with an overload of candidates. We aim at enhancing KELPIE by targeting three goals: reducing the number of candidates, producing explanations at different levels of detail, and improving the effectiveness of the explanations. To accomplish them, we adopt a semantic similarity measure to enhance the filtering of potential explanations, and we focus on a condensed representation of the search space in the form of a quotient graph based on entity types. Three quotient formulations of different granularity are considered to reduce the risk of losing valuable information. We conduct a quantitative and qualitative experimental evaluation of the proposed solutions, using KELPIE as a baseline.

Keywords: Knowledge Graphs · Link Prediction · Explanation

1 Introduction

Knowledge Graphs (KGs) emerged as a tool to represent, navigate and query the growing flood of data by encapsulating knowledge of complex domains into a form accessible to both humans and machines. A KG is a multi-relational graph composed of entities and relations, represented as nodes and edges, respectively. KGs are often integrated with ontologies, that formally define classes and relations, which allow for advanced inference capabilities [18]. Several examples of large KGs exist, including enterprise products, e.g., see [13,30], and various well-known open sources, e.g., [1,5,23].

Despite their effectiveness, working with KGs often suffers from their inherent incompleteness [18], due also to the open-world semantics generally assumed in scenarios that involve them as the result of a complex, incremental and distributed building process. This underpins the importance of tasks like *Link Prediction* (LP) and *triple classification* aiming, respectively, at inferring missing relationships between existing nodes and deciding on the truth of (new) triples.

A. Meroño Peñuela et al. (Eds.): ESWC 2024, LNCS 14664, pp. 180–198, 2024.
https://doi.org/10.1007/978-3-031-60626-7_10

Among the many LP methods grounded on *Machine Learning* (ML) models, those based on *Knowledge Graph Embeddings* (KGEs) have emerged as a prevalent approach, especially for their superior scalability (see [25] for a recent survey). KGEs map elements in the KGs to low-dimensional vector spaces, streamlining complex tasks via linear algebra.

Nevertheless, such models tend to operate as "opaque boxes" whose predictions are difficult to explain, thus undermining their credibility and trust. Indeed, this opacity can be particularly problematic in contexts where understanding the reasons supporting the predictions is critical, such as healthcare or financial decision-making. For example, LP might help to find out potential connections between specific drugs and their side effects [9]. In such cases, not only the predictions but also their underlying rationale is essential, as these may influence decisions about investments on a drug.

To address this opacity, the field of *Explainable Artificial Intelligence* (XAI) [21] has come into the spotlight. XAI aims at making the decisions of ML models more transparent and understandable, enhancing human trust and comprehension across all AI applications. Following [17], XAI methods can be divided into two categories: a) *post-hoc* methods, for addressing specific model outcomes; b) *global interpretability* methods, for offering a holistic understanding of the entire inference process.

We will focus on *post-hoc* methods, which are suitable for scalable LP models based on numerical representations. In the post-hoc setting, given a prediction, the generated explanation typically comprises a specific set of facts that have enabled the inference through the model. A recent, effective, and versatile framework providing such kind of explanation to LP tasks is KELPIE [26]. Specifically, KELPIE provides explanations through three steps: (1) the pre-filter/extraction of a sub-graph designated as the search space, (2) the combination of facts within this sub-graph into candidate explanations, (3) the ranking of such possible explanations. Nevertheless, the pre-filter phase is grounded on the exploitation of a topological measure that could lead to discarding facts that can make up potentially valid explanations. Furthermore, the successive combination of facts for building candidate explanations leads to an extremely large number of possible explanations. Even more so, KELPIE is not able to offer explanations with an adaptable level of detail. Some users might prefer a brief, high-level explanation, while others might require a more detailed account.

Recognizing these limitations, we identified three specific research goals. Firstly, adopting an alternative metric for the pre-filtering phase. Second, decreasing the number of assessments of candidate explanations necessary to identify the optimal ones. Third, to be able to generate explanations at different levels of detail. Overall, as a final goal, we aim at improving the results for the metrics of end-to-end effectiveness.

In agreement with these goals, we formulate our contributions relying on formal ontologies often available in KGs. Firstly, we propose to adopt in the pre-filter/extraction phase a measure of semantic similarity between entities so to focus on facts semantically related to the prediction, thus potentially more suitable as part of an explanation. Secondly, our proposal is to compute a summarization of the search space using different formulations of quotient graphs so to speed up the search, ground it on clusters of semantically related facts and offering explanations at different levels of granularity. We also experimentally prove that these contributions are actually able to improve effectiveness and efficiency.

The rest of this work is organized as follows. Section 2 introduces basic notions that are essential for the paper. Section 3 reviews existing approaches for explaining link predictions. Section 4 provides details on KELPIE, then outlines the proposed contributions. The experimental evaluation of the resulting approach is illustrated in Sect. 5. Finally, Sect. 6 summarizes achievements and limitations, delineating future works.

2   Fundamentals

2.1   Knowledge Graphs and Embedding Models

A KG is a graph-based data structure $\mathcal{G}(\mathcal{V}, \mathcal{E})$ with $\mathcal{V}$ representing a set of nodes, also known as entities, and $\mathcal{E}$ representing a set of edges, labeled with relationships which connect pairs of entities.

In the adopted RDF model, a KG is a collection of triples in the format $\langle s, p, o \rangle$, i.e., *subject*, *predicate*, and *object* where $s, o \in \mathcal{V}$ and $p \in \mathcal{E}$. In RDF, the terms are denoted by the elements of the sets $\mathcal{U}$ (URIs), $\mathcal{B}$ (blank nodes) and $\mathcal{L}$ (literals). Consequently, an RDF graph is a set of triples with: $s \in \mathcal{U} \cup \mathcal{B}, r \in \mathcal{U}$, and $o \in \mathcal{U} \cup \mathcal{B} \cup \mathcal{L}$.

Various models have been proposed for representing KGs in low-dimensional vector spaces, by learning a unique distributed representation (or *embedding*) for each entity and predicate in the KG and considering different representation spaces (e.g., pointwise, complex, discrete, Gaussian, manifold). Here we focus on vector embeddings in the set of real numbers.

In an ML setting, a KG $\mathcal{G}$ may be further split into a training set $\mathcal{G}_{train}$, a validation set $\mathcal{G}_{val}$ and a test set of triples $\mathcal{G}_{test}$. Irrespective of the specific learning approach, these models all represent each entity $x \in \mathcal{V}$ by means of a continuous embedding vector $\mathbf{e}_x \in \mathbb{R}^k$, where $k \in \mathbb{N}$ is a user-defined hyper-parameter. Similarly, each predicate $p \in \mathcal{E}$ is associated to a scoring function $f_p : \mathbb{R}^k \times \mathbb{R}^k \to \mathbb{R}$. For each pair of entities $s, o \in \mathcal{V}$, the score $f_p(\mathbf{e}_s, \mathbf{e}_o)$ is a measure of the plausibility of the statement encoded by $\langle s, p, o \rangle$. The embedding of all entities (and predicates) in $\mathcal{G}$ is learned by minimizing a loss function, often a margin-based one.

2.2   Quotient Graph

In the context of KGs, quotient graphs aim at summarizing the data graph into a higher-level topology [8]. They are based on the concept of a quotient set and equivalence classes. Given a set X and an equivalence relation $\sim$ on it, X is partitioned into disjoint subsets of equivalent elements, the equivalence classes. The *quotient set* $X/\sim$ contains all equivalence classes.

Before diving into the specifics of the quotient graph formation, we introduce the notions of simulation and bisimulation, and we report the definitions as provided in [18]. A simulation is a binary relation from a graph $\mathcal{G}$ to a graph $\mathcal{G}'$ that maintains the existence of a path between connected nodes in $\mathcal{G}$ to $\mathcal{G}'$. Formally:

Definition 1. *A simulation from a graph* $\mathcal{G}(\mathcal{V}, \mathcal{E})$ *to a graph* $\mathcal{G}'(\mathcal{V}', \mathcal{E}')$ *is a relation* $R \subseteq \mathcal{V} \times \mathcal{V}'$ *such that for every edge* $\langle x, y, z \rangle$ *in* $\mathcal{G}$, *if* $x\ R\ x'$ *for some* $x' \in \mathcal{V}'$, *then there exists* $z' \in \mathcal{V}'$ *such that* $z\ R\ z'$ *and* $\langle x', y, z' \rangle$ *is an edge in* $\mathcal{G}'$. $\mathcal{G}'$ *simulates* $\mathcal{G}$ *when a simulation exists from* $\mathcal{G}$ *to* $\mathcal{G}'$.

A stronger, symmetric extension of this relation can be defined as follows:

Definition 2. *A bisimulation between* $\mathcal{G}$ *and* $\mathcal{G}'$ *is a relation that is both a simulation from* $\mathcal{G}$ *to* $\mathcal{G}'$ *and a simulation from* $\mathcal{G}'$ *to* $\mathcal{G}$. *When a bisimulation exists on* $\mathcal{G}$ *and* $\mathcal{G}'$, *they are bisimilar.*

Given a KG $\mathcal{G}(\mathcal{V}, \mathcal{E})$, its *quotient graph* $\mathcal{Q}(\mathcal{V}/\sim, \mathcal{E}/\sim)$ is a graph with node set $\mathcal{V}/\sim$ as the quotient set of $\mathcal{V}$ according to the equivalence relation of choice, and edge set $\mathcal{E}/\sim$, that can be defined in different ways, depending on the desired level of preservation of the structure. In the case of simulation, an edge $\langle X, y, Z \rangle$ in the quotient graph exists if and only if there exists $x \in X$ and $z \in Z$ such that an edge $\langle x, y, z \rangle$ is in the original graph $\mathcal{G}$. Bisimulation, on the other hand, imposes a stronger condition: $\langle X, y, Z \rangle$ is an edge in the quotient graph when, for each $x \in X$, there exists a $z \in Z$ such that $\langle x, y, z \rangle$ is in the input graph.

Bisimulation also involves refining the quotient nodes to split the ones that contain nodes with different outgoing edges in the original graph. To clarify, two nodes belong to the same quotient node if they share identical outgoing edges in the original graph and are in the same equivalence class. This is equivalent to finding the *Relational Stable Coarsest Partition* (RSCP) [15].

To summarize, the steps to compute a quotient graph *bisimilar* to $\mathcal{G}$ are the following: 1) compute $\mathcal{V}/\sim$ as seen before; 2) compute the RSCP to refine $\mathcal{V}/\sim$; 3) add the edges to $\mathcal{E}/\sim$. If an equivalence relation is not available, the RSCP can be computed directly on $\mathcal{V}$.

Several algorithms to compute RSCP are available [14,22], however, such implementations are not suitable for *multi-graphs*, i.e., graphs that allow multiple parallel edges between the same pair of nodes, and therefore KGs. Indeed, these algorithms take into account only one of the multiple edges between the same pair of nodes, since all edges are assumed to be of the same type. To address the issue, the input graph can be pre-processed as formalized in [7]. Specifically, each triple $\langle s, p, o \rangle$ is represented as an unlabeled edge connecting s to the node (p, o).

3   Related Works

In this section, we specifically focus on post-hoc methods for explaining LP. Recent solutions are grounded on the exploitation of a data poisoning technique [33] in order to determine a single fact that, if inserted or eliminated, would most significantly poison the prediction. Another example is CRIAGE [24], which addresses the computational challenges of earlier models [19], but lacks clarity on the model adaptability and focuses on limited cases (solely those facts where the object is either the subject or the object of the prediction).

An objective that is receiving increasing attention and that is targeted in this paper is the development of model-independent explainability solutions. CROSSE [35] and

APPROXSEMANTICCROSSE [10] analyze the KG topology instead of the model's behavior and find the paths from the subject to the object supported by analogous situations in the graph, i.e., with similar paths connecting similar entities. The LP model is used only to exclude relations and entities.

KELPIE [26] explains a prediction by returning a specific set of relevant facts rather than a single fact. To evaluate the relevance of a candidate explanation, KELPIE employs a novel *post-training* process which can be tailored to different KGE models. Similarly to KELPIE, KGEX [2] provide sets of facts as explanations, but employs sub-graph sampling and *Knowledge Distillation* of surrogate models. A complementary approach is proposed in [3], it provides explanation through abductive reasoning on a logical theory learned through a symbolic rule learning approach.

KELPIE represents a very significant advance in the class of model-independent explanation solutions. Nevertheless, it has some drawbacks: (i) in the initial filtering of facts it employs a topological measure which may be a limited insight on the correlation between the training facts and the prediction to explain, (ii) the combination of facts for building candidate explanations leads to a very large number of possible explanations, (iii) the explanations are solely provided as sets of facts, thus not encompassing more general insights such as the classes of the entities. This paper intends to overcome all these limitations.

In parallel to post-hoc explanation solutions, also inherently interpretable LP models are available as for the case of XTRANSE [34], the work in [4], and GNNEX-PLAINER [32] targeting specifically *Graph Neural Networks*. These methods aim at making the model itself interpretable, whereas our goal is to develop a model-agnostic post-hoc solution for explaining LP results.

Also, general-purpose XAI methods have been developed, i.e., solutions that are independent of the task to explain. Such general-purpose approaches can be distinguished between those that identify relevant features of the input data as explanations, called *saliency explanations*, and those that identify training samples as explanations. Saliency-based frameworks (like SHAP [20] grounded on *Shapley values* [28]) are rather popular, but they are not easily adaptable to the LP task since they require interpretable input attributes, which are suitable for images or text, but fall short for LP using graph embeddings, where input samples are numerical. Indeed, in this case, saliency-based methods merely highlight the most significant components of vectors in relation to the outcome, lacking human interpretability. As for the methods identifying training samples as explanations [19], the notion of *Influence Functions* from robust statistics [27] has been exploited, resulting in a very high resource demanding solution.

4   The Proposed Approach

We propose a method for providing post-hoc explanations for LP on KGs exploiting their semantics. Specifically, moving from KELPIE, we aim at the enhancement of its main components. Our contributions capitalize on the exploitation of schema-level information from the shared OWL ontologies adopted by the KGs. The resulting new method remains independent of the KGE model. In the following, we summarize KELPIE in Sect. 4.1, then we delineate the proposed extensions in Sects. 4.2, and 4.3.

4.1 KELPIE

KELPIE [26] stands out in the related works overview as it provides effective explanations including multiple triples, it adapts to any KGE model, and it is supplied with the resources to replicate the experiments. Therefore, we adopt it as the ground of our approach. We now delve into details on how it works. Given a predicted triple $\langle s, p, o \rangle$, a KELPIE explanation consists in a set of training triples featuring s that have enabled to predict the object o through the model. KELPIE, is structured around three main components:

- The Pre-Filter selects the most useful triples featuring s
- The Explanation Builder combines the pre-filtered triples into candidate explanations and identifies the most relevant ones.
- The Relevance Engine computes the relevance of a candidate explanation adopting a ML technique called *post-training*, coined by the authors.

We further detail each component individually.

The Pre-Filter aims at decreasing the complexity for the subsequent stages. Firstly, it extracts the sub-graph $\mathcal{G}^s_{train}$ of all the training triples featuring s as subject or as object. Then, it filters this sub-graph to obtain $\mathcal{F}^s_{train}$ by selecting the top-k most useful triples. The utility measure is based on graph topology. Specifically, for any given $\langle s, q, r \rangle$ (or $\langle r, q, s \rangle$), it computes the length of the shortest path connecting r to the predicted object o, ignoring the direction of the edges.

The Explanation Builder's task is to find a set of triples in $\mathcal{F}^s_{train}$ representing an optimal explanation X^* according to their relevance. It combines the triples in $\mathcal{F}^s_{train}$ into candidate explanations (each denoted as X), and determines whether any X can be accepted as X^* based on its relevance. Identifying X^* is a search problem within a space ($\mathcal{S}$) of candidate explanations of varying lengths, but exhaustive search is impractical. Indeed, Kelpie implements heuristic conditions to prune the search space.

The process is summarized in Algorithm 1. The method first computes the relevance of each triple in $\mathcal{F}^s_{train}$ used as a 1-triple explanation (Line 1). Before exploring any $\mathcal{S}_i$ with, $i > 1$ the algorithm computes the *preliminary relevance* of each explanation as the average relevance of its triples (Line 4). The subset $\mathcal{S}_i$ is then traversed in descending order of preliminary relevance. For each explanation, the Relevance Engine computes its *true relevance* (Line 8) and the algorithm checks if it meets the acceptance criteria (Line 9). The decision on whether to continue exploring the current subset or to move on to the next one is guided by ρ_i, which is defined as the ratio of the relevance of the current explanation to the highest relevance found so far in the subset (Line 14). If ρ_i becomes too small, indicating likely less relevant future explanations, the algorithm decides whether to move on to the next subset, with probability $1 - \rho_i$ (Lines 15–16).

The Relevance Engine adopts two alternative, yet complementary methods, namely *necessary relevance* and *sufficient relevance*. Both methods ground on a 4 steps approach: (a) create s' as a duplicate of s, (b) compute the set $\mathcal{F}^{s'}_{train}$ as $\mathcal{F}^s_{train} - X$ for *necessary relevance* and as $\mathcal{F}^s_{train} \cup X$ for *sufficient relevance*, (c) learn the entity embedding of s' through *post-training*, (d) compute the difference between the scores of the triples $\langle s, p, o \rangle$ and $\langle s', p, o \rangle$. Moreover, for *sufficient relevance*, instead of being applied on s, the process is iterated on the elements of a set C of random entities c for which the model does not lead to the object o for predicting the filler of $\langle c, p, ? \rangle$.

Algorithm 1: Algorithm for identifying the explanation X^*

    Input: the triples in $\mathcal{F}^s_{train}$; the Relevance Engine *engine*;
               the acceptance threshold ξ_0; the explanation size limit i_{max};
    Output: The smallest combination X^* whose relevance exceeds ξ_0;

1   $triple\_to\_relevance \leftarrow \{t : engine.compute([t])$ **for** t **in** $F^s_{train}\}$;
2   **for** $i \leftarrow 2$ **to** i_{max} **do**
3      $\mathcal{S}_i \leftarrow combinations(F^s_{train}, i)$;
4      $pre\_relevances \leftarrow [avg([triple\_to\_relevance[t]$ **for** t **in** $X]) $ **for** X **in** $\mathcal{S}_i]$;
5      $\mathcal{S}_i \leftarrow sort(\mathcal{S}_i, pre\_relevances)$;
6      $best\_relevance \leftarrow None$;
7   **foreach** $X \in \mathcal{S}_i$ **do**
8          $cur\_relevance \leftarrow engine.compute(X)$;
9   **if** $cur\_relevance > \xi_0$ **then**
10   **return** X;
11   **else**
12   **if** $best\_relevance = None$ **or** $cur\_relevance > best\_relevance$ **then**
13              $best\_relevance \leftarrow cur\_relevance$;
14            $\rho_i \leftarrow cur\_relevance/best\_relevance$;
15   **if** $random(0, 1) > \rho_i$ **then**
16   **break**;

4.2 Injecting Semantics in the Pre-Filter

The first new component of our proposed methodology is Semantic Pre-Filter, a refined version of the Pre-Filter. We refine the method for assessing the utility of the triples in $\mathcal{G}^s_{train}$. We recall that in KELPIE the Pre-Filter calculates the utility of a triple $\langle s, q, r \rangle$ (or $\langle r, q, s \rangle$) with respect to the prediction $\langle s, p, o \rangle$ based on a breadth-first search (*bfs*) measuring the length of the shortest non-oriented path connecting r to o. Now, we base this calculation on the path (connecting r to o) with the least cumulative weight instead of the shortest one, making the *bfs* weighted (*wbfs*). Specifically, the weight of an edge connecting r to o is $1 - sim(r, o)$. Such weight is meant as a measure of the semantic similarity between the two connected nodes. This enhancement involves the integration of the approximated semantic similarity measure proposed in [10] as a measure over pairs of entities. Namely, in [10] a function Cl is defined as returning classes to which an entity can be proven to belong to, and its approximated version $\tilde{Cl}$ that simplifies the needed *realization* service and allows bypassing the usage of *retrieval* required in Cl.

    The semantic similarity measure sim is formally defined as a *Jaccard* measure on a couple of entities r, o. Specifically, $sim(r, o) = \frac{|\tilde{Cl}(r) \cap \tilde{Cl}(o)|}{|\tilde{Cl}(r) \cup \tilde{Cl}(o)|}$.

    For instance, consider a very simple KG with the following triples: { $\langle Barack\ Obama, signed, Obamacare \rangle$, $\langle Barack\ Obama, born\ in, Honolulu \rangle$, $\langle Honolulu, located\ in, United\ States \rangle$, $\langle United\ States, signed, Obamacare \rangle$ } and the prediction $\langle Barack\ Obama, nationality, United\ States \rangle$. Both *Honolulu* and *Obamacare* are one-hop distant from *United States*. Nevertheless, *Honolulu* has higher semantic similarity with *United States* than *Obamacare*.

Algorithm 2: Algorithm for identifying X^* in the quotient graph

---

Input: The set $\mathcal{F}_{train}^s$ of training triples; Cl function;
the Relevance Engine object $engine$;
the acceptance threshold ξ_0; the explanation size limit i_{max};
Output: The smallest combination X^* whose relevance exceeds ξ_0;

1   $\mathcal{Q}_{train}^s \leftarrow compute\_quotient\_graph(F_{train}^s, Cl)$;
2   $quot\_to\_orig \leftarrow compute\_mapping(\mathcal{Q}_{train}^s, F_{train}^s)$;
3   $triple\_to\_relevance \leftarrow \{\}$;
4   **foreach** $triple_{quot} \in Q_s^{train}$ **do**
5      $triples_{orig} \leftarrow quot\_to\_orig([triple_{quot}])$;
6      $triple\_to\_relevance[triple_{quot}] \leftarrow engine.compute(triples_{orig})$;

7   **for** $i \leftarrow 2$ **to** i_{max} **do**
8      $\mathcal{S}_i \leftarrow combinations(\mathcal{Q}_{train}^s, i)$;
9      $\mathcal{S}_i \leftarrow preliminary\_sort(\mathcal{S}_i, triple\_to\_relevance)$;
10   **foreach** $X_{quot} \in \mathcal{S}_i$ **do**
11       $X_{orig} \leftarrow quot\_to\_orig(X_{quot})$;
12       $cur\_relevance \leftarrow engine.compute(X_{orig})$;
13       $check\_accept\_threshold(cur\_relevance, \xi_0)$;
14       $update\_best\_relevance(best\_relevance, cur\_relevance)$;
15       $check\_early\_exit(best\_relevance, cur\_relevance)$;

This integration of semantics fundamentally changes how the Pre-Filter assesses the utility of triples within the $\mathcal{G}_{train}^s$ set. The intuition behind this enhancement is to acknowledge the semantic relationships between nodes (entities). Adopting this refined version, $\mathcal{F}_{train}^s$ will eventually contain triples that are not only topologically, but also semantically related to the prediction. This contribution may result in a more accurate extraction of $\mathcal{F}_{train}^s$ enhancing the effectiveness of the explanations; thus addressing our final research goal.

4.3 Injecting Semantics in the Explanation Builder

We propose Quotient Explanation Builder to enhance the Explanation Builder component of KELPIE. In brief, given a prediction $\langle s, p, o \rangle$ to explain, the Explanation Builder combines the pre-filtered training triples ($\mathcal{F}_{train}^s$) into candidate explanations and then identify optimal ones (details in Algorithm 1).

Our enhanced version tackles the goal of limiting the combinatorial explosion when computing candidate explanations by introducing a suitable summarization step of the input subgraph. For the purpose, the method relies on the notion of quotient graph, presented in Subsect. 2.2.

We formalize Quotient Explanation Builder in Algorithm 2. The algorithm requires the same parameters as the original formulation, along with the inclusion of the $\tilde{C}l$ function. Initially, the algorithm computes the quotient graph $\mathcal{Q}_{train}^s$ of $\mathcal{F}_{train}^s$ exploiting $\tilde{C}l$ to determine the equivalence relation (Line 1).

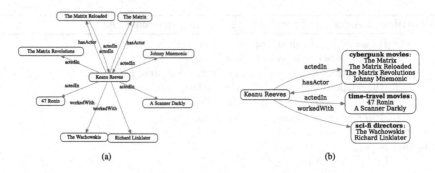

Fig. 1. Sub-graph on *Keanu Reeves* (a) and its (simulation) quotient graph (b)

We refer to a triple in the quotient graph as a *quotient triple* ($triple_{quot}$), and to a triple in the original graph as an *original triple* ($triple_{orig}$). The nodes in the quotient graph are equivalence classes of nodes in the original graph; hence, each quotient triple is mapped to its inherently equivalent original triples (Line 2).

For example, we consider *Keanu Reeves* as the subject s, for which we report $\mathcal{F}^s_{train}$ in Fig. 1a and its quotient graph in Fig. 1b. The quotient triple $\langle s, actedIn,$ *cyberpunk movies*$\rangle$ (where *cyberpunk movies* is a class) is mapped to the original triples: $\langle s, actedIn, The\ Matrix \rangle$, $\langle s, actedIn, The\ Matrix\ Reloaded \rangle$, $\langle s, actedIn,$ *The Matrix Revolutions*$\rangle$, $\langle s, actedIn, Johnny\ Mnemonic \rangle$.

The subsequent steps in Algorithm 2 follow a procedure similar to Algorithm 1, but performing all the computations on $\mathcal{Q}^s_{train}$. It computes the relevance of each quotient triple used as a 1-quotient-triple explanation. It first retrieves the original triples that it corresponds to (Line 5), then it computes the relevance of such original triples using the Relevance Engine (Line 6).

The process then continues by exploring combinations of quotient triples from the smallest size (2) up to a specified limit i_{max} (Line 7). $\mathcal{S}_i$ is sorted in descending order of preliminary relevance, then the algorithm computes the relevance of each X_{quot}. Similarly to the case of 1-quotient-triple explanations, it retrieves the corresponding original triples (Line 11), then it computes their relevance (Line 12). The remaining lines check the heuristic conditions.

The quotient graph condenses the original graph, thus addressing the goal of decreasing the number of candidate explanations. Moreover, quotient graphs contribute to the other goal of generating explanations at varying levels of detail, as both the quotient triples and the entities within the quotient nodes can be output. Finally, the inherent grouping of entities in a quotient graph may enhance the effectiveness of explanations addressing the overall final goal.

A quotient triple represents a collection of similar original triples, so it provides multiple pieces of evidence of the same kind. Hence, a set of quotient triples represents different kinds of evidence, each backed by multiple pieces. This pattern furthers heterogeneity and robustness in the explanation. In essence, these goals hinge on grouping of semantically related triples, an aspect effectively handled by quotient graphs.

Fig. 2. *Bisimulation quotient of* $\mathcal{G}^s_{train}$

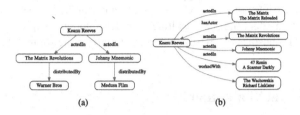

Fig. 3. Fragment of $\mathcal{G}^{s(+)}_{train}$ (a) and its *bisimulation quotient* (b)

The introduced abstraction potentially enhances the method. However, it increases the risk to bypass relevant smaller sets or individual triples. Hence, the use of the quotient graph allows for a trade-off between abstraction and detail.

To optimize the choice, we propose three alternative formulations of the quotient graph: (i) *simulation quotient*, (ii) *bisimulation quotient*, and (iii) *depth-1 bisimulation quotient*. Algorithm 2 is agnostic of the formulation.

The *simulation quotient* computes the quotient graph according to the simulation relation. Being the most abstract formulation, it is the most prone to ignore small sets and individual triples. In Fig. 1b we already provided an example.

In the *bisimulation quotient*, we adopt the bisimulation relation. It splits some equivalence classes to meet the stricter condition on edges, mitigating the risk of the first method. For instance, in Fig. 2 the equivalence class *cyberpunk movies* is split in two nodes according to the presence of the property *hasActor*. This helps when the prediction relies on either the first two movies or the other two. We clarify that, in this formulation, we compute the RSCP of the partition resulting from $\tilde{C}l$ rather than computing it on the whole set of nodes in $\mathcal{F}^s_{train}$, thus maintaining the semantic information provided $\tilde{C}l$. We also recall that to compute RSCP, a pre-processing step is required converting each triple $\langle s, p, o \rangle$ to an unlabeled edge from s to a new node p, o. However, for the new nodes $\tilde{C}l$ is not defined. To address this aspect, we have extended $\tilde{C}l$ so to assign each new node to a separate equivalence class.

So far, information not directly featuring s was not taken into account while looking for local explanations. However, additional information may be employed only to further refine the equivalence classes, leading to the *depth-1 bisimulation quotient*. The algorithm starts by finding the depth-1 sub-graph of s ($\mathcal{G}^{s(+)}_{train}$) which includes neighbors of neighbors of s. In Fig. 3a we exemplify this by reporting a fragment for $s = $ *Keanu Reeves*. Secondly, it computes the *bisimulation quotient* of $\mathcal{G}^{s(+)}_{train}$, as shown in Fig. 3. The equivalence class {*The Matrix Revolutions, Johnny Mnemonic*} is further split in two nodes according to the property *produced by*. This helps when the prediction solely relies either on *The Matrix Revolutions* or on *Johnny Mnemonic*. In this formulation, quotient triples that do not include s will arise. They can be discarded as the graph is expanded only for refining the equivalence classes rather than increasing the set of candidate explanations.

In conclusion, we remark that *bisimulation quotient* and *depth-1 bisimulation quotient* have a finer granularity than *simulation quotient*, but still explanations are more abstract than those produced by KELPIE. The *depth-1 bisimulation quotient*, that is the finest of our formulations can still bypass individual triples, only KELPIE assesses the relevance of all the triples in $\mathcal{F}^s_{train}$. Our assumption is that the bypassed triples are less relevant explanations than the sets resulting from the quotient graphs by virtue of the heterogeneity and robustness of explanations resulting from the combination of quotient triples.

5   Experimental Evaluation

In this section, we illustrate the experiments carried out to assess our contributions with respect to our research goals (stated in Sect. 1). All the code, datasets, and trained models utilized in our study are openly accessible on GitHub[1]. In this section we present our experimental setting, then we provide quantitative and qualitative results.

5.1   Experimental Setting

We performed the experiments on three datasets: DBpedia50 [29] (DB50K), DB100K [12] and YAGO4-20. DB50K and DB100K are both samples of the DBpedia KG. In contrast, we sampled YAGO4-20 from the YAGO4 [23] KG by extracting triples about entities involved in at least 20 triples and then filtering out triples with literal objects. We report statistics on the datasets in Table 1.

The common aspect of these datasets is that along with the RDF triples they include or can be integrated with OWL (specifically OWL2-DL) statements including class assertions, and other schema axioms concerning classes and relationships. We exploited the HERMIT [16] reasoner offline to materialize the implicit class assertions in the KGs. Next, we implemented $\tilde{C}l$ as a simple lookup of class assertions to execute our method without any other adjustment.

We highlight that for DB50k and DB100K, only the RDF triples are available off the shelf. We retrieved the class assertions with custom SPARQL queries to the DBpedia endpoint[2], while we employed the schema axioms provided by the full DBpedia[3]. The resulting KGs DB50K and DB100K proved to be inconsistent, while YAGO4-20 turned out to contain unsatisfiable classes. We manually repaired the KGs (see Appendix A) as such problems hinder the use of the reasoner.

KELPIE and consequently our extension, supports any LP model based on embeddings; therefore, we performed our experiments on TRANSE [6], CONVE [11] and COMPLEX [31] as KELPIE adopts the same ones and these represent three prevalent families: Geometric Models, Deep Learning Models, and Tensor Decomposition Models. Table 2 reports the LP performance of each model that we measured on each dataset

---

[1] https://github.com/rbarile17/kelpiePP.

[2] https://dbpedia.org/sparql.

[3] https://databus.dbpedia.org/ontologies/dbpedia.org/ontology--DEV.

Table 1. Statistics of the datasets

	Entities	Relations	Train triples	Valid triples	Test triples
DB50K	24620	351	32194	123	2095
DB100K	98776	464	587688	49172	49114
YAGO4-20	96910	70	555182	69398	69398

Table 2. Performance of the LP models

	DB50K		DB100K		YAGO4-20	
	H@1	MRR	H@1	MRR	H@1	MRR
TRANSE	0.288	0.380	0.101	0.211	0.081	0.135
CONVE	0.366	0.410	0.285	0.360	0.163	0.213
COMPLEX	0.407	0.464	0.359	0.430	0.195	0.242

in terms of typical measures, namely Mean Reciprocal Rank (MRR) and Hits-at-1 ($H@1$) in their filtered variant.

We evaluate the approach adopting both *necessary* and *sufficient* relevance (details in Subsect. 4.1). In both scenarios, we select a set P of 100 correct test predictions randomly, for each model. Then, we adopt the methodology utilized by CRIAGE, and subsequently by KELPIE, to assess if we meet our goal of optimizing the effectiveness. Specifically, in the *necessary* scenario, after extracting explanations for all predictions in P, we remove their triples and retrain the model. Since the original model correctly led to those predictions, their initial $H@1$ and MRR are both 1.0; however, if the extracted explanations are indeed necessary, the inference through the retrained model should fail to lead to the predictions in P. Therefore, the effectiveness is the decrease in $H@1$ and MRR over P.

Conversely, in the *sufficient* scenario, for each prediction $\langle s, p, o \rangle \in P$, we draw a set C of 10 random entities c and extract sufficient explanations that should lead the model to predict $\langle c, p, o \rangle$. We define P_C as the set of these hypothetical $10 \times 100 = 1000$ predictions $\langle c, p, o \rangle$. As the original model did not lead to the predictions in P_C by design, their original $H@1$ and MRR are approximately null. However, if the extracted explanations are indeed sufficient, the retrained model should lead to the predictions in P_C. Therefore, the effectiveness is the increase in the $H@1$ and MRR over P_C. In both scenarios, the variation metrics are denoted as $\Delta H@1$ and ΔMRR. Moreover, as for our first research goal, we aim at improving the efficiency by decreasing the number of candidate explanations. Hence, we also measure the number of Relevance Engine invocations (indicated by #r). In Appendix B we report details on the hyper-parameters used for training the models, for the *post-training* in the extraction of the explanations, and for the re-training in the evaluation of the explanations.

5.2 Quantitative Evaluation

In Table 3 we report the outcomes of the experiments on KELPIE with the various extensions in terms of the effectiveness and efficiency metrics. Firstly, we measure the impact

Table 3. Results of the experimental evaluation

Model	Method	Necessary									Sufficient								
		DB50K			DB100K			YAGO4-20			DB50K			DB100K			YAGO4-20		
		#r	ΔH@1	ΔMRR	#r	ΔH@1	ΔMRR	#r	ΔH@1	ΔMRR	#r	ΔH@1	ΔMRR	#r	ΔH@1	ΔMRR	#r	ΔH@1	ΔMRR
TransE	KELPIE	1234	-0.93	**-0.867**	2568	**-0.69**	-0.546	2614	-0.48	-0.356	1086	0.521	0.580	3109	0.317	0.443	6872	0.260	0.365
	wbfs	1241	**-0.94**↑	-0.857	2420	-0.64	-0.511	2771	-0.49↑	-0.357↑	1097	0.499	0.562	3065	**0.323**↑	0.441	6953	0.299↑	0.391↑
	sim	510	-0.93	-0.856	1953	-0.63	-0.518	**1437**	**-0.57**↑	**-0.450**↑	362	0.449	0.515	2464	0.298	0.426	3819	**0.377**↑	**0.475**↑
	bisim	752	-0.91	-0.815	1954	-0.58	-0.481	1470	-0.57↑	-0.433↑	622	0.508	0.568	**2345**	0.287	0.422	3789	0.335↑	0.441↑
	bisim_d1	925	-0.91	-0.836	2555	-0.61	-0.481	2461	-0.46	-0.332	804	**0.528**↑	**0.587**↑	3210	0.320↑	0.440	6858	0.270	0.390↑
	wbfs + sim	**502**	**-0.94**↑	-0.861	1860	-0.68	**-0.566**↑	1458	-0.56↑	-0.443↑	361	0.475	0.531	2465	0.301	0.432	**3774**	0.324↑	0.434↑
	wbfs + bisim	762	-0.90	-0.818	2037	-0.65	-0.551	1474	-0.55↑	-0.439↑	632	0.482	0.556	2517	0.321↑	**0.446**↑	4035	0.343↑	0.434↑
	wbfs + bisim_d1	938	-0.91	-0.824	2555	-0.65	-0.533	2422	-0.46	-0.333	813	0.499	0.565	3288	0.307	0.429	6239	0.278	0.387↑
ConvE	KELPIE	818	-0.42	-0.385	1590	-0.60	-0.483	5349	-0.39	-0.263	1177	0.629	**0.648**	6090	0.141	0.184	9550	0.349	0.378
	wbfs	817	-0.44↑	-0.397↑	1558	**-0.91**↑	**-0.833**↑	5481	-0.36	-0.256	1177	0.602	0.605	5485	0.140	0.201↑	9514	0.345	0.381↑
	sim	267	**-0.50**↑	**-0.463**↑	1083	-0.84↑	-0.771↑	3360	-0.41↑	-0.308↑	**520**	0.610	0.599	**4380**	0.254↑	0.333↑	**4953**	0.386↑	0.399↑
	bisim	341	-0.48↑	-0.427↑	1229	-0.64↑	-0.481	**3269**	**-0.43**↑	**-0.322**↑	636	0.615	0.607	5230	0.381↑	0.444↑	5077	0.387↑	0.396↑
	bisim_d1	518	-0.45↑	-0.390↑	1891	-0.80↑	-0.697↑	5258	-0.38	-0.282↑	864	0.626	0.622	5158	**0.423**↑	**0.468**↑	8315	0.343	0.365
	wbfs + sim	**258**	-0.42	-0.389↑	**1047**	-0.79↑	-0.716↑	3373	-0.39	-0.294↑	530	0.565	0.576	5173	0.403↑	0.467↑	5013	0.379↑	**0.403**↑
	wbfs + bisim	350	-0.41	-0.375	1701	-0.85↑	-0.764↑	3362	-0.41↑	-0.308↑	634	**0.635**↑	0.631	4580	0.247↑	0.321↑	5086	0.349	0.366
	wbfs + bisim_d1	518	-0.46↑	-0.396↑	1919	-0.81↑	-0.727↑	5030	-0.39	-0.277↑	1026	**0.658**↑	0.639	5717	0.386↑	0.433↑	9072	0.347	0.368
ComplEx	KELPIE	626	**-0.98**	**-0.944**	2682	-0.82	-0.723	3882	-0.65	-0.499	261	0.795	0.720	1263	0.658	0.705	1310	**0.543**	**0.519**
	wbfs	626	**-0.98**	**-0.944**	2760	-0.84↑	-0.726↑	3939	-0.59	-0.461	261	0.795	0.720	1263	0.659↑	**0.708**↑	1315	0.520	0.503
	sim	549	-0.96	-0.895	1783	-0.80	-0.684	**1515**	-0.62	-0.510	219	0.777	0.716	774	0.526	0.624	**576**	0.475	0.463
	bisim	571	-0.90	-0.872	1952	-0.85↑	**-0.733**↑	1606	-0.62	-0.507	225	0.772	0.705	897	0.581	0.666	595	0.461	0.449
	bisim_d1	617	-0.98	-0.921	2419	-0.81	-0.693	2930	-0.61	-0.505	259	**0.804**↑	**0.726**↑	1228	0.657	0.697	1153	0.494	0.476
	wbfs + sim	548	-0.97	-0.935	**1758**	-0.80	-0.710	1527	-0.64	**-0.551**↑	217	0.790	0.723↑	740	0.499	0.598	578	0.498	0.482
	wbfs + bisim	573	-0.90	-0.872	1910	-0.81	-0.704	1584	**-0.66**↑	-0.534↑	222	0.779	0.718	857	0.556	0.646	597	0.471	0.463
	wbfs + bisim_d1	617	-0.98	-0.921	2506	-0.83↑	-0.722	2900	-0.57	-0.471	259	0.804↑	0.726↑	1228	**0.660**↑	0.699	1148	0.481	0.472

of the Semantic Pre-Filter (*wbfs*). Moreover, we gauge the effects of the Quotient Explanation Builder in its various formulations: *sim (simulation quotient)*, *bisim (bisimulation quotient)*, and *bisim_d1 (depth-1 bisimulation quotient)*. Finally, we assess how KELPIE performs when equipped with both contributions reporting *wbfs + sim*, *wbfs + bisim*, and *wbfs + bisim_d1*. In all experiments, we set the Pre-Filter k threshold to 20 as in KELPIE.

Across almost all configurations of scenario, model, and dataset, the Quotient Explanation Builder improves the effectiveness through at least one of the alternative formulations. Exceptions were observed in the *necessary* scenario with TRANSE on DB50K and DB100K, the *sufficient* scenario with CONVE on DB50K, in the *necessary* scenario+ with COMPLEX on DB50K, and in the *sufficient* scenario with COMPLEX on YAGO4-20. The Semantic Pre-Filter improves its effectiveness either when adopted independently or when combined with the Quotient Explanation Builder. It has no benefit in the *sufficient* scenario with TRANSE on DB50K, in the *necessary* scenario with CONVE on YAGO4-20, in the *necessary* scenario with COMPLEX on DB50K, and in both scenarios with COMPLEX on YAGO4-20. We posit that the cases of limited performance can be attributed to the selected predictions. Specifically, many predictions may have a subject s associated with a limited number of triples in G_{train}^s, thus leading to a limited search space. Explaining such predictions may be challenging also for KELPIE itself as it looks for explanations in G_{train}^s where few potentially effective explanations are available when explaining such predictions.

Particularly, for predictions associated with less than k (Pre-Filter parameter) triples the Pre-Filter selects all the triples in G_{train}^s; hence, any pre-filter is equivalent except for the order of triples. Furthermore, in instances of graphs characterized by 2 or 3

Table 4. Examples of explanations with Quotient Explanation Builder and plain KELPIE

Simulation quotient	
quotient triples	**entities**
⟨Movie, actor, Gabourey Sidibe⟩	Casting By, Top Five, Seven Psychopaths,White Bird in a Blizzard, Tower Heist, Precious
⟨TVSeries, actor, Gabourey Sidibe⟩	Empire (2015 TV Series)
Depth-1 bisimulation quotient	
quotient triples	**entities**
⟨Movie - subset 1, actor, Gabourey Sidibe⟩	Top Five
⟨Movie - subset 2, actor, Gabourey Sidibe⟩	White Bird in a Blizzard
⟨TVSeries, actor, Gabourey Sidibe⟩	Empire (2015 TV Series)
KELPIE	
⟨Empire (2015 TV Series), actor, Gabourey Sidibe⟩	
⟨Top Five, actor, Gabourey Sidibe⟩	
⟨White Bird in a Blizzard, actor, Gabourey Sidibe⟩	

triples, the compressive potential of quotient graphs is restricted, as the resulting quotient graph precisely mirrors the original structure.

For instance, for DB50K 98 out of 100 predictions with TRANSE are associated with less than 20 triples, with 91 predictions associated with less than 5 triples. We focused on streamlining the search in cases of very extensive search spaces, rather than expanding the search space when it is limited. Indeed, for predictions on highly connected entities for which the space of candidate explanations becomes very extensive, our approach enhances KELPIE.

As regards # r, the impact of the Semantic Pre-Filter appears nearly negligible. In contrast, the Quotient Explanation Builder is intended to optimize # r, indeed the lowest value was observed in the cases of *sim* or *wbfs + sim* in all configurations. Next values are those reported for *bisim*, *bisim_d1*, and finally for KELPIE, that showed the lowest efficiency, thus aligning with the quotient formulations that feature incremental granularity, with *simulation quotient* being the most abstract. In only two cases, the lowest value is reported for *bisim*, specifically in the *necessary* scenario with CONVE on YAGO4-20 and in the *sufficient* scenario with TRANSE on DB100K. These cases are likely due to meeting exceptionally early the heuristic conditions in KELPIE.

5.3 Qualitative Evaluation

In this part, we show a typical example of explanation output by the proposed method, focusing on the impact of the Quotient Explanation Builder.More specifically, we report *necessary explanations* for the prediction ⟨*Gabourey Sidibe*, *nationality*, *United States*⟩ found in the experiments with COMPLEX on YAGO4-20.

In Table 4 we show the explanation computed using the *simulation quotient* and the *depth-1 bisimulation quotient* and the baseline KELPIE. We omit *bisimulation quotient*, as for this prediction it provides the same explanation of *simulation quotient*. The *simulation quotient* provides as explanation both *quotient triples* (first column), and specific entities (second column) supporting our third research goal. Indeed, the *quotient triples* represent relationships between a class expression and the subject of the prediction; then, the specific entities within such a class are explicated. For the given prediction, the first form suggests that the model predicted the correct *nationality* because in the training phase it had seen works like movies and TV series starring *Gabourey Sidibe* as *actor* which is useful to understand the inference made through the model with a general insight based on classes. The second form indicates specific works, which is useful to inspect specific instances and verify their commonalities. Indeed, all the works have the *United States* as country of origin or are related to a producer or actor who, in turn, is related to the *United States*. In the *depth-1 bisimulation quotient* the quotient triples involve two specific subsets of movies, as in this formulation the equivalence classes in the quotient can be split into multiple nodes according to outgoing edges. Indeed, the two sets of movies feature different actors.

The baseline explanation is grounded on the same rationale, retrieving works related to the *United States*. However, it provides a narrower array of evidence. The insight is more precise and more easily comprehensible given its brevity, but limited in terms of coverage. We claim that, despite its length, the explanation with the *simulation quotient* is still intelligible because the effort needed to understand a single triple is approximately the same needed for a *quotient triple* with the corresponding entities. Thus, our approach seems able to provide richer explanations with little extra effort required for their interpretation.

6   Conclusions

We introduced a novel approach for explaining predictions made on KGs that enhances the existing framework KELPIE. Specifically, we identified three research goals: decreasing the number of candidates, providing explanations on different levels of detail, and improving explanation effectiveness. We employed a semantic similarity measure to focus the process on triples semantically related to the prediction. Furthermore, we employed quotient graphs to compact the search space for explanations. We experimentally proved that our solutions furthers the goals on three datasets endowed with semantic information and three representative LP models. We also qualitatively assessed the impact of the quotient graph with respect to the third goal.

A natural progression of this work involves incorporating additional knowledge on a semantic level, e.g., about relationships. Moreover, we plan to explore other graph summarization techniques, such as those in [8]. Furthermore, we aim at improving the explanations for the predictions associated with a limited number of triples. Finally, we may involve users in the evaluation.

Acknowledgments. This work was partially supported by project *FAIR - Future AI Research* (PE00000013), spoke 6 - Symbiotic AI (https://future-ai-research.it/), under the NRRP MUR program funded by the NextGenerationEU and by project *HypeKG - Hybrid Prediction and Explanation with Knowledge Graphs* (H53D23003700006), under PRIN 2022 program funded by MUR.

Disclosure of Interests. The authors declare to have no competing interests.

Table 5. Hyper-parameters of the models

	DB50K	DB100K	YAGO4-20
TRANSE	D: 256, p: 2, Ep: 65, Lr: 0.0017, B: 2048, γ: 5, N: 5	D: 256, p: 2, Ep: 133, Lr: 0.0036, B: 2048, γ: 5, N: 5	D: 128, p: 2, Ep: 59, Lr: 0.0098, B: 2048, γ: 10, N: 5
CONVE	D: 200, $Drop$: {in: 0, h: 0, $feat$: 0}, Ep: 69, Lr: 0.018, B: 512	D: 200, $Drop$: {in: 0, h: 0.2, $feat$: 0}, Ep: 109, Lr: 0.0432, B: 512	D: 200, $Drop$: {in: 0.2, h: 0.1, $feat$: 0.3}, Ep: 512, Lr: 0.0427, B: 512
COMPLEX	D: 200, Ep: 43, Lr: 0.043, B: 512	D: 200, Ep: 83, Lr: 0.0814, B: 512	D: 200, Ep: 88, Lr: 0.0425, B: 512

A   Appendix: Repair of Inconsistent and Unsatisfiable Ontologies

We integrated the datasets DB100K and DB50K with OWL schema axioms retrieved from various sources. The resulting ontologies were inconsistent; hence, we manually repaired them. Moreover, YAGO4-20 turned out to contain unsatisfiable classes. Specifically, we identified the causes of such problems by running the explanation facility of the reasoner. In this appendix, we report some insights on the adjustments that we performed to make DB100K consistent and YAGO4-20 satisfiable, hence "reasonable". In our GitHub repository, we report all the adaptations.

For DB100K, we modified the types for certain instances. For instance, our SPARQL query to retrieve the class assertions returned *Politician*, and *TimePeriod* for several entities. Such types led to inconsistencies as *Politician* is a subclass of *Person* which, in turn, is disjoint with *TimePeriod*. We modified these entities, keeping only the type *Politician*.

We also modified schema axioms in certain cases to preserve several triples, which otherwise would have led to inconsistencies. For instance, we modified the range of the property *location*. We recall that the range of a relationship p specify the classes whose instances can occur as object in a triple with predicate p. We changed the range from *Place* to *Place*⊔*Company*. Through this adjustment, we accommodate also triples having *location* as predicate and an instance of *Company* as object.

In other cases, we needed to remove triples. For instance, we removed the triple ⟨*Subramanian_Swamy, region, Economics*⟩. These triples caused an inconsistency because the type of *Economics* is, *University* which in turn is a descendant of *Agent* in the class hierarchy. However, the range of *region* is *Place* which is disjoint with *Agent*.

For YAGO4-20, we removed certain *subClassOf* axioms. For instance, the class *Districts_of_Slovakia* was a subclass of *AdministrativeArea*, *Product*, and *CreativeWork*. The class *Districts_of_Slovakia* was unsatisfiable as *AdministrativeArea* is subclass of *Localization* which is disjoint with *CreativeWork*. We modified it keeping only *AdministrativeArea* and *Product* as super-classes.

B   Appendix: Hyper-parameters

In this appendix, we report in Table 5 the hyper-parameters that we adopted to train each model on each dataset. Furthermore, we employed the same set of hyper-parameters to execute the *post-training* in the extraction of the explanations and to retrain the models in the evaluation of the explanations.

Note that:

- D is the embedding dimension, in the models that we adopted entity and relation embeddings always have same dimension
- p is the exponent of the p-norm
- Lr is the learning rate
- B is the batch size
- Ep is the number of epochs
- γ is the margin in the Pairwise Ranking Loss
- N is the number of negative triples generated for each positive triple
- ω is the size of the convolutional kernels
- $Drop$ is the training dropout rate, specifically:
  - in is the input dropout
  - h is the dropout applied after a hidden layer
  - $feat$ is the feature dropout

We adopted random search to find the values of the hyper-parameters. Exceptions are given by B and Ep. For B we adopted the value leading to optimize execution times and parallelism. For Ep we adopted early stopping with 1000 as maximum number of epochs and 5 as patience threshold during the training of the models, and we reported the epoch on which the training stopped. Hence, we used such value as number of epochs in the *post-training* and in the evaluation. Furthermore, as in KELPIE, for TRANSE we adopted the learning rate (Lr) values in Table 5 during training and evaluation, but for the *post-training* we used a different value. For TRANSE the batch size (B) is particularly large (2048) and usually exceeds by far the number of triples featuring an entity. This affects *post-training* because in any *post-training* epoch the entity would only benefit from one optimization step. We easily balanced this by increasing the Lr to 0.01.

References

1. Auer, S., Bizer, C., Kobilarov, G., Lehmann, J., Cyganiak, R., Ives, Z.: DBpedia: a nucleus for a web of open data. In: Aberer, K., et al. (eds.) ISWC 2007. LNCS, vol. 4825, pp. 722–735. Springer, Heidelberg (2007). https://doi.org/10.1007/978-3-540-76298-0_52
2. Baltatzis, V., Costabello, L.: Kgex: explaining knowledge graph embeddings via subgraph sampling and knowledge distillation. arXiv preprint arXiv:2310.01065 (2023)
3. Betz, P., Meilicke, C., Stuckenschmidt, H.: Adversarial explanations for knowledge graph embeddings. In: IJCAI-22. International Joint Conferences on Artificial Intelligence Organization, pp. 2820–2826 (2022). https://doi.org/10.24963/ijcai.2022/391
4. Bhowmik, R., de Melo, G.: Explainable link prediction for emerging entities in knowledge graphs. In: Pan, J.Z., et al. (eds.) ISWC 2020. LNCS, vol. 12506, pp. 39–55. Springer, Cham (2020). https://doi.org/10.1007/978-3-030-62419-4_3

5. Bollacker, K., Cook, R., Tufts, P.: Freebase: a shared database of structured general human knowledge. In: AAAI 2007, pp. 1962–1963. AAAI Press (2007). https://doi.org/10.5555/1619797.1619981
6. Bordes, A., Usunier, N., Garcia-Duran, A., Weston, J., Yakhnenko, O.: Translating embeddings for modeling multi-relational data. In: Advances in Neural Information Processing Systems, vol. 26 (2013). https://doi.org/10.5555/2999792.2999923
7. Buneman, P., Staworko, S.: RDF graph alignment with bisimulation. Proc. VLDB Endow. **9**(12), 1149–1160 (2016). https://doi.org/10.14778/2994509.2994531
8. Čebirić, Š, et al.: Summarizing semantic graphs: a survey. VLDB J. **28**(3), 295–327 (2019). https://doi.org/10.1007/s00778-018-0528-3
9. Cohen, S., et al.: Drug repurposing using link prediction on knowledge graphs with applications to non-volatile memory. In: Benito, R.M., Cherifi, C., Cherifi, H., Moro, E., Rocha, L.M., Sales-Pardo, M. (eds.) COMPLEX NETWORKS 2021. Studies in Computational Intelligence, vol. 1073, pp. 742–753. Springer, Cham (2022). https://doi.org/10.1007/978-3-030-93413-2_61
10. d'Amato, C., Masella, P., Fanizzi, N.: An approach based on semantic similarity to explaining link predictions on knowledge graphs. In: WI-IAT 2021, pp. 170–177 (2021). https://doi.org/10.1145/3486622.3493956
11. Dettmers, T., Minervini, P., Stenetorp, P., Riedel, S.: Convolutional 2D knowledge graph embeddings. In: AAAI 2018. AAAI Press (2018). https://doi.org/10.5555/3504035.3504256
12. Ding, B., Wang, Q., Wang, B., Guo, L.: Improving knowledge graph embedding using simple constraints. In: ACL 2018, pp. 110–121. ACL (2018). https://doi.org/10.18653/v1/P18-1011
13. Dong, X.L.: Building a broad knowledge graph for products. In: IEEE-ICDE 2019, pp. 25–25. IEEE (2019). https://doi.org/10.1109/ICDE.2019.00010
14. Dovier, A., Piazza, C., Policriti, A.: A fast bisimulation algorithm. In: Berry, G., Comon, H., Finkel, A. (eds.) CAV 2001. LNCS, vol. 2102, pp. 79–90. Springer, Heidelberg (2001). https://doi.org/10.1007/3-540-44585-4_8
15. Gentilini, R., Piazza, C., Policriti, A.: From bisimulation to simulation: coarsest partition problems. J. Autom. Reason. **31**, 73–103 (2003). https://doi.org/10.1023/A:1027328830731
16. Glimm, B., Horrocks, I., Motik, B., Stoilos, G., Wang, Z.: HermiT: an OWL 2 reasoner. J. Autom. Reason. **53**(3), 245–269 (2014). https://doi.org/10.1007/s10817-014-9305-1
17. Guidotti, R., Monreale, A., Ruggieri, S., Turini, F., Giannotti, F., Pedreschi, D.: A survey of methods for explaining black box models. ACM Comput. Surv. **51**(5), 1–42 (2018). https://doi.org/10.1145/3236009
18. Hogan, A., et al.: Knowledge Graphs. No. 22 in Synthesis Lectures on Data, Semantics, and Knowledge, Springer, Switzerland (2021). https://doi.org/10.2200/S01125ED1V01Y202109DSK022, https://kgbook.org/
19. Koh, P.W., Liang, P.: Understanding black-box predictions via influence functions. In: ICML 2017, pp. 1885–1894. PMLR (2017). https://doi.org/10.5555/3305381.3305576
20. Lundberg, S.M., Lee, S.I.: A unified approach to interpreting model predictions. In: Advances in Neural Information Processing Systems, vol. 30 (2017). https://doi.org/10.5555/3295222.3295230
21. Monroe, D.: AI, explain yourself. Commun. ACM **61**(11), 11–13 (2018). https://doi.org/10.1145/3276742
22. Paige, R., Tarjan, R.E.: Three partition refinement algorithms. SIAM J. Comput. **16**(6), 973–989 (1987). https://doi.org/10.1137/0216062
23. Pellissier Tanon, T., Weikum, G., Suchanek, F.: YAGO 4: a reason-able knowledge base. In: Harth, A., et al. (eds.) ESWC 2020. LNCS, vol. 12123, pp. 583–596. Springer, Cham (2020). https://doi.org/10.1007/978-3-030-49461-2_34
24. Pezeshkpour, P., Tian, Y., Singh, S.: Investigating robustness and interpretability of link prediction via adversarial modifications. arXiv preprint arXiv:1905.00563 (2019)

25. Rossi, A., Barbosa, D., Firmani, D., Matinata, A., Merialdo, P.: Knowledge graph embedding for link prediction: a comparative analysis. ACM Trans. Knowl. Discovery Data (TKDD) 15(2), 1–49 (2021). https://doi.org/10.1145/3424672

26. Rossi, A., Firmani, D., Merialdo, P., Teofili, T.: Explaining link prediction systems based on knowledge graph embeddings. In: SIGMOD 2022, pp. 2062–2075 (2022). https://doi.org/10.1145/3514221.3517887

27. Rousseeuw, P.J., Hampel, F.R., Ronchetti, E.M., Stahel, W.A.: Robust Statistics: The Approach Based on Influence Functions. Wiley, Hoboken (2011). https://doi.org/10.1002/9781118186435

28. Shapley, L.S.: A value for n-person games. In: Contributions to the Theory of Games II (1953). Princeton University Press (1997). https://doi.org/10.1515/9781400829156-012

29. Shi, B., Weninger, T.: Open-world knowledge graph completion. In: AAAI 2018, vol. 32 (2018). https://doi.org/10.1609/aaai.v32i1.11535

30. Singhal, A.: Introducing the knowledge graph: things, not strings. https://blog.google/products/search/introducing-knowledge-graph-things-not

31. Trouillon, T., Welbl, J., Riedel, S., Gaussier, É., Bouchard, G.: Complex embeddings for simple link prediction. In: ICML 2016, pp. 2071–2080. PMLR (2016). https://doi.org/10.5555/3045390.3045609

32. Ying, Z., Bourgeois, D., You, J., Zitnik, M., Leskovec, J.: GNNExplainer: generating explanations for graph neural networks. In: Advances in Neural Information Processing Systems, vol. 32 (2019). https://dl.acm.org/doi/10.5555/3454287.3455116

33. Zhang, H., et al.: Data poisoning attack against knowledge graph embedding. In: International Joint Conference on Artificial Intelligence, pp. 4853–4859 (2019). https://doi.org/10.24963/ijcai.2019/674

34. Zhang, W., Deng, S., Wang, H., Chen, Q., Zhang, W., Chen, H.: Xtranse: explainable knowledge graph embedding for link prediction with lifestyles in e-commerce. In: Wang, X., Lisi, F., Xiao, G., Botoeva, E. (eds.) JIST 2019. LNCS, vol. 1157, pp. 78–87. Springer, Singapore (2020). https://doi.org/10.1007/978-981-15-3412-6_8

35. Zhang, W., Paudel, B., Zhang, W., Bernstein, A., Chen, H.: Interaction embeddings for prediction and explanation in knowledge graphs. In: WSDM 2019, pp. 96–104 (2019). https://doi.org/10.1145/3289600.3291014

Large Language Models for Scientific Question Answering: An Extensive Analysis of the SciQA Benchmark

Jens Lehmann[1,2]([✉]) [ID], Antonello Meloni[3] [ID], Enrico Motta[4] [ID],
Francesco Osborne[4,5] [ID], Diego Reforgiato Recupero[3] [ID],
Angelo Antonio Salatino[4] [ID], and Sahar Vahdati[1] [ID]

[1] ScaDS.AI - TU Dresden, Dresden, Germany
`jens.lehmann@tu-dresden.de`
[2] Amazon, Munich, Germany
[3] Department of Mathematics and Computer Science,
University of Cagliari, Cagliari, Italy
[4] Knowledge Media Institute, The Open University, Milton Keynes, UK
[5] Department of Business and Law, University of Milano-Bicocca, Milan, Italy

Abstract. The SciQA benchmark for scientific question answering aims
to represent a challenging task for next-generation question-answering
systems on which vanilla large language models fail. In this article, we
provide an analysis of the performance of language models on this benchmark including prompting and fine-tuning techniques to adapt them to
the SciQA task. We show that both fine-tuning and prompting techniques
with intelligent few-shot selection allow us to obtain excellent results on
the SciQA benchmark. We discuss the valuable lessons and common
error categories, and outline their implications on how to optimise large
language models for question answering over knowledge graphs.

Keywords: Knowledge graphs · Question answering · Language
models. · Fine-tuning · Few-shot learning

1 Introduction

Knowledge graphs have gained popularity over the last decades as fact storage systems and, more recently, as retrieval backends for large language models
(LLMs) [18,30]. Most knowledge graphs are currently relatively simple semantic
structures and question-answering (QA) benchmarks have been designed mostly
for such structures and encyclopedic knowledge graphs such as DBpedia [25]
and Wikidata [39]. The SciQA ORKG benchmark [2] focuses on a new direction by leveraging the Open Research Knowledge Graph[1](ORKG) [36], which
models scholarly artifacts and is meant to pose a challenge for next-generation
QA systems. Auer et al. [2] underline this by pointing out that the performance

---

[1] Open Research Knowledge Graph - https://orkg.org.

A. Meroño Peñuela et al. (Eds.): ESWC 2024, LNCS 14664, pp. 199–217, 2024.
https://doi.org/10.1007/978-3-031-60626-7_11

of knowledge graph question answering (KGQA) systems as well as ChatGPT (GPT-3.5 backend) on the benchmark is rather low.

The main goal of this paper is to dive deeper into analysing the LLM performance on the SciQA. In particular, we studied the accuracy of different language models on the SciQA ORKG benchmark and whether simple training or prompting approaches are already sufficient to address the challenges or whether more sophisticated techniques are required. The most common approaches to improving the performance of a language model on a given task are appropriate prompting and fine-tuning. We investigated the impact of both strategies on different language models. If a language model can be prompted to solve the task, then this indicates that the model itself has the inherent capability to solve the problem but requires further context to direct its next-token prediction optimisation towards solving the problem [14,41]. We were, therefore, particularly interested in this direction and analysed the impact of zero- and few-shot learning techniques on SciQA performance. For few-shot learning, we investigated different example selection strategies leveraging semantic similarity [28], diversity [26], and entropy criteria [19]. To ensure reproducibility, we released a repository with the codebase and the prompts used in the experiments[2].

The result of this study gives insights into how to optimise language models for SciQA, and more in general, for scientific question answering. Our findings demonstrate that different combinations of LLMs, both fine-tuned and utilizing few-shot learning, achieve remarkable outcomes on the benchmark. Notably, the best-performing model was the fine-tuned version of T5-base, a relatively small language model. The fine-tuned version of GPT-2-large also performed well. This suggests that low-resource models, when meticulously fine-tuned, can match or even surpass the performance of larger, more resource-intensive models in complex tasks like scientific question answering. GPT-3.5 also produced excellent results using a few-shot approach based on the semantic similarity between the question and the samples in the training set.

In summary, the contributions of this paper are the following:

1. Performance analysis of four language models for SciQA using three methodologies (zero-shot learning, few-shot learning, and fine-tuning).
2. Analysis of seven alternative approaches to select samples for few-shot learning for LLM-based KGQA.
3. Provision of several insights which indicate that optimising LLMs with appropriate prompting and fine-tuning techniques is sufficient to obtain excellent results (>94.1% exact query match accuracy and >97.5% F1-score).

The remainder of this paper is structured as follows. Section 2 discusses related works on knowledge graph question answering and LLM few-shot optimizations. The SciQA benchmark and the LLM models are described in Sect. 3. Section 4 outlines the experiments and the metrics used for evaluation. Section 5 reports the results. Section 6 provides an analysis of the types of errors made by the models while Sect. 7 discusses the limitations of our study. Finally, Sect. 8 ends the paper with concluding remarks and future directions.

---

[2] Codebase and prompts - https://github.com/NIMI-research/SciQA-LLM.

2   Related Work

Semantic Parsing for Knowledge Graph Question Answering. Semantic parsing [21] is the process of converting an utterance in natural language into a query with a format that machines can understand and by using a specific formal language. This conversion is considered successful when the query accurately represents the original intent of the natural language input. Various strategies have been proposed for semantic parsing, encompassing Combinatory Categorial Grammars (CCG), rule-based systems, and neural-network-based methods [3,12,34,35]. The work introduced in [12] addresses semantic parsing using neural network methods for knowledge graphs. Recent methods of semantic parsing for KGQA often utilize sequence-to-sequence models, such as those described by [34]. In these models, an encoder transforms a natural language utterance into an intermediate representation. Subsequently, a decoder processes this representation to generate a logical form.

A type of neural network architecture that combines the features of both sequence-to-sequence models and pointer networks is called pointer generator networks [3]. These networks enable the replication of specific input segments, such as data values, in the output. Although these methods show high accuracy when trained with large amounts of high-quality data, the difficulties in obtaining such data and the significant effort needed for updates lead to an increased interest in training semantic parsers with fewer datasets. The work in [35] explores an unsupervised approach for semantic parsing which shows promising results. While promising, these methods have scalability issues for real scenarios. Additionally, they present challenges in terms of easy adjustability or tunability for practical applications. LLMs such as GPT-2 [32], GPT-3.5 [9], and Dolly [15], which builds upon the Pythia models [6] show potential in semantic parsing with zero or minimal examples, especially following improvements to mitigate inference costs. However, LLMs do not inherently support KG-specific parsing without access to entities in the underlying knowledge graph. Consequently, they require augmentation with a method for entity retrieval. A work that exploited LLMs for KGQA semantic parsing is described in [24], where authors claimed that training data requirements are substantially reduced when using controlled natural languages as targets for KGQA. Some conversational agents also use language models to map questions to a set of predefined templates that facilitate the generation of queries [29].

The developers of the SciQA benchmark hosted the Scholarly QALD Challenge as a satellite event of the 22nd International Semantic Web Conference [4]. This resulted in the development of some dedicated methods for SciQA [20,31,37], which usually involve additional components for incorporating a representation of the ORKG ontological schema or detecting domain-specific entities. In contrast, our study seeks to evaluate the performance of conventional LLMs when augmented with fine-tuning and prompting techniques.

LLM Optimisation. In this paper, we focus on two standard optimization techniques for LLMs: few-shot learning and fine-tuning. Few-shot optimization in LLMs is a technique where the models are given one or more samples when performing a certain task. This contrasts with the zero-shot approach, where LLMs function without any training data [22]. There are several ways of optimizing the examples for few-shot learning. Kumagai et al. [23] proposed an approach that employs a feature selector and a decoder: the selector uses concrete random variables to choose relevant features for the decoder to reconstruct the original features of unseen instances. The sampling method in Bansal et al. [5] enables optimization-based meta-learning across diverse tasks with varying numbers of classes. Levy et al. [26] study how to select a diverse set of samples in in-context learning. Similarly, Lin et al. [27] studied the few-shot learning capabilities of LLMs across a broad spectrum of tasks with a diverse selection of examples. The recent Cross Entropy Difference method [19] also demonstrated promising results, especially in semantic parsing. Another optimization method is fine-tuning, a technique where a pre-trained LLM is further trained on a specific task or dataset to improve its performance for that particular domain [40]. Fine-tuning is particularly valuable when the labeled data for the target task are limited, as it maximises the utility of pre-existing knowledge in the model [17].

3 Background

This section describes the SciQA benchmark and introduces the four models employed in our experiments.

3.1 Dataset

We adopted as a benchmark the recent SciQA dataset[3]. SciQA includes 2,565 pairs, each composed of a natural language question and the equivalent SPARQL query. The SPARQL queries are specifically designed to extract information from the Open Research Knowledge Graph (ORKG) [36]. This KG presently consists of 170,000 resources that detail the research contributions of nearly 15,000 scholarly articles spanning 709 research fields.

SciQA is a mix of manually crafted and automatically generated pairs of questions and queries. Specifically, 100 pairs have been manually crafted by a team of researchers from different countries and institutions. First, they selected a number of research fields and relevant studies on ORKG to narrow down the data scope. Next, they formulated natural language questions of various types, such as single comparison, True/False, and aggregation questions. Each question was then associated with a corresponding SPARQL query and relevant metadata (e.g., type, query shape). The entire process underwent multiple peer reviews. The authors sought diverse perspectives and consulted 21 domain experts involved in ORKG curation grants to ensure the relevance and importance of the questions. Next, from these initial 100 pairs, the authors identified eight query templates and used them to automatically produce an additional 2,465 pairs using the GPT-3 model [9]. Specifically, they defined a set of

---

[3] SciQA dataset - https://huggingface.co/datasets/orkg/SciQA.

eight templates and 32 questions, with one manual question and three GPT-3-generated variations for each template. The query templates were then filled with all possible entities from ORKG and the resulting queries were randomly assigned to one of the four questions. The results of these queries were collected, and metadata for the final question set was extracted. The addition of automatically generated questions expanded the SciQA dataset to 2,565 questions. We report an example of a question and query pair in Example 1.

The benchmark divides the dataset into three parts: 70% for training (1795 samples), 10% for validation (257 samples), and 20% for testing (513 samples). Table 1 reports relevant statistics for each set. The unbalanced distribution of the template frequency across the datasets is noteworthy. Further details about the SciQA dataset and its construction are available in [2].

Table 1. For the training, validation, and test splits of the SciQA dataset, we show from the left to the right, respectively, the number of human-generated questions, the number of automatically generated questions, the percentage of human-generated questions (H), and the percentage of automatically generated questions for each template T_i.

	#Hum. Gen.	#Aut. Gen.	H	T1	T2	T3	T4	T5	T6	T7	T8
Training	66	1729	3.7%	13.5%	13.3%	12.9%	13.3%	18.1%	2.7%	22.3%	0.3%
Validation	13	244	5.1%	14.0%	15.6%	14.8%	9.7%	16.3%	3.5%	21.0%	0%
Test	21	492	4.1%	12.5%	12.7%	14.2%	15.4%	18.5%	2.9%	19.5%	0.2%

3.2 Approaches

We employed four LLMs in our experiments: i) T5-base [33], ii) GPT-2-large [32], iii) Dolly-v2-3b [15], and iv) GPT-3.5 Turbo [9]. These models are all based on the transformer architecture, which leverages the attention mechanism to learn how words and sentences relate to each other [38]. We chose these models due to their significant variability in dimension (220M to 175B parameters) and structure (both encoder-decoder and decoder-only), enabling us to evaluate various options on the benchmark. In the following, we will briefly illustrate their main characteristics.

The **T5-base** model features an encoder-decoder structure with 220 million parameters and a context capacity of 512 tokens. It was trained on the C4 dataset, a vast collection of text totaling around 750 GB of web pages. The model is trained using a mask-filling task, where 15% of the tokens are randomly masked. The **GPT-2-large** model uses a decoder-only architecture with 774 million parameters. It is capable of handling a maximum of 1024 tokens encompassing both input and output. It was trained on the WebText dataset, which includes approximately 40 GB of text sourced from 45 million web pages. The **Dolly-v2-3b** is a decoder architecture with 2.8 billion parameters and a token context capacity of 2048. This model was developed by Databricks through the fine-tuning of EleutherAI's Pythia-12b[4]. Specifically, Dolly-v2-3b underwent fine-tuning using a dataset comprising approximately 15,000 instruction-response pairs. Finally, the **GPT-3.5 Turbo** model features a decoder-only architecture with 175 billion parameters and supports a context of up to 4,096 tokens. This model builds upon the architecture and pre-training data of GPT-3, having been trained on roughly 45 terabytes of text from diverse sources, including books, websites, social media, and news articles. GPT-3.5 Turbo underwent further fine-tuning with data and tasks pertinent to chat applications, covering areas like dialogues, question-and-answer sessions, command-response interactions, and code generation and execution.

4   Methodology

The primary objective of the study presented in this paper was to evaluate the performance of various types of LLMs on the Sci-QA datasets when applying common optimization techniques, such as few-shot learning and fine-tuning. To achieve this, we methodically conducted experiments using 7 distinct few-shot learning approaches as well as fine-tuning. We omitted certain experiments that were either unfeasible due to the context window limitations of some language models or too demanding computationally, specifically the fine-tuning of GPT-3.5 and Dolly. We also conducted an error analysis (reported in Sect. 6) to determine the strengths and weaknesses of the leading models. Contrary to other studies [20,31,37], we did not apply any specialized adjustments to improve the performance of these methods for this specific dataset. This section details the experiments, along with the metrics and setup utilised in our study.

4.1   Experiments

We focused on three approaches for optimizing the LLMs: fine-tuning (FT), zero-shot learning (ZSL), and few-shot learning (FSL). For the FT, we fine-tuned the model on the training set before applying it to the test set. For the ZSL, the test set was processed directly using a basic prompt without any examples. Specifically, we tried for each model different types of prompts, ultimately selecting the most effective prompt for each model based on performance. Finally, for the FSL, the prompt was enhanced by providing a number S of samples extracted from the training set. Given a question of the test set, we evaluated seven methods from the literature to select the most relevant samples for each question:

- 1) **Random.** S samples are randomly chosen for each test set element.
- 2) **Similarity** [28]. The samples are ranked in descending order based on their semantic similarity to the question associated with the corresponding test set element, and the top S samples are chosen.

---

[4] Pythia-12b - https://huggingface.co/EleutherAI/pythia-12b.

- **Diversity of template** [26].
  - **3)** *Test A* (All diverse templates): The samples are arranged in a descending sequence according to their semantic similarity to the question. The top S samples, using all different templates, are selected.
  - **4)** *Test B* (Same template for all): The samples are arranged in a descending sequence according to their semantic similarity to the question. Subsequently, the top S samples that utilize the same template as the first sample are selected.
- **Entropy** [19]. We implemented three different techniques derived from [19]. For each template, we first select the 8 shortest (in terms of string size) samples of the training set and then:
  - **5)** *Same_Templ*: We select the most similar question in the training set and identify its template. Next, we select the sample using the same template from the set of 8 samples;
  - **6)** *Ran_Templ*: We select a random sample from the set of 8 samples;
  - **7)** *Low_Perp*: We generate the SPARQL query for each of the 8 input templates and then compute the perplexity for each of them. We then select the sample which produced the lowest perplexity.

In the following, we will illustrate the experiments conducted on the four models: T5-base, GPT-2, Dolly-v2-3b, and GPT-3.5-Turbo.

T5-Base Model. The T5-base model underwent only fine-tuning. ZSL and FSL were not applicable due to the limited size of the prompt for this model and its pre-training on specific tasks, such as translating from English to German or summarizing text. Consequently, when prompted for English to SPARQL translation tasks, the model tended to provide a German translation of the question instead. In the FT process, the T5 model was assigned the task of translating a specified English natural language question into a SPARQL query. The prefix used for this purpose was *"Translate English to SPARQL:"*.

The model was fine-tuned with 1,795 labeled samples from the training set over 20 epochs. Each sample consisted of a short prefix and a natural language question, serving as the input request to the model, and was paired with the corresponding SPARQL query, which formed the expected response. Example 4 illustrates the type of data utilized in the fine-tuning process of the T5-base model.

GPT-2-large Model. The primary limitation of the GPT-2 model was its restricted context size, which constrained the use of lengthy prompts and multiple samples in few-shot learning (FSL). Despite this limitation, we successfully conducted the full suite of experiments. Specifically, we performed the following experiments:

- ZSL, using the prompt: *"input (English text): [nl question] \n output (Sparql query):"*.

- FSL with 1, 2, and 3 samples, due to the 1024 token context restriction, which did not permit the inclusion of more samples. In the following, we report an example using 1-short learning:
  "input (English text): [the sample question in natural language]"
  "output (Sparql query): [the sample SPARQL query]"
  "input (English text): [the question in natural language of the underlying test set sample]"
  "output (SPARQL query):"
  We tried all the methods described in the previous section (similarity, diversity, and entropy) to select relevant samples.
- FT: The GPT-2-large model was fine-tuned using a text file containing all question and query pairs from the training set. The special token $<|endoftext|>$ was inserted before each pair, serving as a signal for the model to cease output generation. The fine-tuning process was carried out over 20 epochs.

Dolly-v2-3b Model. Due to the substantial dimensions of the Dolly model, fine-tuning posed significant challenges. Consequently, we focused on ZSL and FSL. Specifically, we performed the following experiments:

- ZSL, using the prompt: *"Translate to a SPARQL query the following English question:"*.
- FSL: We utilized 1, 3, 5, and 7 examples for few-shot learning. The model's token limit of 2,048 restricted our input to a maximum of seven samples. We employed the same prompt format as used with the GPT-2 model. Similar to the previous approach, we applied the full range of previously discussed metrics to select various samples for testing.

GPT-3.5-Turbo. Similarly to Dolly, fine-tuning was potentially resource-intensive due to the model's large size. Therefore, we concentrated on ZSL and FSL instead. In summary, we performed the following experiments:

- ZSL, using the prompt: *"What is the SPARQL query for:"*.
- FSL: To ensure comparability with the other models, we conducted experiments using 1, 3, 5, and 7 samples. We applied the complete set of methodologies for sample selection, with one exception. We did not test the Entropy - *Low_Perp* methodology because it was not possible to compute the perplexity with OpenAI API. The samples were provided using the following template:
  "input (English text): [the sample question in natural language]"
  "output (SPARQL query): [the sample SPARQL query]"
  "What is the SPARQL query for: [the question in natural language]"

4.2 Experimental Setup

We adopted an extensive set of metrics for evaluating the performance of the models over SciQA: Precision, Recall, F1 Score, Bleu 4, Bleu Cumulative, Rouge

1, and Rouge 2. The precision is calculated as the ratio of common tokens to the total number of tokens in the predicted output. Similarly, recall is computed by dividing the number of common tokens by the total number of tokens in the ground truth. The F1-score is computed as $2 \times \frac{prec \times rec}{prec + rec}$. The BLEU score evaluates the quality of predicted text, referred to as the candidate, by comparing it to a set of references. Representing a precision-based measure, the BLEU score ranges from 0 to 1, with a higher value indicating better prediction quality. A value above 0.3 is generally considered a good score. Cumulative BLEU is determined by adding the n-gram accuracy scores from all reference translations and then taking the geometric mean. Finally, Rouge-n measures the number of matching n-grams between the model-generated text and a human-produced reference. We refer to Chauhan and Daniel [13] for a comprehensive review of these metrics.

The setup we have used for the approaches is the following. T5-base has been first fine-tuned using HuggingFace *Seq2SeqTrainer*[5] with the following parameters: learning_rate = 2e−5, weight_decay = 0.01, fp16 = True. Then, we used the method *generate* of the class *AutoModelForSeq2SeqLM*[6] from the Transformers[7] library with the following setting: max_new_tokens = 512, do_sample = True, top_k = 30, top_p = 0.95. GPT-2-large was fine-tuned using HuggingFace *Trainer*[8] using all the default parameters. GPT-2-large was utilized through the *pipeline*[9] abstraction from the Transformers (see footnote 7) library and setting the following parameters: task = text-generation, max_new_tokens = 400, return_full_text = False. Dolly-v2-3b was also employed using the *pipeline* abstraction from the Transformers (see footnote 7) library and the following parameters: torch_dtype = torch.bfloat16, trust_remote_code = True. Finally, GPT-3.5 Turbo was used through the OPENAI API[10] via the *create* method of the class *ChatCompletion*[11] from the *openai*[12] Python library. The parameters were set as follows: system role: *"You are a translator from natural language to SPARQL using the provided examples."*, temperature = 0.5, top_p = 0.95, frequency_penalty = 0, presence_penalty = 0, stop = None, timeout = 30.

---

[5] Seq2SeqTrainer - https://huggingface.co/docs/transformers/v4.35.1/en/main_classes/trainer#transformers.Seq2SeqTrainer.

[6] AutoModelForSeq2SeqLM - https://huggingface.co/docs/transformers/v4.35.1/en/model_doc/auto#transformers.AutoModelForSeq2SeqLM.

[7] Transformers - https://huggingface.co/docs/transformers/v4.35.1/en/index.

[8] Trainer - https://huggingface.co/docs/transformers/v4.35.1/en/main_classes/trainer

[9] Pipeline class - https://huggingface.co/docs/transformers/v4.35.1/en/main_classes/pipelines

[10] OPENAI API - https://openai.com/blog/openai-api.

[11] ChatCompletion -https://platform.openai.com/docs/guides/text-generation/chat-completions-api

[12] OpenAI Libraries - https://platform.openai.com/docs/libraries

5   Results and Discussion

Tables 2, 3, 4, and 5 report the complete set of experiments on the four models, covering all the metrics introduced in Sect. 4.2. Table 6 presents a comparative summary of each configuration, focusing specifically on F1-scores and exact matches.

For the sake of space, we will primarily focus on each model's top performance. The highest F1 score was achieved by the fine-tuned T5 model, with a score of 0.9751. This was closely followed by GPT-3.5, which attained a score of 0.9736 using similarity and a 7-sample few-shot approach, and the fine-tuned GPT-2 (0.9669). Dolly's best performance, yielded using a 1-sample few-shot method with the sample selected via similarity, resulted in a score of 0.8792. Although this is the lowest among the compared models, it is still fairly good.

Table 2. T5-base model - Performance when applying the fine-tuning strategy.

Strat.	Prec.	Rec.	F1 Score	Blue 4	Bleu C	Rouge 1	Rouge 2	Exact matches
FT	0.9767	0.9760	0.9751	0.9597	0.9631	0.9790	0.9683	483

Table 3. GPT2-large model - Performance based on the strategy (Strat.) and the criteria for FSL sample selection (C): Random, Similarity, Diversity, Entropy. S is the number of samples. In bold the best results for each metric.

Strat.	C	Test name	S	Prec.	Rec.	F1 Score	Blue 4	Bleu C	Rouge 1	Rouge 2	Exact matches
FT				**0.9693**	**0.9669**	**0.9669**	**0.9462**	**0.9504**	**0.9708**	**0.9580**	430
ZSL				0.0464	0.1579	0.0653	0.0004	0.0009	0.0932	0.0087	0
FSL	Rand.		1	0.1119	0.3001	0.1499	0.0191	0.02950	0.1826	0.0692	0
			2	0.1230	0.2968	0.1580	0.0255	0.0372	0.1976	0.0807	0
			3	0.1336	0.3195	0.1730	0.0303	0.0449	0.2287	0.1003	0
	Simil.		1	0.1477	0.4877	0.2076	0.0685	0.0822	0.2535	0.1640	0
			2	0.2054	0.5112	0.2719	0.1205	0.1355	0.3237	0.2454	1
			3	0.2366	0.5692	0.3107	0.1502	0.1670	0.3767	0.3077	2
	Dive.	Test A	3	0.1680	0.4422	0.2215	0.0519	0.0721	0.2879	0.1573	0
		Test B	3	0.2205	0.5636	0.2988	0.1496	0.1635	0.3555	0.2926	1
	Entro.	Same_Templ	1	0.1567	0.3903	0.2029	0.0541	0.0696	0.2383	0.1373	0
		Ran_Templ	1	0.1105	0.2778	0.1421	0.0180	0.0280	0.1798	0.0675	0
		Low_Perp	1	0.2408	0.4360	0.2788	0.0924	0.1263	0.3427	0.2302	0

The scenario is similar when considering exact matches. Once again, the top three performing approaches are the fine-tuned T5 model (scoring 483/513, 94.1%), GPT-3.5 employing the similarity-based 7-sample few-shot approach

Table 4. Dolly-v2-3b model - Performance based on the strategy (Strat.) and the criteria for FSL sample selection (C): Random, Similarity, Diversity, Entropy. S is the number of samples. In bold the best results for each metric.

Strat.	C	Test name	S	Prec.	Rec.	F1 Score	Blue 4	Bleu C	Rouge 1	Rouge 2	Exact matches
ZSL				0.1911	0.0993	0.1087	0.0033	0.0062	0.1734	0.0358	0
FSL	Random		1	0.5976	0.5973	0.5659	0.2799	0.3456	0.6269	0.4559	27
			3	0.5830	0.6676	0.5900	0.3484	0.4038	0.6435	0.5131	31
			5	0.6075	0.7147	0.6242	0.3934	0.4450	0.6847	0.5575	51
			7	0.6358	0.7423	0.6576	0.4460	0.4947	0.7153	0.6001	69
	Similarity		1	**0.8742**	0.9088	**0.8792**	**0.7728**	**0.8015**	**0.8861**	**0.8494**	167
			3	0.7979	0.9163	0.8304	0.7204	0.7432	0.8775	0.8470	182
			5	0.7845	**0.9238**	0.8242	0.7102	0.7322	0.8711	0.8407	180
			7	0.7681	0.9057	0.8052	0.6911	0.7113	0.8515	0.8215	181
	Diversity	Test A	3	0.6621	0.8240	0.7000	0.4587	0.5138	0.7766	0.6565	43
			5	0.6350	0.8151	0.6825	0.4329	0.4912	0.7702	0.6327	39
			7	0.6316	0.7941	0.6729	0.4246	0.4836	0.7529	0.6116	46
		Test B	3	0.7623	0.9068	0.8025	0.6926	0.7139	0.8533	0.8254	171
			5	0.7793	0.9223	0.8181	0.7148	0.7346	0.8647	0.8411	201
			7	0.7961	0.9122	0.8261	0.7279	0.7456	0.8647	0.8398	**212**
	Entr.	Same_Templ	1	0.4866	0.8089	0.5734	0.3662	0.4005	0.5741	0.5474	2
		Ran_Templ	1	0.3789	0.6107	0.4402	0.1604	0.2155	0.4811	0.3288	0
		Low_Perp	1	0.5854	0.8647	0.6757	0.4499	0.5013	0.7262	0.6201	1

Table 5. GPT-3.5-Turbo model - performance based on the strategy (Strat.) and the criteria for FSL sample selection (C): Random, Similarity, Diversity, Entropy. S is the number of samples. In bold the best results for each metric.

Strat.	C	Test name	S	Prec.	Rec.	F1 Score	Blue 4	Bleu C	Rouge 1	Rouge 2	Exact matches
ZSL				0.4963	0.1931	0.2632	0.0039	0.0088	0.4010	0.1019	0
FSL	Random		1	0.8162	0.6962	0.7362	0.3954	0.4828	0.8194	0.6107	45
			3	0.8707	0.8052	0.8259	0.5914	0.6558	0.8817	0.7543	113
			5	0.9024	0.8499	0.8675	0.6813	0.7343	0.9073	0.8141	165
			7	0.9226	0.8726	0.8905	0.7356	0.7799	0.9227	0.8476	189
	Similarity		1	0.9509	0.9305	0.9368	0.8655	0.8833	0.9505	0.9137	356
			3	0.9730	0.9635	0.9667	0.9378	0.9439	0.9750	0.9558	451
			5	0.9759	0.9685	0.9709	0.9481	0.9521	0.9772	0.9609	464
			7	**0.9768**	**0.9727**	**0.9736**	**0.9534**	**0.9571**	**0.9788**	**0.9638**	475
	Diversity	Test A	3	0.9693	0.9156	0.9378	0.8613	0.8784	0.9619	0.9232	315
			5	0.9730	0.9209	0.9428	0.8699	0.8875	0.9642	0.9290	328
			7	0.9688	0.9158	0.9375	0.8599	0.8783	0.9610	0.9230	313
		Test B	3	0.9678	0.9492	0.9561	0.9057	0.9181	0.9672	0.9391	412
			5	0.9683	0.9502	0.9566	0.9076	0.9192	0.9681	0.9401	417
			7	0.9673	0.9501	0.9562	0.9061	0.9180	0.9671	0.9386	422
	Ent.	Same_Templ	1	0.9303	0.8781	0.8988	0.7759	0.8119	0.9252	0.8631	205
		Ran_Templ	1	0.8268	0.6345	0.7016	0.3222	0.4070	0.7946	0.5651	26

(92.6%), and the fine-tuned GPT-2 (83.8%). However, in contrast to the previous findings, the performance of the Dolly model is consistently subpar across all configurations, with the best outcome achieved using a 7-sample few-shot approach (Diversity - Test B, always using the same template), yielding a score of only 41.3% (212/513). The low performance of Dolly may be due to its exclusive reliance on FSL and ZST, as it has not undergone fine-tuning due to its significant size. The fact that the best version of Dolly uses Diversity - Test B may depend on the fact that using a substantial number of examples following a consistent template enables Dolly to produce comparatively better queries.

When considering all the other metrics introduced in Sect. 4.2 and reported in Tables 2, 3, 4 and 5, the trends are very similar. The top three methods in terms of performance for Rouge 1, Rouge 2, Bleu 4, and Bleu C are again the fine-tuned T5, GPT 3.5 with 7-sample few-shot, and the fine-tuned GPT-2.

These results reveal several noteworthy insights. First, despite the challenges presented by the SciQA benchmark and the generally low performance in zero-shot learning observed among all models (no exact matches and <26% F1), the highest-performing solutions that employed fine-tuning and few-shot learning exhibited excellent results. This suggests that as LLMs advance and our understanding of them deepens, more challenging benchmarks may be soon required.

Second, the T5 model, although the smallest model by a large margin (220M parameters), surpassed all the larger models and even robust commercial solutions like GPT-3.5 (175B) when fine-tuned with relevant data. This supports the idea that appropriately tailored datasets can enable the development of efficient, low-resource models that perform comparably to more computationally intensive options. The financial and sustainability implications of this finding warrant further exploration.

Third, our analysis identified semantic similarity as the most effective approach for selecting samples in few-shot learning tasks. This method consistently produced the highest performance across all models. However, it was observed that different models react differently to changes in sample size. For example, GPT 3.5 showed improved performance with larger sample sizes when using similarity-based selection. In contrast, Dolly's performance declined, suggesting that an increase in examples might confuse this model. These phenomena merit additional investigation in future studies.

6   Error Analysis

We performed a comprehensive analysis of erroneous queries, defined as those not matching the expected response. The analysis focused on the errors produced by T5-base (30 errors), which was the top-performing model, and GPT-3.5 (38 errors), which was the second-best performer.

We identified five unique error categories. Table 7 details the definitions of these categories and the respective proportions of erroneous queries for both T5 and GPT-3.5. Each query can be associated with multiple categories due to the presence of more than one type of error. Therefore, the sum of all these categories does not necessarily equal to 100%.

Table 6. Summary of F1-scores and exact matches (in parenthesis) based on the different strategies (Strat.), and the various Criteria (C): Random, Similarity, Diversity, Entropy. S is the number of samples. In bold the best results for each model.

Strat.	C	Test name	S	T5-base	GPT2-large	Dolly-v2-3b	GPT-3.5-turbo
ZSL					0.0653 (0)	0.1087 (0)	0.2632 (0)
FT				**0.9751 (483)**	**0.9669 (430)**		
FSL	Random		1		0.2005 (0)	0.5659 (27)	0.7362 (45)
			3		0.2187 (0)	0.5900 (31)	0.8259 (113)
			5			0.6242 (51)	0.8675 (165)
			7			0.6576 (69)	0.8905 (189)
	Similarity		1		0.2718 (0)	**0.8792** (167)	0.9368 (356)
			3		0.4051 (2)	0.8304 (182)	0.9667 (451)
			5			0.8242 (180)	0.9709 (464)
			7			0.8052 (181)	**0.9736 (475)**
	Diversity	Test A	3		0.2215 (0)	0.7000 (43)	0.9378 (315)
			5			0.6525 (39)	0.9428 (328)
			7			0.6729 (46)	0.9375 (313)
		Test B	3		0.2988 (1)	0.8025 (171)	0.9561 (412)
			5			0.8181 (201)	0.9566 (417)
			7			0.8261 **(212)**	0.9562 (422)
	Ent.	Same_Templ	1		0.2029 (0)	0.5734 (2)	0.8988 (205)
		Ran_Templ	1		0.1421 (0)	0.4402 (0)	0.7016 (26)
		Low_Perp	1		0.2788 (0)	0.6757 (1)	

The comparison of incorrect results between T5 and GPT-3.5, reveals a total of 21 overlapping queries. They included 10 instances of error type 1 (misspelled entity), 11 of type 2 (wrong entity type), 9 of error type 3 (wrong predicate), and 4 of error type 5 (semantic error). Despite these overlaps, T5 and GPT-3.5 exhibit significant differences across several categories of errors, underscoring their distinct strengths and weaknesses. Indeed, they only exhibit similar behavior with respect to the *misspelled entity* category (60.0% and 60.5%, respectively), which explains the high number of errors of type 1 in the overlapping mistakes. Both models occasionally alter the names of entities, often using synonyms. An example of this is provided in Example 3, where the entity *linear-chain CRFs* was modified to *label-chain CRFs*.

The primary issue with T5 pertains to syntactic errors, accounting for 40.0% of errors compared to only 5.2% in GPT-3.5. This suggests that T5 has a relatively weaker grasp of SPARQL syntax compared to GPT-3.5. Example 4 reports a typical case in which T5 generate a syntactically incorrect query. In all other error categories, T5 demonstrates notably fewer mistakes than GPT-3.5.

GPT-3.5 is characterized by a range of issues, most notably a semantic error rate that is more than double that of T5 (57.8% vs 26.6%). This indicates that while GPT-3.5 typically generates syntactically accurate queries, it tends to misinterpret their meaning more often than T5. Additionally, GPT-3.5 is more prone to incorrectly generating the wrong entity type or predicate. These problems may partly stem from the inclusion of SPARQL queries from Wikidata and DBpedia in its training data. Consequently, instead of accurately translating queries, GPT-3.5 may hallucinate and erroneously insert SPARQL query fragments that refer to Wikidata and other large knowledge graphs. Example 5 shows a typical case in which GPT-3.5 hallucinates a fictitious predicate (orkgp:P2067), probably influenced by the Wikidata predicate *Property:P2067*. This analysis underscores the challenges LLMs face when attempting to use specialized domain ontologies, which may diverge from standard predicate and type conventions. Consequently, a less complex and more straightforward model, such as T5, may at times produce better results.

Table 7. Categories of errors and their frequencies for GPT-3.5 and T5.

Error Category	Category Definition	GPT-3.5	T5
(1) Misspelled entity	The query misspells an entity name (e.g., "robotic navigation" instead of "robot navigation")	60.5%	60.0%
(2) Wrong entity type	The query uses an entity of a different type (e.g., the query searching an entity of type Energy Source instead of one of type Paper)	52.6%	36.6%
(3) Wrong predicate	The query uses an incorrect predicate in the query (e.g., orkgp:P7046 has been replaced with orkgp:HAS_METHOD)	76.3%	36.6%
(4) SPARQL syntactic error	The query cannot be syntactically parsed by a SPARQL compiler	5.2%	40.0%
(5) Semantic error	The query does not reflect the meaning of the question	57.8%	26.6%

7   Limitations

This paper focuses on the SciQA benchmark, since it is meant to be a challenge for next-generation QA systems. However, when analyzing the findings, it is important to consider some limitations and factors regarding the general applicability of the proposed solutions. First, SciQA includes numerous questions that have been automatically generated from initial seed questions. Despite the complexity of these queries, LLMs may be capable of identifying the some inherent patterns. Therefore, the results shown here do not necessarily carry over to other benchmarks, which were entirely handcrafted, e.g., [24] indicates that semantic parsing is still a challenging task for LLMs. Additionally, success on the SciQA is heavily reliant on the appropriate application of literals, unlike other QA benchmarks where the accuracy of entities plays a more crucial role. While SciQA was just recently released, we would also like to point out that the precise training schedules for OpenAI models are unknown. Therefore, there exists a small yet plausible chance that the GPT-3.5 model may have encountered information from the SciQA benchmark during its training phase.

A further limitation is that we focused on query similarity metrics on the generated query string as opposed to execution metrics for computational reasons. For example, for our exact match metric, there is a risk that an LLM-generated query is rated as incorrect even if retrieves the correct results and the query formulation is aligned with the intent of the question, i.e., we could underestimate model performance. However, the manual examination of the sample of 68 errors produced by T5 and GPT-3.5 reported in Sect. 6 found that all queries marked as erroneous would not have yielded the correct results upon execution.

8 Conclusions

This paper presented an in-depth examination of the performance of LLMs on the SciQA benchmark. We evaluated four distinct language models, employing three methodologies: zero-shot learning, few-shot learning, and fine-tuning. Additionally, we explored seven different strategies for choosing examples in few-shot learning. Our findings demonstrate that optimizing LLMS with appropriate prompting and fine-tuning techniques produces outstanding results on this benchmark, with the best model yielding a >94.1% exact query match accuracy and >97.5% F1-score. This performance highlights the potential need for more challenging benchmarks in this space.

Interestingly, the most effective results across nearly all metrics were achieved by the fine-tuned T5 model, which is relatively small (220M parameters), followed by GPT-3.5 using a seven-sample few-shot, and the fine-tuned GPT-2. This suggests that low-resource models, if carefully fine-tuned, can rival the performance of larger, more resource-intensive models even on a complex task such as scientific question answering. Finally, the analysis shows that semantic similarity is the most effective method for sample selection in few-shot learning for this specific task.

In our future research, we plan to broaden our examination by conducting comprehensive experiments on the capability of LLMs to perform knowledge graph question answering across various domains. We are also investigating the potential of LLMs in performing related tasks, such as generating scientific hypotheses [8], classifying research articles [11], recommending citations [10], and producing literature reviews [7]. Finally, we plan to apply the valuable insights gained from this study to develop a more challenging benchmark for scientific question answering. Specifically, we plan to produce a new resource based on large-scale knowledge graphs in this space such as the Academia/Industry DynAmics Knowledge Graph (AIDA-KG) [1] and the Computer Science Knowledge Graph (CS-KG) [16].

A Appendix - Examples

Example 1 Question: What are the titles and IDs of research papers that include a benchmark for the DDI extraction 2013 corpus dataset?

```
SELECT DISTINCT ?paper ?paper_lbl
WHERE { ?dataset a orkgc:Dataset; rdfs:label ?dataset_lbl.
   FILTER (str(?dataset_lbl) = ''DDI extraction 2013 corpus'')
     ?benchmark orkgp:HAS_DATASET ?dataset.
     ?cont orkgp:HAS_BENCHMARK ?benchmark.
     ?paper orkgp:P31 ?cont; rdfs:label ?paper_lbl.
}
```

Example 2 Type of data used in the fine-tuning process of the T5-base model.

Request:

"Translate English to SPARQL: Which model has achieved the highest Accuracy score on the ARC (Challenge) benchmark dataset?"

Expected output:

```
SELECT DISTINCT ?model ?model_lbl
  WHERE { ?metric a orkgc:Metric; rdfs:label ?metric_lbl.
  FILTER (str( ?metric_lbl) = ''Accuracy") { SELECT ?model ?model_lbl
  WHERE { ?dataset a orkgc:Dataset; rdfs:label ?dataset_lbl.
  FILTER (str( ?dataset_lbl) = ''ARC (Challenge)")
  ?benchmark orkgp:HAS_DATASET ?dataset; orkgp:HAS_EVALUATION ?eval.
  ?eval orkgp:HAS_VALUE ?value; orkgp:HAS_METRIC ?metric.
  ?cont orkgp:HAS_BENCHMARK ?benchmark; orkgp:HAS_MODEL ?model.
  ?model rdfs:label ?model_lbl. } ORDER BY DESC( ?value) LIMIT 1 } }
```

Example 3 Example of *misspelled entity* generated by T5. In blue the original entity and in red the misspelled one. Question: *List the code links in papers that use the linear-chain CRFs model in any benchmark?*

Original query:

```
SELECT DISTINCT ?code
WHERE {
    ?model a orkgc:Model;
        rdfs:label ?model_lbl.
    FILTER (str( ?model_lbl) =
    "linear-chain CRFs")
    ?benchmark orkgp:HAS_DATASET ?dataset.
    ?cont orkgp:HAS_BENCHMARK ?benchmark.
    ?cont orkgp:HAS_MODEL ?model;
        orkgp:HAS_SOURCE_CODE ?code.
}
```

Generated output:

```
SELECT DISTINCT ?code
WHERE {
    ?model a orkgc:Model;
        rdfs:label ?model_lbl.
    FILTER (str( ?model_lbl) =
    "label-chain CRFs")
    ?benchmark orkgp:HAS_DATASET ?dataset,
    ?cont orkgp:HAS_BENCHMARK ?benchmark.
    ?cont orkgp:HAS_MODEL ?model;
        orkgp:HAS_SOURCE_CODE ?code.
}
```

Example 4 Example of *SPARQL syntactic error* generated by T5. In red the unmatching part. Question: *What is Raman spectroscopy?*

Original query:

```
SELECT ?definitions
WHERE {
    ?terms orkgp:P24009 ?definitions.
    ?terms rdfs:label ?terms_labels.
    FILTER(REGEX( ?terms_labels,
    "Raman spectroscopy"))
}
```

Generated output:

```
SELECT ?rmdfs:label ?rmdfs:label
?rmdfs:label ?rmdfs:label ?rmdfs:label
?rmdfs:label ?rmdfs:label ?rmdfs:label
?rmdfs:label ?rmdfs:label ?rmdfs:label
?rmdfs:label. }
```

Example 5 Example of *wrong predicate* error generated by GPT-3.5. In red the hallucinated predicate. Question: *What quantity of iron oxide was discovered on Elorza crater?*

Original query:

```
SELECT ?properties_values,
        ?property_description
WHERE {
    ?papers rdf:type orkgc:Paper.
    ?papers rdfs:label ?papers_labels.
    ?papers orkgp:P31 ?contrib.
    ?contrib ?properties ?properties_values.
    ?properties rdfs:label ?properties_labels.
    ?properties orkgp:description ?property_description.
    FILTER(REGEX( ?properties_labels, "FeO"))
    FILTER(REGEX(?papers_labels, "Elorza crater", "i"))
}
```

Generated output:

```
SELECT ?quantity
WHERE {
    ?crater rdf:type orkgc:Crater.
    ?crater rdfs:label "Elorza crater".
    ?crater orkgp:P31 ?iron_oxide_discoveries.
    ?iron_oxide_discoveries orkgp:P2067 ?quantity.
}
```

References

1. Angioni, S., Salatino, A., Osborne, F., Recupero, D.R., Motta, E.: AIDA: a knowledge graph about research dynamics in academia and industry. Quant. Sci. Stud. **2**(4), 1356–1398 (2021)

2. Auer, S., et al.: The SciQA scientific question answering benchmark for scholarly knowledge. Sci. Rep. **13**(1), 7240 (2023). https://doi.org/10.1038/s41598-023-33607-z

3. Babu, G.A., Badugu, S.: A survey on automatic text summarisation. In: Reddy, A.B., Nagini, S., Balas, V.E., Raju, K.S. (eds.) Proceedings of Third International Conference on Advances in Computer Engineering and Communication Systems. LNNS, vol. 612, pp. 679–689. Springer, Singapore (2023). https://doi.org/10.1007/978-981-19-9228-5_58

4. Banerjee, D., Usbeck, R., Mihindukulasooriya, N., Singh, G., Mutharaju, R., Kapanipathi, P. (eds.): Joint Proceedings of Scholarly QALD 2023 and SemREC 2023 Co-located with 22nd International Semantic Web Conference ISWC 2023, Athens, Greece, 6–10 November 2023, CEUR Workshop Proceedings, vol. 3592. CEUR-WS.org (2023), https://ceur-ws.org/Vol-3592

5. Bansal, T., Jha, R., McCallum, A.: Learning to few-shot learn across diverse natural language classification tasks. In: Proceedings of the 28th International Conference on Computational Linguistics, pp. 5108–5123 (2020)

6. Biderman, S., et al.: Pythia: a suite for analyzing large language models across training and scaling. In: International Conference on Machine Learning, pp. 2397–2430. PMLR (2023)

7. Bolanos, F., Salatino, A., Osborne, F., Motta, E.: Artificial intelligence for literature reviews: opportunities and challenges. arXiv preprint arXiv:2402.08565 (2024)

8. Borrego, A., et al.: Completing scientific facts in knowledge graphs of research concepts. IEEE Access **10**, 125867–125880 (2022)

9. Brown, T.B., et al.: Language models are few-shot learners (2020)

10. Buscaldi, D., Dessí, D., Motta, E., Murgia, M., Osborne, F., Recupero, D.R.: Citation prediction by leveraging transformers and natural language processing heuristics. Inf. Process. Manage. **61**(1), 103583 (2024)

11. Cadeddu, A., et al.: A comparative analysis of knowledge injection strategies for large language models in the scholarly domain. Eng. Appl. Artif. Intell. **133**, 108166 (2024)

12. Chakraborty, N., Lukovnikov, D., Maheshwari, G., Trivedi, P., Lehmann, J., Fischer, A.: Introduction to neural network-based question answering over knowledge graphs. Wiley Interdisc. Rev.: Data Min. Knowl. Discov. **11**(3), e1389 (2021)

13. Chauhan, S., Daniel, P.: A comprehensive survey on various fully automatic machine translation evaluation metrics. Neural Process. Lett. **55**, 12663–12717 (2022). https://doi.org/10.1007/s11063-022-10835-4

14. Chen, Y., Kang, H., Zhai, V., Li, L., Singh, R., Raj, B.: Token prediction as implicit classification to identify LLM-generated text. arXiv preprint arXiv:2311.08723 (2023)

15. Conover, M., et al.: Free dolly: introducing the world's first truly open instruction-tuned LLM (2023). https://www.databricks.com/blog/2023/04/12/dolly-first-open-commercially-viable-instruction-tuned-llm

16. Dessí, D., Osborne, F., Reforgiato Recupero, D., Buscaldi, D., Motta, E.: CS-KG: a large-scale knowledge graph of research entities and claims in computer science. In: Sattler, U., et al. (eds.) ISWC 2022. LNCS, vol. 13489, pp. 678–696. Springer, Cham (2022). https://doi.org/10.1007/978-3-031-19433-7_39

17. Fu, Z., Yang, H., So, A.M.C., Lam, W., Bing, L., Collier, N.: On the effectiveness of parameter-efficient fine-tuning (2022)

18. Hogan, A., et al.: Knowledge graphs. ACM Comput. Surv. (CSUR) **54**(4), 1–37 (2021)

19. Iter, D., et al.: In-context demonstration selection with cross entropy difference. arXiv preprint arXiv:2305.14726 (2023)

20. Jiang, L., Yan, X., Usbeck, R.: A structure and content prompt-based method for knowledge graph question answering over scholarly data. CEUR Workshop Proceedings, vol. 3592 (2023). https://ceur-ws.org/Vol-3592/paper3.pdf

21. Kamath, A., Das, R.: A survey on semantic parsing. arXiv preprint arXiv:1812.00978 (2018)

22. Kojima, T., Gu, S.S., Reid, M., Matsuo, Y., Iwasawa, Y.: Large language models are zero-shot reasoners (2023)

23. Kumagai, A., Iwata, T., Fujiwara, Y.: Few-shot learning for unsupervised feature selection. arXiv preprint arXiv:2107.00816 (2021)

24. Lehmann, J., Gattogi, P., Bhandiwad, D., Ferré, S., Vahdati, S.: Language models as controlled natural language semantic parsers for knowledge graph question answering. In: European Conference on Artificial Intelligence (ECAI), vol. 372, pp. 1348–1356. IOS Press (2023)

25. Lehmann, J., et al.: DBpedia-a large-scale, multilingual knowledge base extracted from Wikipedia. Semant. Web **6**(2), 167–195 (2015)

26. Levy, I., Bogin, B., Berant, J.: Diverse demonstrations improve in-context compositional generalization. arXiv preprint arXiv:2212.06800 (2022)

27. Lin, X.V., et al.: Few-shot learning with multilingual generative language models. In: Proceedings of the 2022 Conference on Empirical Methods in Natural Language Processing, pp. 9019–9052 (2022)

28. Liu, J., Shen, D., Zhang, Y., Dolan, B., Carin, L., Chen, W.: What makes good in-context examples for GPT-3? arXiv preprint arXiv:2101.06804 (2021)

29. Meloni, A., et al.: AIDA-Bot 2.0: enhancing conversational agents with knowledge graphs for analysing the research landscape. In: Payne, T.R., et al. (eds.) ISWC 2023. LNCS, vol. 14266, pp. 400–418. Springer, Cham (2023). https://doi.org/10.1007/978-3-031-47243-5_22

30. Peng, C., Xia, F., Naseriparsa, M., Osborne, F.: Knowledge graphs: opportunities and challenges. Artif. Intell. Rev. 1–32 (2023)

31. Pliukhin, D., Radyush, D., Kovriguina, L., Mouromtsev, D.: Improving subgraph extraction algorithms for one-shot SPARQL query generation with large language models. In: Scholarly-QALD-23: Scholarly QALD Challenge at The 22nd International Semantic Web Conference (ISWC 2023), Athens, Greece. vol. 3592, pp. 1–10 (2023). https://ceur-ws.org/Vol-3592/paper6.pdf

32. Radford, A., et al.: Language models are unsupervised multitask learners. OpenAI Blog **1**(8), 9 (2019)

33. Raffel, C., et al.: Exploring the limits of transfer learning with a unified text-to-text transformer. J. Mach. Learn. Res. **21**(1), 1–67 (2020)

34. Rongali, S., Soldaini, L., Monti, E., Hamza, W.: Don't parse, generate! A sequence to sequence architecture for task-oriented semantic parsing. In: Proceedings of The Web Conference 2020, pp. 2962–2968 (2020)

35. Rony, M.R.A.H., Chaudhuri, D., Usbeck, R., Lehmann, J.: Tree-KGQA: an unsupervised approach for question answering over knowledge graphs. IEEE Access **10**, 50467–50478 (2022)

36. Stocker, M., et al.: Fair scientific information with the open research knowledge graph. FAIR Connect **1**, 19–21 (2023). https://doi.org/10.3233/FC-221513

37. Taffa, T.A., Usbeck, R.: Leveraging LLMs in scholarly knowledge graph question answering. In: Scholarly-QALD-23: Scholarly QALD Challenge at the 22nd International Semantic Web Conference (ISWC 2023), Athens, Greece, vol. 3592, pp. 1–10 (2023). https://ceur-ws.org/Vol-3592/paper5.pdf

38. Vaswani, A., et al.: Attention is all you need (2023)

39. Vrandečić, D., Krötzsch, M.: Wikidata: a free collaborative knowledgebase. Commun. ACM **57**(10), 78–85 (2014)

40. Wei, J., et al.: Finetuned language models are zero-shot learners (2022)

41. Zhao, S., Dang, J., Grover, A.: Group preference optimization: Few-shot alignment of large language models. arXiv preprint arXiv:2310.11523 (2023)

Efficient Evaluation of Conjunctive Regular Path Queries Using Multi-way Joins

Nikolaos Karalis[✉][iD], Alexander Bigerl[iD], Liss Heidrich[iD],
Mohamed Ahmed Sherif[iD], and Axel-Cyrille Ngonga Ngomo[iD]

DICE group, Department of Computer Science, Paderborn University,
Paderborn, Germany
{nikolaos.karalis,alexander.bigerl,liss.heidrich,
mohamed.sherif,axel.ngonga}@uni-paderborn.de

Abstract. Recent analyses of real-world queries show that a prominent
type of queries is that of conjunctive regular path queries. Despite the
increasing popularity of this type of queries, only limited efforts have
been invested in their efficient evaluation. Motivated by recent results
on the efficiency of worst-case optimal multi-way join algorithms for the
evaluation of conjunctive queries, we present a novel multi-way join algorithm for the efficient evaluation of conjunctive regular path queries. The
hallmark of our algorithm is the evaluation of the regular path queries
found in conjunctive regular path queries using multi-way joins. This
enables the exploitation of regular path queries in the planning steps of
the proposed algorithm, which is crucial for the algorithm's efficiency, as
shown by the results of our detailed evaluation using the Wikidata-based
benchmark WDBench. The results of this evaluation also show that our
approach achieves a value of query mixes per hour that is 4.3 higher than
the state of the art and that it outperforms all of the competing graph
storage solutions in almost 70% of the benchmark's queries.

Keywords: knowledge graphs · conjunctive regular path queries ·
multi-way joins

1 Introduction

The ability to express queries requesting paths of arbitrary length between entities of a knowledge graph, also known as *(two-way) regular path queries* (RPQs)
[3,10], is a unique characteristic of graph query languages [18]. Finding paths
of arbitrary length is crucial for many applications on knowledge graphs, such
as path-based fact checking [27] and class expression learning [14]. As a result,
many recent works have focused on developing approaches for the efficient evaluation of RPQs (e.g., [4,6,30]). However, most of these works do not take the
fact that RPQs are usually part of more complex queries into account [1]. In
fact, the results of two detailed studies of Wikidata's query logs [11,21] show
that *conjunctive two-way regular path queries* (C2RPQs), which are conjunctive

© The Author(s), under exclusive license to Springer Nature Switzerland AG 2024
A. Meroño Peñuela et al. (Eds.): ESWC 2024, LNCS 14664, pp. 218–235, 2024.
https://doi.org/10.1007/978-3-031-60626-7_12

queries extended with regular path queries [11], have received a lot of attention recently and are often used in practice.

Recently, Cucumides et al. [13] performed a theoretical analysis of the evaluation of C2RPQs using worst-case optimal multi-way join algorithms. Worst-case optimal multi-way join algorithms [22] are a recent advancement in query processing and have achieved state-of-the-art performance in evaluating conjunctive queries. Among other contributions, Cucumides et al. proposed the algorithm GenericJoinCRPQ that evaluates C2RPQs using multi-way joins. However, as previous theoretical analyses of such algorithms, they do not discuss the involvement of RPQs in aspects of multi-way joins that are crucial for their efficiency in practice. First, they assume a given variable ordering. Second, given a particular variable, they do not discuss the order in which operations should be carried out. For example, provided the SPARQL graph pattern {?x <p1> <o1> . ?x <p2> <o2> . ?x <p3>* <o3> .}, should the RPQ be evaluated before or after the join operation between the first two triple patterns? To the best of our knowledge, there have not been any implementations of multi-way join algorithms for the evaluation of C2RPQs.

In this work, we hence focus on presenting a novel multi-way join algorithm for the evaluation of C2RPQs over RDF graphs using SPARQL. The main characteristic of our approach is that it evaluates RPQs using multi-way joins. This evaluation (1) enables the integration of RPQs in multi-way join plans; our approach is generic and can be integrated in any system supporting multi-way joins, (2) allows for the evaluation of C2RPQs without the need for materializing the results of RPQs—which can be large—and (3) enables the accurate size estimation of RPQs, as it allows for their evaluation up to an arbitrary depth, and thus allows their inclusion in optimization steps of the evaluation, such as the variable ordering process. The proposed algorithm is implemented in a state-of-the-art triple store supporting worst-case optimal multi-way joins. We evaluate our approach using WDBench [2]. We compare the performance of our approach against the performance of multiple state-of-the-art graph storage solutions. The results of this comparison suggest that our solution is on average 4.3 times faster than the second best system. We also carry out a detailed evaluation of different execution strategies. Its results show the importance of including RPQs in the optimization steps of the proposed algorithm. Like a number of approaches for the evaluation of conjunctive queries based on worst-case optimal joins [5,8,19], the performance of our approach depends on the order of variables and the order in which operations are carried out for a particular variable.

The rest of the paper is structured as follows. In Sect. 2, we provide background knowledge on the topics that are covered in this paper. We discuss related works in Sect. 3. We present our approach for the evaluation of C2RPQs using multi-way joins in Sect. 4. Our experimental results are presented in Sect. 5. In Sect. 6, we conclude and discuss possible future research directions.

2   Preliminaries

Below, we cover the features of SPARQL that are relevant to this work and briefly summarize worst-case optimal multi-way joins and GenericJoinCRPQ [13].

2.1   RDF and SPARQL

The semantics of SPARQL have been extensively covered in previous works (e.g.,
[20,24–26]). Here, following standard notation, we recapitulate the semantics and
properties of the features of the language that are used later in this work. More
specifically, we cover *basic graph patterns*, *union graph patterns* and *property path
patterns*. The formal definitions presented below rely on the following pairwise
disjoint sets. Let $\mathbf{I}$ be an infinite set of IRIs, $\mathbf{B}$ an infinite set of blank nodes, and
$\mathbf{L}$ an infinite set of literals. Furthermore, let $\mathbf{V}$ be an infinite set of variables. An
RDF graph is a set of *subject-predicate-object* triples and is formally defined as
$G = \{(s, p, o) \mid s \in (\mathbf{I} \cup \mathbf{B}), p \in \mathbf{I}, o \in (\mathbf{I} \cup \mathbf{B} \cup \mathbf{L})\}$.

Basic and Union Graph Patterns [19,24]. A basic graph pattern (BGP) is
a set of triples and is formally defined as $P = \{(s, p, o) \mid s \in (\mathbf{I} \cup \mathbf{L} \cup \mathbf{V}), p \in
(\mathbf{I} \cup \mathbf{V}), o \in (\mathbf{I} \cup \mathbf{L} \cup \mathbf{V})\}$. An element of P is called a triple pattern and is denoted
as *tp*. As in [19,24], we do not consider blank nodes in triple patterns because
they behave as variables. The set of variables of a triple pattern is denoted
as *var(tp)*. BGPs are conjunctive queries and their semantics are defined using
mappings. A mapping is a partial function assigning RDF terms (i.e., IRIs, blank
nodes, or literals) to variables and is formally defined as $\mu : \mathbf{V} \rightarrow (\mathbf{I} \cup \mathbf{B} \cup \mathbf{L})$.
The set of variables, over which a mapping μ is defined, is called the domain of μ
and is denoted as $dom(\mu)$. Provided a triple pattern *tp*, $\mu(tp)$ denotes the RDF
triple obtained by replacing every variable in *var(tp)* with its corresponding
value in μ. Two mappings μ_1 and μ_2 are compatible, if and only if for every
variable $?v \in dom(\mu_1) \cap dom(\mu_2)$ holds that $\mu_1(?v) = \mu_2(?v)$. The join and
union operations between two sets of mappings Ω_1 and Ω_2 are defined as:

$$\Omega_1 \bowtie \Omega_2 = \{\mu_1 \cup \mu_2 \mid \mu_1 \in \Omega_1, \mu_2 \in \Omega_2 \text{ and } \mu_1, \mu_2 \text{ are compatible}\} \text{ and}$$
$$\Omega_1 \cup \Omega_2 = \{\mu \mid \mu \in \Omega_1 \text{ or } \mu \in \Omega_2\}.$$

A triple pattern *tp* and a BGP P are evaluated over an RDF graph G as follows:

$$[\![tp]\!]_G = \{\mu \mid dom(\mu) = var(tp) \text{ and } \mu(tp) \in G\} \text{ and}$$
$$[\![P]\!]_G = [\![tp_1, \ldots, tp_n]\!]_G = [\![tp_1]\!]_G \bowtie \ldots \bowtie [\![tp_n]\!]_G.$$

In SPARQL, braces allow us to form different graph patterns. Both BGPs and
triple patterns are graph patterns. The concatenation (conjunction) and union of
two graph patterns P_1 and P_2 are denoted as $(P_1 \; AND \; P_2)$ and $(P_1 \; UNION \; P_2)$,
respectively, and they are evaluated as follows:

$$[\![P_1 \; AND \; P_2]\!]_G = [\![P_1]\!]_G \bowtie [\![P_2]\!]_G \text{ and } [\![P_1 \; UNION \; P_2]\!]_G = [\![P_1]\!]_G \cup [\![P_2]\!]_G.$$

A graph pattern P is in *UNION* normal form if it is in the form $(P_1 \; UNION \; \ldots$
$UNION \; P_n)$ and each P_i, for $1 \leq i \leq n$, is *UNION*-free. A graph pattern is
UNION-free, if it does not contain any union graph patterns.

Theorem 1 (Existence of UNION normal form [24]**).** *Every graph pattern
P using the AND and UNION operators is equivalent to a graph pattern P',
which is in UNION normal form.*

Property Paths Patterns [20]. In SPARQL, two-way RPQs are expressed as property path patterns. A property path pattern is a triple $r = (s, e, o)$, where $s, o \in (\mathbf{I} \cup \mathbf{L} \cup \mathbf{V})$ and e is constructed by the following grammar:

$$e := \alpha \mid e^- \mid e \cdot e \mid e + e \mid e^+ \mid e^* \mid e? \,, \alpha \in \mathbf{I}.$$

The set of variables of a property path pattern r is denoted as $var(r)$. For simplicity, we assume that property paths do not have the same variable in both the subject and object position. The implementation of our approach (Sect. 4) supports RPQs having the same variable in the subject and object position. Note that we omit the rules of negated property sets from the grammar. As in [1], we do not consider negated property sets in this work. Expressions of the shape $e_1 \cdot e_2$ and $e_1 + e_2$ can be rewritten as joins and unions, respectively (see Example 1) [17,25]. Consequently, property path patterns using only the first four rules of the grammar can be rewritten to equivalent graph patterns consisting only of basic and union graph patterns and be evaluated as shown above. Property paths using the transitive closure operators $*$ or $+$ are evaluated under set semantics. The expression $e?$ is a special case of e^*; it returns solutions for paths of length 0 and length 1. From this point on, we will use the term RPQ only for property paths using any of the $+$, $*$, or $?$ operators, as the remaining expressions can be rewritten as graph patterns without property paths.

Example 1. The queries provided below are semantically equivalent.

$Q1$: SELECT ?s WHERE {?s (<e1>/(<e2>/<e3>)+)|(<e4>/<e5>) <o>}

$Q2$: SELECT ?s WHERE {{?s <e1> ?t . ?t (<e2>/<e3>)+ <o>} UNION
{?s <e4> ?v . ?v <e5> <o>}}

Conjunctive Two-Way Regular Path Queries. In SPARQL, a C2RPQ is a pattern that only uses triple patterns, the operator *AND*, and RPQs [11]. In $Q2$ of Example 1, the first graph pattern of the *UNION* is a C2RPQ consisting of a triple pattern and an RPQ, whereas the second graph pattern is a BGP. As the property path patterns that we consider in this work use only the *AND* and *UNION* operators, graph patterns comprised of basic, union and property path patterns can be rewritten to a graph pattern in *UNION* normal form [25]. Our approach presented below deals with graph patterns that are in *UNION* normal form, where each union operand is either a C2RPQ or a BGP (e.g., $Q2$). Our implementation applies the *UNION* normal form to queries, while parsing them.

2.2 Worst-Case Optimal Multi-way Joins

Worst-case optimal multi-way join algorithms [22] satisfy the AGM bound [7], i.e., their runtime complexity is bounded by the worst-case size of the result of the input query [19]. Since their recent introduction, they have gained a lot of attention and, in particular, have achieved state-of-the-art performance when evaluating graph pattern queries (e.g., [5,8,15,19]). Unlike pair-wise joins that carry out join operations on two join operands at a time, worst-case optimal

Algorithm 1. Generic Join for Basic Graph Patterns

1: // Generator function: execution is resumed after a **yield** operation
2: **function** GENERICJOIN(P, G, X)   ▷ P: BGP, G: RDF Graph, X: Mapping
3:     **if** all variables are resolved **then yield** X and **return**
4:     $?x \leftarrow$ select an unresolved variable from X
5:     $K \leftarrow \bigcap_{tp \in P | ?x \in var(tp)} \{\mu(?x) \mid \mu \in [\![tp]\!]_G\}$
6:     **for all** $k \in K$ **do**
7:         $X(?x) \leftarrow k$; $P' \leftarrow$ assign k to all occurrences of $?x$ in P
8:         **yield all** GENERICJOIN(P', G, X) ▷ after yielding proceeds with the next k

---

multi-way join algorithms evaluate queries recursively on a per variable basis [15, 19]. This evaluation method does not store any intermediate results and allows for mappings to be written to the result incrementally. A worst-case optimal multi-way join algorithm based on Generic Join [23] is shown in Algorithm 1. In practice, the performance of Generic Join is mostly affected by the order of variables (line 4) [19] and the set intersection, which finds the possible values of the selected variable (line 5). In fact, the set intersection should be guided by the triple pattern with the smallest set of values for the selected variable [29]. Last, indices (e.g., [5,8]) also play an important role in the efficiency of worst-case optimal multi-way join algorithms [15].

2.3 GenericJoinCRPQ

As mentioned, Cucumides et al. [13] perform a theoretical analysis of the evaluation of C2RPQs using worst-case optimal algorithms. In their work, the authors obtain size bounds for several classes of C2RPQs and show that there are C2RPQs that cannot be evaluated by a worst-case optimal algorithm. Despite their latter finding, the authors devise algorithms based on existing worst-case optimal algorithms. One of these algorithms is GenericJoinCRPQ, which is an extension of Generic Join (Algorithm 1). More specifically, GenericJoinCRPQ also considers the RPQs of the graph pattern that have the selected variable $?x$ for the set intersection (line 5). Additionally, GenericJoinCRPQ materializes the RPQs that have the selected variable $?x$, once $?x$ is replaced with a value k (line 7). In subsequent steps, materialized RPQs are treated as triple patterns. However, the materialization of RPQs can be avoided by evaluating them with multi-way joins. As the authors study the complexity of GenericJoinCRPQ theoretically, they assume a given variable ordering (line 4). As discussed, the variable ordering is crucial for the performance of multi-way join algorithms. Hence, it is necessary to consider RPQs in the variable selection process. Furthermore, the authors assume an arbitrary order for the set intersection (line 5). In practice, should a triple pattern or an RPQ guide the set intersection? Last, the authors argue that the running time of GenericJoinCRPQ might end up being too high when there are multiple recursive steps evaluating RPQs. Our experimental results show that, even in such cases, the evaluation of C2RPQs with multi-way joins outperforms existing solutions by considering RPQs in planning steps.

3   Related Work

Multiple algorithms have been proposed for the evaluation of RPQs. A type of approaches that has received a lot of attention is that based on finite automata [6,28,30] and recently, an approach based on matrix algebra was introduced [4]. As existing SPARQL engines (e.g., Blazegraph and Virtuoso), our work falls into the category of approaches that use existing relational operations for the evaluation of RPQs [1]. However, to the best of our knowledge, our work is the first to use multi-way joins for the evaluation of RPQs. In addition to the specialized evaluation algorithms, specialized indices have also been proposed for the efficient evaluation of RPQs (e.g., [6,16]). We refer the reader to [6] for a more detailed review of the literature on the evaluation of RPQs.

For the efficient evaluation of C2RPQs, the works in the literature mostly focus on obtaining accurate cardinality estimations of RPQs that ultimately lead to good orderings of two-way joins [1]. A recent work in this direction is presented in [1]. In [1], the authors propose a cost model for RPQs and an approach based on random walks for estimating the size of RPQs that do not have any nested transitive closures (i.e., for path expressions α^+ and α^*). Given an RPQ, a fixed number of random walks evaluate the RPQ up to a specified depth and the RPQ's estimated cost is ultimately equal to the sum of the number of results returned by each random walk. A shortcoming of this approach is that random walks use bag semantics instead of set semantics, which, as per the authors, might lead to overestimated cardinalities in dense graphs. Our approach presented below proposes an end-to-end evaluation methodology for C2RPQs based on multi-way joins and follows set semantics for estimating the size of RPQs.

4   Evaluating C2RPQs with Multi-way Joins

In this section, we present our approach for the evaluation of C2RPQs using multi-way joins. Note that the proposed algorithm is not worst-case optimal [13]. The main characteristic of our approach is the evaluation of RPQs found in C2RPQs using multi-way joins. The evaluation of RPQs using multi-way joins offers multiple benefits. First, it allows for the easy integration of RPQs into multi-way join plans. Our approach is generic and can be adopted by any system supporting multi-way joins. Second, it does not require the materialization of RPQs. As in multi-way joins for conjunctive queries, we generate the results of RPQs incrementally. However, as we discuss later, there are cases where the materialization of RPQs leads to an improved performance. Third, it allows for the accurate estimation of the size of RPQs up to an arbitrary depth. The accurate size estimation of RPQs, enables their inclusion in planning steps of the algorithm, which is crucial for the efficient evaluation of C2RPQs [1].

4.1   Evaluation of RPQs

The SPARQL standard [17] defines the function ALP for the evaluation of transitive closures (i.e., path patterns using the * or + operators). Consider an RPQ

$r = (s, \alpha^+, ?o)$, with $s \notin \mathbf{V}$. At depth 1, ALP evaluates the triple pattern $tp_1 = (s, \alpha, ?o)$. At depth 2, ALP evaluates the triple pattern $tp_2 = (u, \alpha, ?o)$ for every term $u \in \{\mu(?o) \mid \mu \in [\![tp_1]\!]_G\}$. In general, for $i > 1$, ALP uses the terms generated at depth $(i - 1)$—that have not been visited already—to evaluate a triple pattern tp_i at depth i. This is generalized to BGPs and unions of BGPs by replacing α^+ with e^+ in r, provided e does not have any nested transitive closures. Here, the main observation is that a transitive closure is evaluated by a recursive procedure, which, in turn, evaluates a graph pattern P that does not have any RPQs at each recursive step [1]. We leverage this observation and use multi-way joins to evaluate P. If P is a BGP, we simply use Generic Join. If P is a union graph pattern, it can be rewritten to an equivalent pattern P' in *UNION* normal form (Sect. 2.1, Theorem 1). The BGPs of P' are then evaluated independently and in a serialized manner. The UNION normal form might lead to a large number of joins; however, this is not common in practice. For nested transitive closures, ALP needs to evaluate a C2RPQ or a union of C2RPQs at each step. In such cases, we use Algorithm 3 (Sect. 4.2).

Algorithm 2 presents the evaluation of RPQs using multi-way joins. The starting point of the evaluation is EvalRPQ. Based on the number of variables that the provided RPQ has, EvalRPQ calls the appropriate function for its evaluation. EvalRPQ_TV (TV stands for term and variable) is based on the ALP function and is called for RPQs that have only one variable (line 5). Here, we assume that the object of the RPQ is the variable. The function works in the same manner for the case of the subject being the variable. As their names suggest, the list to_visit and the set visited keep track of the nodes of the input RDF graph that need to be visited and have already been visited, respectively. EvalRPQ_TV checks first if the RPQ should return paths of length 0 (lines 14–17). This is the case for the * and ? operators. As discussed above, an RPQ that does not have any nested RPQs can be rewritten to a graph pattern that does not have any property path patterns after removing its transitive closure operator. The resulting graph pattern P' (line 18) is used to evaluate paths of length greater than zero (lines 20–29). The while-loop runs until there are no more nodes left to be visited. The term in P' corresponding to the original term of the RPQ's subject position is replaced every time with the node that needs to be visited (line 23). The updated P' is then evaluated by Generic Join (line 24). Nodes that have already been visited are discarded (line 26). The RPQ's mapping is updated for each distinct node and returned as a result (line 29). If the max_depth is not exceeded, the nodes returned by Generic Join are pushed into to_visit. For the ? operator, max_depth is set to one. For the + and * operators, it is set to the largest possible integer value.

We omit the pseudocode for EvalRPQ_VV (VV stands for variable and variable) and EvalRPQ_TT (TT stands for term and term) due to space considerations. As in the SPARQL standard [17], EvalRPQ_VV assigns every node of the input graph to the subject position of the RPQ and calls EvalRPQ_TV. In fact, for the + operator, it restricts the visited nodes using the first IRI of the property path expression. EvalRPQ_TT treats either the subject or the object

Algorithm 2. Evaluation of RPQs with Multi-way Joins

1:   // The pseudocode for EVALRPQ_TT and EVALRPQ_VV is omitted for brevity
2:   // As in the SPARQL standard [17], it relies on EVALRPQ_TV
3:   **function** EVALRPQ(r, G, X)   ▷ r: RPQ, G: RDF Graph X: Mapping
4:       **if** $|var(r)| = 1$ **then**
5:           **yield all** EVALRPQ_TV(r, G, X)
6:       **else if** $|var(r)| = 0$ **then**
7:           **if** EVALRPQ_TT(r, G, X) is **true then yield** X
8:       **else if** $|var(r)| = 2$ **then**
9:           **yield all** EVALRPQ_VV(r, G, X)
10: **function** EVALRPQ_TV(r, G, X)
11:      // For brevity, we only cover the case of the subject being known
12:      to_visit $\leftarrow$ [(subject, 0)]   ▷ List of (term, depth) pairs
13:      visited $\leftarrow$ {}   ▷ Set of terms
14:      **if** paths of length 0 need to be returned **then**   ▷ * or ? operator
15:          X(object_var) $\leftarrow$ subject   ▷ mapping is updated
16:          **yield** X ▷ The mapping is yielded
17:          insert subject into visited   ▷ keep track of visited terms
18:      $P' \leftarrow$ graph pattern corresponding to r without the transitive closure
19:      $X' \leftarrow$ empty solution mapping for P
20:      **while** to_visit is not empty **do**
21:          (term, depth) $\leftarrow$ last item from to_visit
22:          remove the last item from to_visit
23:          replace the value of subject with the value of term in P'
24:          **for all** GENERICJOIN(P', G, X') **do**
25:              object $\leftarrow$ X'(object_var)
26:              **if** visited contains object **then continue**
27:              X(object_var) $\leftarrow$ object ; insert object into visited
28:              **yield** X ▷ Output the updated mapping of the RPQ r
29:              **if** depth+1 < max_depth **then** push (object, depth+1) into to_visit
---

as a variable and then calls EvalRPQ_TV. If EvalRPQ_TV yields the term that was replaced by the variable, it returns true; otherwise it returns false. To avoid clutter, we assume that P' is in *UNION* normal form and that the BGPs comprising P' are evaluated one after another by Generic Join (line 26). To deal with nested RPQs, we use Algorithm 3 (Sect. 4.2) instead of Generic Join.

4.2   Evaluation of C2RPQs

Algorithm 3 presents our approach for the evaluation of C2RPQs. One of the main characteristics of our algorithm is that it does not consider in any part of the evaluation RPQs that have two variables. Such RPQs are considered once their subject or object position is bounded to a particular value in one of the recursive steps of the algorithm. The motivation behind this choice is twofold. First, RPQs having only one variable are evaluated more efficiently; EvalRPQ does not have to iterate over unnecessary nodes of the provided graph. Second,

Algorithm 3. Evaluation of C2RPQs with Multi-way Joins

1: **function** EVALC2RPQ(Q, G, X)   ▷ Q: C2RPQ, G: RDF Graph X: Mapping
2:    **if** there are no RPQs in Q **then**
3:        **yield all** GENERICJOIN(Q, G, X) and **return**

4:    **for all** RPQs $r \in Q, |var(r)| = 0$ **do**   ▷ Evaluate RPQs having no variables
5:        **if** EVALRPQ_TT(r, Q, X) is **false then return**
6:        **else** remove r from Q

7:    **if** Q is empty **then yield** X and **return**   ▷ A solution is found
8:    $?x \leftarrow$ select an unresolved variable from X ▷ Uses RPQs and triple patterns
9:    $q \leftarrow$ PRIORPQ(Q, $?x$)   ▷ Check if an RPQ should guide the set intersection
10:   **if** q is a triple pattern **then**   ▷ Set intersection only between triple patterns
11:       $K \leftarrow \bigcap_{tp \in Q | ?x \in var(tp)} \{\mu(?x) \mid \mu \in \llbracket tp \rrbracket_G\}$ ▷ Set intersection guided by q
12:       **for all** $k \in K$ **do**
13:           $X(?x) \leftarrow k$; $Q' \leftarrow$ assign k to all occurrences of $?x$ in Q
14:           **yield all** EVALC2RPQ(Q', G, X) and **return**

15:   **for all** EVALRPQ(q, G, X) **do**   ▷ q is an RPQ, set intersection guided by q
16:       **if** $X(?x) \notin \{\mu(?x) \mid \mu \in \llbracket tp \rrbracket_G\}$ for any $tp \in Q, ?x \in var(tp)$ **then continue**
17:       $Q' \leftarrow$ assign k to all occurrences of $?x$ in Q and remove q from Q
18:       **yield all** EVALC2RPQ(Q', G, X)

19: **function** PRIORPQ(Q, $?x$)   ▷ See Sect. 4.3, Set Intersection
20:   // Checks if the set intersection should be guided by an RPQ or a triple pattern
21:   // Considers all RPQs r, for which $?x \in var(r)$ and $|var(r)| = 1$

---

we are able to acquire more accurate size estimates for the remaining variable. RPQs having two variables are evaluated only if they do not participate in any join operations with triple patterns or are part of a cross product. We do not cover such cases, as they cannot be optimized.

The core of our algorithm is the function EvalC2RPQ. If the input query is not a C2RPQ, EvalC2RPQ simply calls Generic Join (lines 2–3). EvalC2RPQ prioritizes RPQs that do not have any variables, as they are not subject to any further changes. (line 4–6). If the evaluation of all RPQs having no variables returns true, EvalC2RPQ proceeds with the evaluation of the remaining C2RPQ. As in Generic Join, EvalC2RPQ first selects the variable to be resolved (line 8). Here, C2RPQ considers triple patterns and RPQs having only one variable. The variable selection strategies are detailed in Sect. 4.3. Once a variable is selected, EvalC2RPQ uses PrioRPQ to find out whether an RPQ or a triple pattern should guide the set intersection (line 9). Again, when it comes to RPQs, PrioRPQ considers only RPQs having one variable. If a triple pattern is selected, EvalRPQ behaves as Generic Join. If an RPQ is selected, it is evaluated using EvalRPQ (lines 15–18). For every value $X(?x)$ returned by EvalRPQ, EvalC2RPQ checks if $X(?x)$ is found in the evaluation of all triple patterns tp, with $?x \in var(tp)$ (line 16). Values that are not found in any of the evaluations, are discarded. RPQs having only the selected variable that are not evaluated in this recursive step will be evaluated in the subsequent step (lines 4–6). For example, if there are two RPQs that can be evaluated for the selected variable, at least one of them will be evaluated in the following recursive step.

4.3   Query Planning and Optimizations

Size Estimation of RPQs. Estimating the size of RPQs (i.e., the number of solutions they return) enables their consideration in the planning steps of the algorithm. Recall that in our algorithm, in planning steps, we consider only RPQs that have one variable. To estimate the size of such RPQs, we evaluate them up to a specified depth. This is possible by assigning a particular value to max_depth in EvalRPQ_TV (Algorithm 2). Note that the returned estimation is equal to the number of solutions found until the provided depth. Following [1], we evaluate RPQs up to depth 5 to estimate their size. Henceforth, we will refer to this estimation as *default estimation*. For RPQs having property path expressions that consist of a single term (e.g., $(s, a^+, ?o)$, with $s \notin \mathbf{V}$), we introduce the *shallow estimation*. In such cases, the shallow estimation is equal to the evaluation's size of the triple pattern that results after removing the transitive closure operator (i.e., $(s, a, ?o)$, with $s \notin \mathbf{V}$). The size of such triple patterns are provided in constant time by indices used for multi-way joins [5,8]. As discussed below, if possible, we use the shallow estimation to avoid using the default estimation, which is more computationally expensive. Last, to avoid computing the estimation of an RPQ for a particular subject or object multiple times, we cache the estimated size for each subject and object.

Set Intersection. In Generic Join, the triple pattern that has the smallest cardinality for the selected variable should guide the search for finding the variable's possible values [29]. For C2RPQs, we also need to consider the RPQs that have the selected variable. In EvalC2RPQ (line 9, Algorithm 3), we first find the minimum cardinality of the selected variable among the triple patterns (see Variable Ordering). Then, for each RPQ, we first calculate its shallow estimation, if possible. If the shallow estimation is greater than the minimum cardinality, we do not calculate the default estimation, as the RPQ will not guide the set intersection. If the shallow estimation is lower than the minimum cardinality, we calculate the default estimation to get a more accurate estimation and update the minimum cardinality accordingly. In the end, the RPQ or the triple pattern having the minimum cardinality for the selected variable guides the set intersection.

Variable Ordering. The order in which variables are resolved is imperative for the efficiency of multi-way joins [19]. The order can be static [5,19] or dynamic [8]. In the first case, the variable ordering of a query is determined before the query's evaluation. In the second case, the variable to be resolved is selected at each recursive step. As described above, the proposed algorithm considers only RPQs with one variable at planning steps. With a static variable ordering, the algorithm would have to completely disregard RPQs that have two variables at the beginning of a query's evaluation. For this reason, our algorithm uses a dynamic variable ordering. During the evaluation of C2RPQs, as variables are recursively resolved, RPQs that start with two variables end up at some point having only one variable and hence, can be considered for the variable ordering.

For selecting a variable at each recursive step, we experiment with two strategies. The first strategy was proposed in [8] and at each recursive step, it selects the variable that has the largest guaranteed *reduction* of the search space spanned by Cartesian products of the triple patterns' solutions. We refer to this strategy as *reduction factor*. The second strategy is based on the one proposed in [5] and at each step, it selects the variable that has the *minimum cardinality* (i.e., the variable that is estimated to have the smallest set of possible values). For both strategies, the size of RPQs is estimated as described in the set intersection.

Materialization of RPQs. By using multi-way joins, our approach is able to evaluate C2RPQs without having to materialize the results of RPQs. However, there are cases where the materialization of RPQs improves the performance of the proposed algorithm. Consider the SPARQL query `SELECT * WHERE { ?x <p1> ?z . ?z <p2> ?y . ?y <p3>+ <o> }`. If $?y$ is not the first variable to be evaluated by EvalC2RPQ, the RPQ might end up being evaluated multiple times in intermediate recursive steps (line 15, Algorithm 3), while always yielding the same results. To avoid unnecessary computations, we materialize the results of such RPQs, i.e., of RPQs that have one variable *before the start of the query evaluation* and are evaluated in intermediate recursive steps of EvalC2RPQ. Note that in [13], all RPQs are materialized.

5   Experimental Results

In this section, we present the performance evaluation of the proposed algorithm, which we have implemented in the tensor-based triple store Tentris [8,9]. We refer to our implementation as TentrisRPQ. For the evaluation of TentrisRPQ, we used the the recently introduced benchmark WDBench [2]. WDBench consists of a real-world dataset based on Wikidata and queries that are extracted from Wikidata's query logs. The experiments presented below were carried out on a Debian 10 server with an AMD EPYC 7282 CPU, 256GB RAM and a 2TB Samsung 970 EVO Plus SSD. Supplementary material—including datasets, binaries, queries, scripts, and configurations—is available online.[1]

5.1   Systems and Experimental Setup

We compared the performance of TentrisRPQ against the performance of the following well-established triple stores: (i) Blazegraph 2.1.6.RC, (ii) Fuseki 4.9.0, (iii) GraphDB 10.3.3 (free version), and (iv) Virtuoso 7.2.10. In our experiments, we also included the graph database MilleniumDB[2] [28]. In MilleniumDB, BGPs

---

[1] https://github.com/dice-group/c2rpqs-benchmark.

[2] https://github.com/MillenniumDB/MillenniumDB, commit: 442e650.

are evaluated by a worst-case optimal multi-way join algorithm and RPQs are evaluated following an automaton-based approach. When it comes to C2RPQs, in MilleniumDB, "paths are pushed to the end of join plans" [28] and hence are not considered in multi-way join plans. We also carried out an evaluation of different execution strategies using multiple versions of TentrisRPQ.

Our experiments were carried over HTTP using the benchmark execution framework IGUANA 3.3.3 [12]. For the set of queries that we used in this work, we created a stress test that was executed by every system four consecutive times, with the first execution serving as a warm-up run. The query timeout was set to three minutes. The timeout was set in the systems' respective configurations. As in previous works [9], we measured the performance of the systems using the following metrics: (i) QPS, i.e., the number of queries executed per second, (ii) pAvgQPS, i.e., the penalized average QPS and (iii) QMPH, i.e., the number of query mixes executed per hour. Queries that failed (e.g., timed out or returned an error code) are penalized with a runtime of three minutes.

5.2 Datasets and Queries

The dataset of WDBench is an extract of Wikidata containing 1.26B triples (92.4M distinct subjects, 8.6K distinct predicates, and 305M distinct objects). In this work, we focus on the set of C2RPQs provided by WDBench[3]. As in [1], we did not consider queries with cross products and queries not using any of the +, *, or ? operators. To ensure the fair comparison of the benchmarked systems, from the remaining queries, we only kept those queries, for which the systems that were able to evaluate them before the specified timeout returned the same number of solutions and bindings. For example, we had to discard the queries returning more than 2^{20} results, which is Virtuoso's hard limit [2,6,8]. To alleviate this issue, some works in the literature (e.g., [2,28]) restrict the number of solutions using SPARQL's LIMIT. However, the use of LIMIT without ordering (ORDER BY in SPARQL) does not guarantee that the benchmarked systems return the same results. In addition, systems supporting multi-way joins— including ours—have an inherent advantage when LIMIT is used without ordering, as they do not have to compute the full result set; they can terminate once they have output the requested number of solutions. For these reasons, instead of using LIMIT, we opted for discarding queries as described above. Ultimately, we used 305 queries in our experiments, which, as in previous works (e.g., [1,28]), were evaluated under set semantics (i.e., SPARQL's DISTINCT is used). Note that no queries were discarded because of TentrisRPQ, as it returns the same amount of solutions with at least one other triple store in all queries.

5.3 Evaluation of Execution Strategies

To get insights on the impact that our proposed query planning and optimizations have on the evaluation of C2RPQs, we compared different execution strategies using different versions of TentrisRPQ, which are presented below.

---

[3] https://github.com/MillenniumDB/WDBench/blob/master/Queries/c2rpqs.txt.

Table 1. The comparison of the different execution strategies. The column failed reports the number of queries for which the corresponding system failed (e.g., timed out) at least once.

	Warm Runs			Cold Run		
	QMPH	pAvgQPS	failed	QMPH	pAvgQPS	failed
TentrisRPQ^{FMC}	10.167	571.654	0	6.433	241.101	0
TentrisRPQ^{FRF}	1.304	587.925	5	1.195	222.749	5
TentrisRPQ^{RPC}	3.127	576.814	3	2.701	212.881	3
TentrisRPQ^{RP}	1.323	586.958	7	1.257	217.213	6
TentrisRPQ^{JP}	0.311	78.298	36	0.301	32.791	38

TentrisRPQ^{JP} (Join Prioritization). This version always prioritizes join operations, which leads to RPQs being evaluated only after their variables have been resolved. RPQs are not considered in the variable selection process.

TentrisRPQ^{RP} (RPQ Prioritization). This version always prioritizes RPQs with one variable. At each recursive step, an RPQ (alongside its remaining variable) is chosen to be evaluated using the shallow estimation. Note that the set intersection of a variable is always guided by an RPQ. If the shallow estimation is not applicable to any of the RPQs, we select the first available RPQ.

TentrisRPQ^{RPC} (RPQ Prioritization and Materialization). This version follows the same evaluation process as the one presented above and, in addition, materializes RPQs as discussed in Sect. 4.3.

TentrisRPQ^{FRF} (Full Version with Reduction Factor). This version fully implements the approaches presented in Sect. 4. For selecting variables, it follows the strategy based on the reduction factor. In fact, all of the previously presented versions of TentrisRPQ use the reduction factor strategy (Sect. 4.3) for selecting variables, which is the strategy supported by Tentris.

TentrisRPQ^{FMC} (Full Version with Minimum Cardinality). This version also fully implements the proposed approaches, but for selecting variables, it uses the minimum cardinality strategy (Sect. 4.3). The minimum cardinality strategy is not supported by Tentris; it is only part of TentrisRPQ.

The performance of each execution strategy is presented in Table 1 and Fig. 1a. TentrisRPQ^{JP} achieves the lowest QMPH and pAvgQPS. The remaining versions of TentrisRPQ achieve similar pAvgQPS values, with TentrisRPQ^{FMC} achieving the best median penalized QPS followed by TentrisRPQ^{FRF}. Regarding the QMPH, TentrisRPQ^{FMC} achieves the best performance, which is 3.2 higher than the second best reported performance (TentrisRPQ^{RPC}).

These results show the importance of considering RPQs in the planning steps of the proposed algorithm. TentrisRPQ^{FMC}—which is the full version of Tentris-RPQ and uses the minimum cardinality estimation—does not time out in any of the queries and achieves the highest QMPH value, outperforming all of the three

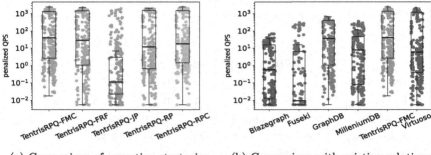

(a) Comparison of execution strategies   (b) Comparison with existing solutions

Fig. 1. Performance of the benchmarked systems, including the different execution strategies of TentrisRPQ, in terms of penalized QPS (warm runs).

versions of TentrisRPQ that prioritize a particular execution strategy. This is not the case for the second full version of TentrisRPQ, namely TentrisRPQ^{FRF}, that uses the reduction factor estimation for selecting variables. The difference between the QMPH values reported by TentrisRPQ^{FMC} and TentrisRPQ^{FRF} highlights once again the importance of selecting a good variable ordering. The reduction factor strategy leads TentrisRPQ^{FRF} to time out in some queries, which has a negative impact on its QMPH value. However, after closely examining the results, we found queries for which TentrisRPQ^{FRF} reports a more than ten times higher QPS. In these queries (e.g., the queries 53 and 186 in our supplementary material), the minimum cardinality prioritizes variables that do not participate in join operations and have values with large multiplicities. As a result, the set of possible values of the join variables that are found in triple patterns with the prioritized variables are not restricted enough. This leads to the conclusion that, as part of our future work, we need to devise a variable selection strategy that combines both of the strategies used in this work. Last, the results of TentrisRPQ^{RP} and TentrisRPQ^{RPC} show that the proposed materialization of RPQs improves the performance of our algorithm.

5.4 Comparison with Existing Solutions

To compare TentrisRPQ with existing storage solutions, we use TentrisRPQ^{FMC}. The comparison's results are shown in Table 2 and Fig. 1b. TentrisRPQ^{FMC} achieves the highest QMPH value, which is 4.3 times higher than the second best value (GraphDB). TentrisRPQ^{FMC} also achieves the highest pAvgQPS and median QPS. Virtuoso and GraphDB achieve the second best values for pAvgQPS and median QPS, respectively.

TentrisRPQ^{FMC} outperforms all of the competing systems in 213 out of 305 queries. More specifically, it achieves at least two and five times higher penalized QPS than every other system in 81 and 27 queries, respectively. To find the shortcomings of our approach, we focused on the queries for which

Table 2. The comparison of different systems. The column failed reports the number of queries for which the corresponding system failed (e.g., timed out) at least once.

	Warm Runs			Cold Run		
	QMPH	pAvgQPS	failed	QMPH	pAvgQPS	failed
Blazegraph	0.257	8.315	49	0.256	7.733	48
Fuseki	0.129	19.672	145	0.128	12.932	145
GraphDB	2.322	184.851	1	2.110	80.443	1
MilleniumDB	0.754	48.034	9	0.728	39.395	10
TentrisRPQ^{FMC}	10.167	571.654	0	6.433	241.101	0
Virtuoso	1.104	450.330	10	1.098	124.812	10

TentrisRPQ^{FMC} achieves at least 2 times lower penalized QPS than any other system; this is the case for 45 queries. In these queries, TentrisRPQ^{FMC} is mostly outperformed by the competing systems due to a bad variable ordering or due to the computational overhead introduced by the default size estimation of RPQs. This is justified by the fact that, in the majority of these queries, another version of TentrisRPQ achieves the overall best performance. For example, in most of the queries where MilleniumDB outperforms TentrisRPQ^{FMC}, TentrisRPQ^{JP} performs better than or similar to MilleniumDB. Note that TentrisRPQ^{JP} and MilleniumDB follow similar execution strategies.

6   Conclusion and Future Work

We presented an approach for the efficient evaluation of C2RPQs using multi-way joins. By evaluating RPQs found in C2RPQs using multi-way joins, the proposed algorithm is able to exploit RPQs in the planning steps of multi-way joins, which, as demonstrated by our detailed evaluation of different execution strategies, is crucial for its performance. The experimental results of our evaluation using WDBench [2] show that our approach outperforms the state of the art.

As part of our future work, we plan to improve the variable selection process of the proposed algorithm by combining the strategies based on the reduction factor and minimum cardinality. We also plan to reduce the computational overhead of our default estimation by terminating the estimation once the active minimum cardinality has been exceeded. We have already used our approach for the evaluation of negated property sets and plan to further study their efficient evaluation. However, we have observed a lack of benchmark queries using negated property sets, which makes their optimization challenging.

Acknowledgments. This work has received funding from the European Union's Horizon Europe research and innovation programme under grant agreement No 101070305 and the Ministry of Culture and Science of North Rhine-Westphalia (MKW NRW) within the project SAIL under the grant no NW21-059D.

References

1. Aimonier-Davat, J., Skaf-Molli, H., Molli, P., Dang, M.H., Nédelec, B.: Join ordering of SPARQL property path queries. In: Pesquita, C., et al. (eds.) ESWC 2023. LNCS, vol. 13870, pp. 38–54. Springer, Cham (2023). https://doi.org/10.1007/978-3-031-33455-9_3

2. Angles, R., Aranda, C.B., Hogan, A., Rojas, C., Vrgoc, D.: WDBENCH: a Wikidata graph query benchmark. In: Sattler, U., et al. (eds.) ISWC 2022. LNCS, vol. 13489, pp. 714–731. Springer, Cham (2022). https://doi.org/10.1007/978-3-031-19433-7_41

3. Angles, R., Arenas, M., Barceló, P., Hogan, A., Reutter, J.L., Vrgoc, D.: Foundations of modern query languages for graph databases. ACM Comput. Surv. **50**(5), 68:1–68:40 (2017). https://doi.org/10.1145/3104031

4. Arroyuelo, D., Gómez-Brandón, A., Navarro, G.: Evaluating regular path queries on compressed adjacency matrices. In: Nardini, F.M., Pisanti, N., Venturini, R. (eds.) SPIRE 2023. LNCS, vol. 14240, pp. 35–48. Springer, Cham (2023). https://doi.org/10.1007/978-3-031-43980-3_4

5. Arroyuelo, D., Hogan, A., Navarro, G., Reutter, J.L., Rojas-Ledesma, J., Soto, A.: Worst-case optimal graph joins in almost no space. In: Li, G., Li, Z., Idreos, S., Srivastava, D. (eds.) SIGMOD 2021: International Conference on Management of Data, Virtual Event, China, 20–25 June 2021, pp. 102–114. ACM (2021). https://doi.org/10.1145/3448016.3457256

6. Arroyuelo, D., Hogan, A., Navarro, G., Rojas-Ledesma, J.: Time- and space-efficient regular path queries. In: 38th IEEE International Conference on Data Engineering, ICDE 2022, Kuala Lumpur, Malaysia, 9–12 May 2022, pp. 3091–3105. IEEE (2022). https://doi.org/10.1109/ICDE53745.2022.00277

7. Atserias, A., Grohe, M., Marx, D.: Size bounds and query plans for relational joins. In: 49th Annual IEEE Symposium on Foundations of Computer Science, FOCS 2008, 25–28 October 2008, Philadelphia, PA, USA, pp. 739–748. IEEE Computer Society (2008). https://doi.org/10.1109/FOCS.2008.43

8. Bigerl, A., Conrads, F., Behning, C., Sherif, M.A., Saleem, M., Ngonga Ngomo, A.-C.: Tentris – a tensor-based triple store. In: Pan, J.Z., et al. (eds.) ISWC 2020, Part I. LNCS, vol. 12506, pp. 56–73. Springer, Cham (2020). https://doi.org/10.1007/978-3-030-62419-4_4

9. Bigerl, A., Conrads, L., Behning, C., Saleem, M., Ngomo, A.N.: Hashing the hypertrie: space- and time-efficient indexing for SPARQL in tensors. In: Sattler, U., et al. (eds.) ISWC 2022. LNCS, vol. 13489, pp. 57–73. Springer, Cham (2022). https://doi.org/10.1007/978-3-031-19433-7_4

10. Bonifati, A., Fletcher, G.H.L., Voigt, H., Yakovets, N.: Querying Graphs. Synthesis Lectures on Data Management. Morgan & Claypool Publishers (2018). https://doi.org/10.2200/S00873ED1V01Y201808DTM051

11. Bonifati, A., Martens, W., Timm, T.: Navigating the maze of wikidata query logs. In: Liu, L., et al. (eds.) The World Wide Web Conference, WWW 2019, San Francisco, CA, USA, 13–17 May 2019, pp. 127–138. ACM (2019). https://doi.org/10.1145/3308558.3313472

12. Conrads, F., Lehmann, J., Saleem, M., Ngomo, A.N.: Benchmarking RDF storage solutions with IGUANA. In: Nikitina, N., Song, D., Fokoue, A., Haase, P. (eds.) Proceedings of the ISWC 2017 Posters & Demonstrations and Industry Tracks co-located with 16th International Semantic Web Conference (ISWC 2017), Vienna, Austria, 23–25 October 2017. CEUR Workshop Proceedings, vol. 1963. CEUR-WS.org (2017). https://ceur-ws.org/Vol-1963/paper621.pdf

13. Cucumides, T., Reutter, J.L., Vrgoc, D.: Size bounds and algorithms for conjunctive regular path queries. In: Geerts, F., Vandevoort, B. (eds.) 26th International Conference on Database Theory, ICDT 2023, 28–31 March 2023, Ioannina, Greece. LIPIcs, vol. 255, pp. 13:1–13:17. Schloss Dagstuhl - Leibniz-Zentrum für Informatik (2023). https://doi.org/10.4230/LIPIcs.ICDT.2023.13

14. Demir, C., Ngomo, A.N.: Neuro-symbolic class expression learning. In: Proceedings of the Thirty-Second International Joint Conference on Artificial Intelligence, IJCAI 2023, 19–25 August 2023, Macao, SAR, China, pp. 3624–3632. ijcai.org (2023). https://doi.org/10.24963/ijcai.2023/403

15. Freitag, M.J., Bandle, M., Schmidt, T., Kemper, A., Neumann, T.: Adopting worst-case optimal joins in relational database systems. Proc. VLDB Endow. **13**(11), 1891–1904 (2020). http://www.vldb.org/pvldb/vol13/p1891-freitag.pdf

16. Gubichev, A., Bedathur, S.J., Seufert, S.: Sparqling kleene: fast property paths in RDF-3X. In: Boncz, P.A., Neumann, T. (eds.) First International Workshop on Graph Data Management Experiences and Systems, GRADES 2013, co-located with SIGMOD/PODS 2013, New York, NY, USA, 24 June 2013, p. 14. CWI/ACM (2013). http://event.cwi.nl/grades2013/14-gubichev.pdf

17. Harris, S., Seaborne, A.: SPARQL 1.1 query language (2013). https://www.w3.org/TR/2013/REC-sparql11-query-20130321/. Accessed 21 Nov 2023

18. Hogan, A., et al.: Knowledge graphs. ACM Comput. Surv. 71:1–71:37 (2021)

19. Hogan, A., Riveros, C., Rojas, C., Soto, A.: A worst-case optimal join algorithm for SPARQL. In: Ghidini, C., et al. (eds.) ISWC 2019, Part I. LNCS, vol. 11778, pp. 258–275. Springer, Cham (2019). https://doi.org/10.1007/978-3-030-30793-6_15

20. Kostylev, E.V., Reutter, J.L., Romero, M., Vrgoč, D.: SPARQL with property paths. In: Arenas, M., et al. (eds.) ISWC 2015, Part I. LNCS, vol. 9366, pp. 3–18. Springer, Cham (2015). https://doi.org/10.1007/978-3-319-25007-6_1

21. Malyshev, S., Krötzsch, M., González, L., Gonsior, J., Bielefeldt, A.: Getting the most out of Wikidata: semantic technology usage in Wikipedia's knowledge graph. In: randečić, D., et al. (eds.) ISWC 2018, Part II. LNCS, vol. 11137, pp. 376–394. Springer, Cham (2018). https://doi.org/10.1007/978-3-030-00668-6_23

22. Ngo, H.Q., Porat, E., Ré, C., Rudra, A.: Worst-case optimal join algorithms. J. ACM **65**(3), 16:1–16:40 (2018). https://doi.org/10.1145/3180143

23. Ngo, H.Q., Ré, C., Rudra, A.: Skew strikes back: new developments in the theory of join algorithms. SIGMOD Rec. **42**(4), 5–16 (2013), https://doi.org/10.1145/2590989.2590991

24. Pérez, J., Arenas, M., Gutierrez, C.: Semantics and complexity of SPARQL. ACM Trans. Database Syst. **34**(3), 16:1–16:45 (2009). https://doi.org/10.1145/1567274.1567278

25. Salas, J., Hogan, A.: Semantics and canonicalisation of SPARQL. Semant. Web **13**(5), 829–893 (2022). https://doi.org/10.3233/SW-212871

26. Schmidt, M., Meier, M., Lausen, G.: Foundations of SPARQL query optimization. In: Segoufin, L. (ed.) Proceedings of the Database Theory - ICDT 2010, 13th International Conference, Lausanne, Switzerland, 23–25 March 2010. ACM International Conference Proceeding Series, pp. 4–33. ACM (2010), https://doi.org/10.1145/1804669.1804675

27. Syed, Z.H., Röder, M., Ngomo, A.-C.N.: Unsupervised discovery of corroborative paths for fact validation. In: Ghidini, C., et al. (eds.) ISWC 2019, Part I. LNCS, vol. 11778, pp. 630–646. Springer, Cham (2019). https://doi.org/10.1007/978-3-030-30793-6_36

28. Vrgoč, D., et al.: MillenniumDB: an open-source graph database system. Data Intell. 1–39 (2023). https://doi.org/10.1162/dint_a_00209

29. Wang, Y.R., Willsey, M., Suciu, D.: Free join: unifying worst-case optimal and traditional joins. Proc. ACM Manag. Data 1(2), 150:1–150:23 (2023). https://doi.org/10.1145/3589295
30. Yakovets, N., Godfrey, P., Gryz, J.: Query planning for evaluating SPARQL property paths. In: Özcan, F., Koutrika, G., Madden, S. (eds.) Proceedings of the 2016 International Conference on Management of Data, SIGMOD Conference 2016, San Francisco, CA, USA, 26 June–01 July 2016, pp. 1875–1889. ACM (2016). https://doi.org/10.1145/2882903.2882944

Can Contrastive Learning Refine Embeddings

Lihui Liu[1] , Jinha Kim[2(✉)] , and Vidit Bansal[2]

[1] University of Illinois at Urbana-Champaign, Champaign, IL 61820, USA
lihui2@illinois.edu
[2] Amazon, Seattle, WA 98109, USA
{jinhak,bansalv}@amazon.com

Abstract. Recent advancements in contrastive learning have revolutionized self-supervised representation learning and achieved state-of-the-art performance on benchmark tasks. While most existing methods focus on applying contrastive learning on input data modalities like images, natural language sentences, or networks, they overlook the potential of utilizing output from previously trained encoders. In this paper, we introduce SIMSKIP, a novel contrastive learning framework that specifically refines the input embeddings for downstream tasks. Unlike traditional unsupervised learning approaches, SIMSKIP takes advantage of the output embedding of encoder models as its input. Through theoretical analysis, we provide evidence that applying SIMSKIP does not lead to larger upper bounds on downstream task errors than that of the original embedding which is SIMSKIP's input. Experiment results on various open datasets demonstrate that the embedding by SIMSKIP improves the performance on downstream tasks.

1 Introduction

Embedding symbolic data such as text, graphs, and multi-relational data has become a key approach in machine learning and AI [24]. The learned embeddings can be utilized in various applications. For instance, in NLP, word embeddings generated by WORD2VEC [23] or BERT [4] have been employed in tasks like question answering and machine translation. In the field of graph learning, embeddings of graphs like NODE2VEC [7] and DEEPWALK [28] have been used for node classification and link prediction in social networks. Similarly, in computer vision, image embeddings such as ResNet [9] can be used for image classification.

Despite the progress in representation learning, learning effective embeddings remains a challenging problem. Deep learning models often require a large amount of labeled training data, which can be costly and limit their applicability. Additionally, the learned embeddings often perform well on one task but not on others.

Contrastive learning has the advantage of being able to learn representations without label information, thus saving a significant amount of human effort and resources that would have been used for data labeling. The fundamental idea of contrastive learning is to bring together an anchor and a "positive" sample in the embedding space while pushing apart the anchor from many "negative" samples [3]. As there are no labels available,

---

L. Liu—Work conducted while the author was an intern at Amazon.

a positive pair often consists of data augmentations of the sample, and negative pairs are formed by the anchor and randomly chosen samples from the minibatch [3]. Although the concept is simple, recent research has shown that contrastive learning methods can achieve comparable results to supervised methods [13,29].

Given the success of contrastive learning, a logical question to ask is whether using the output of another embedding model as input to contrastive learning can further refine the embedding space and make it perform better for downstream tasks. This is the question we aim to answer in this paper. We propose a new approach, called SIMSKIP, that takes the output embedding of another model as input and applies contrastive learning on it. Our proposed method aims to fine-tune the input embedding space, making it more robust for downstream tasks. We theoretically prove that after applying SIMSKIP on the input embedding space, for a downstream task, the error upper bound of the new learned fine-tuned embedding will not be larger than that of the original embedding space. We conduct extensive experiments on various datasets and downstream tasks to evaluate the performance of our proposed approach and compare it with other state-of-the-art methods. The results show that the proposed SIMSKIP can refine the input embedding space and achieve better performance on downstream tasks.

In summary, the main contributions of this paper are:

- **Problem Definition.** To the best of our knowledge, we are the first to propose and investigate the use of contrastive learning to improve the robustness of embedding spaces.
- **Algorithm** We propose a skip-connection-based contrastive learning model, SIM-SKIP, and theoretically prove that it can reduce the error upper bound of downstream tasks.
- **Empirical Evaluations**. We conduct extensive experiments on several real-world datasets and various downstream tasks. The results of our experiments demonstrate the effectiveness of SIMSKIP.

2   Preliminaries and Problem Definition

2.1   Contrastive Learning

Contrastive learning aims to learn effective representations by pulling semantically similar samples together and pushing dissimilar samples apart [6]. In self-supervised setting as such contrastive learning, constructing positive and negative pairs from unlabeld dataset through data augmentation is critical. For example, in visual representations, an effective approach is to generate two augmented images from one input image and use them as the positive pair, while other images in the same mini-batch are treated as negative pairs of the input image. There are several different data augmentation methods such as cropping, flipping, distortion, and rotation [3]. In node representations in graphs, one idea is to use the neighborhood of the given node as positive pairs, while nodes that are farther away are treated as negative pairs. For graph-level representations, operations such as node deletion and edge deletion can be used to generate positive augmentations of the input graph.

After building positive and negative pairs, neural network-based encoders are used to learn representation vectors from augmented data examples. Various network architectures such as ResNet [3] for images and BERT [6] for text can be used. The output representation vectors of the encoders are used as the final embedding of the input data. To learn an effective embedding space discriminating positive and negative pairs, a simpler neural network called projector is stacked on top of a encoder and the contrastive loss is applied against the projector output. A commonly used projector is an MLP with one or two layers, which is simple to implement.

In training, first, a random sample of N examples is taken for a mini-batch. Then, N pairs are constructed from N samples through data augmentation, which lead $2N$ examples total in the mini-batch. N augmented pair of an input data point are treated as the positive pair in the mini-batch. For each augmented positive pair (i, j), the remaining $2N - 2$ example are used to construct negative examples (i, k). The commonly used contrastive learning loss is

$$l_{i,j} = -\log \frac{\exp(\text{sim}(z_i, z_j)/\tau)}{\sum_{k=1}^{2N} \mathbf{I}_{k \neq i,j}\exp(\text{sim}(z_i, z_k)/\tau)} \tag{1}$$

where τ is the temperature, sim is a similarity function such as the cosine similarity, $z_i(= p(f(x_i)))$ is the output of the projector p which takes the output of the encoder f [3].

2.2 Problem Statement

In this paper, we focus on investigating whether contrastive learning can refine the embeddings for downstream tasks. Given an input dataset $D = \{d_i\}_1^N$, an arbitrary embedding function $h()$, and its output embedding $\mathcal{X}$, where $x_i \in \mathcal{X}$ is the embedding of data point d_i ($x_i = h(d_i)$), our goal is to design a new embedding function f such that $f(h(d_i))$ performs no worse than $h(d_i)$ given an arbitrary downstream task T.

3   Method

In the previous sections, we outlined the concept of unsupervised contrastive learning. In this section, we will delve into the specifics of using SIMSKIP that refines pre-existing embeddings.

3.1   Contrastive Learning Limitation

The architecture of contrastive learning ensures that augmentations of the same data point are close to each other in the embedding space. However, this alone does not guarantee that the learned embeddings are suitable for downstream tasks. As shown in Fig. 1, assuming we have eight input embedding points that belong to two different classes, red and blue. When adding Gaussian noise to the original embedding points to create their augmentations, the augmented positive points are represented by the circles around the points on the left side of Fig. 1. When there are two different contrastive learning encoders, f_1 and f_2, they will map all augmentations of the same data point to the embedding points close to that of the original data point in the contrastive embedding space. If the two augmentations are denoted as x_{i1} and x_{i2}, $\text{sim}(f_1(x_{i1}), f_1(x_{i2}))$ will be close to 1 (the same applies to f_2). Since all the other augmentation examples are treated as negative examples, it is clear that the contrastive loss of $f_1(x)$ will be very similar as that of $f_2(x)$ when they map the original data points as shown in the right side of Fig. 1.

Even though the contrastive loss of f_1 and f_2 are very similar, the performance of the downstream classification task may differ between the two embedding spaces. For example, f_1 separates the red and blue points into distinct clusters, which makes it easy for the downstream classification task to accurately classify them. However, in the embedding space created by f_2, the red and blue points are mixed together, which results in poor performance for the downstream classification task.

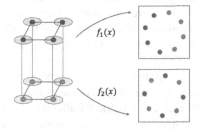

Fig. 1. The problem of existing unsupervised contrastive learning

3.2 SIMSKIP Details

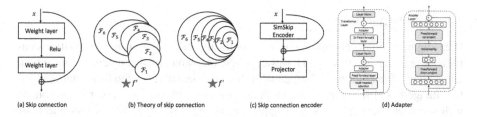

Fig. 2. The Skip Connection. The picture of Adapter is from [10].

To address this problem, we introduce skip connection contrastive learning. This method is similar in principle to ResNet [9] as illustrated in Fig. 2. The idea behind skip connections is that it retains the expressiveness of the original network. A specific network architecture defines a class of functions F that it can represent. Suppose f' is the optimal function we aim to find. If it is within $\mathcal{F}$, we are in good shape. However, it is often the case that it is not. Therefore, our goal is to find the best approximation of f' within $\mathcal{F}$.

A naive way to achieve this is to increase the depth and width of the neural network. By adding more layers and neurons, the network can represent a new class of functions $\mathcal{F}'$, which is more expressive than $\mathcal{F}$. In general, we expect that $f_{\mathcal{F}'}$ would be better than $f_{\mathcal{F}}$, as a more expressive function class should be able to capture more complex patterns in the data. However, this may not be the case. In fact, increasing the depth and width of the network can lead to a worse $f_{\mathcal{F}'}$, as illustrated by Fig. 2(b). In this example, even though $\mathcal{F}_6$ is larger than $\mathcal{F}_3$, its optimal approximation is farther from the optimal function f'.

To solve this problem, Kaiming et al. in [9] proposes to use skip-connection to avoid the aforementioned issue from the non-nested function classes. The idea of skip-connection is that it can create nested function classes where $\mathcal{F}_1 \subseteq ... \subseteq \mathcal{F}_6$ as shown on the right of Fig. 2(b). Because the larger function classes contain the smaller ones, it can guarantee that increasing them strictly increases the expressive power of the network. For deep neural networks, if we can train the newly-added layer into an identity function $f(x) = x$, the new model will be as effective as the original model. As the new model may get a better solution to fit the training dataset, the added layer might make it easier to reduce training errors.

Building on the idea of incorporating skip connections, we propose a model named SIMSKIP that utilizes contrastive learning to refine embedding based on the original input embedding. The architecture of SIMSKIP is illustrated in Fig. 2(c). SIMSKIP consists of two components: a skip connection based encoder and a projector. The detail of the encoder can be found in Fig. 3.

The projector is a Multi-layer Perceptron (MLP) with one hidden layer, represented as $W_2\sigma(W_1x)$, where σ is a ReLU non-linearity. By incorporating skip connections, the expressive power of the network (contrastive learning encoder) is increased. Therefore, the new learned embedding should perform at least as well as the original embedding in downstream tasks.

3.3 Data Augmentation

Data augmentation is commonly used in contrastive learning to generate positive samples for a given data point. However, when the input to the model is the output embedding of another model, traditional data augmentation methods are not applicable. Image-based techniques like cropping, resizing, cut-out, and color distortion, as well as Sobel filtering, can only be applied to images [3]. Other methods such as node deletion and edge deletion for graphs are also not suitable for this purpose. Designing an effective data augmentation strategy is critical for contrastive learning methods.

Inspired by [30], in this paper, we use two types of data augmentation to embedding output of an encoder network – masking and Gaussian noise.

A - Random Masking. Random masking is applied to the input embedding. Specifically, given an input embedding $\mathbf{e}_i \in R^d$, a random vector $M \in \{0, 1\}^d$ is created where 0 indicates that the element will be masked and 1 indicates no change. The number of 0 s in M is drawn from a Bernoulli distribution $B(1 - \alpha)$, where α is a hyperparameter. The output after applying random masking is $\mathbf{e}_i \circ M$, where $\circ$ represents element-wise multiplication.

B - Gaussian Noise. When adding Gaussian noise to the input embedding $\mathbf{e}_i \in R^d$, a random vector $\epsilon \sim N(0, \mathbf{I})$ is first sampled from a multi-variable Gaussian distribution, where $\epsilon \in R^d$ and each element in ϵ is drawn from a Gaussian distribution with zero mean and unit variance. The output after adding the Gaussian noise is $\mathbf{e}_i + \delta \circ \epsilon$, where δ is a hyper-parameter.

3.4 Theoretical Proof

In this section, we theoretically demonstrate why SIMSKIP may refine it's input embedding. Here, 'refine' means that the embedding which SIMSKIP produces has no worse downstream performance than that of the original embedding which is SIMSKIP's input. We initially establish the upper bound for the loss in any downstream task within the context of contrastive unsupervised learning, as demonstrated in [31]. Then, we prove that using a skip connection-based network as the contrastive learning encoder can achieve a smaller or equal loss upper bound for downstream classification tasks compared to using original input embedding directly.

A - Preliminary. Let $\mathcal{X}$ denote the set of all possible data points. Let $f_1(x)$ represent an arbitrary neural network that takes x as its input. Then, a neural network with skip connection f_2 can be denoted as

$$f_2(x) = f_1(x) + x = f_1(x) + f_I \tag{2}$$

where $f_I = x$.

B - Downstream Task Loss for Contrastive Unsupervised Learning. In this section, we present an upper bound for the loss of a supervised downstream task which uses representation learned by any contrastive learning, as originally shown in [31].

In unsupervised learning, given a contrastive encoder f, the primary objective is to make ensure that the embeddings of the positive pair (x^+, x), generated by the function f, are close to each other, while the embeddings of the negative pair (x^-, x), generated by the same function, are far away from each other. Contrastive learning assumes access to similar data in the form of pairs (x, x^+) that come from a distribution D_{sim} as well as k i.i.d. negative samples $x_1^-, x_2^-, ..., x_k^-$ from a distribution D_{neg} that are presumably unrelated to x. Learning is done over $\mathcal{F}$, a class of representation functions $f : \mathcal{X} \longrightarrow R^d$ where f is the embedding function. The quality of the representation function f (contrastive encoder) is evaluated by its performance on a multi-class classification task $T \in \mathcal{T}$ using linear classification. A multi-class classifier for $T \in \mathcal{T}$ is a function g whose output coordinates are indexed by the classes c in task $T \in \mathcal{T}$. For example, in SimCLR [3], the encoder is denoted as f and a linear classifier is used as the projector. So the whole framework of SimCLR can be expressed as $g(x) = wf(x)$. The loss considered in [31] is the logistic loss $l(v) = \log_2(1 + \sum_i \exp(-v_i))$ for $v \in R^d$. Then the supervised loss of the downstream task classifier g is

$$L_{sup}(T, g) = E[l(\{g(x)_c - g(x)_{c'}\}_{c \neq c'})] \tag{3}$$

where c and c' are different classes. For simplicity, we use $L_{sup}(f)$ to denote the downstream loss of the model with function f which satisfies $L_{sup}(T, g)$ and $g(x) = wf(x)$.

We outline the objective of contrastive learning: k denotes number of negative samples used for training. The unsupervised loss can be defined as

$$L_{un}(f) = E[l(\{f(x)^T(f(x^+) - f(x^-)\}_{i=1}^k)] \tag{4}$$

After training, suppose $\hat{f}$ is the function which can minimizes the empirical unsupervised loss and we denote its corresponding loss for supervised downstream task as $L_{sup}(\hat{f})$. According to the theorem 4.1 in [31], we have

$$L_{sup}(\hat{f}) <= \alpha L_{un}(f) + \eta Gen_M + \epsilon \tag{5}$$

where Gen is the generalization error which is defined as

$$Gen_M = O(R\sqrt{k}\frac{R_s(\mathcal{F})}{M} + (R^2 + logk)\sqrt{\frac{log\frac{1}{\epsilon}}{M}}) \tag{6}$$

and M is the sample size, $R_s(\mathcal{F})$ is the Rademacher average of $\mathcal{F}$ [31], $\mathcal{F}$ is the function space defined by f, and R is a constant which satisfies $||f(x)|| <= R$ for any x. This shows that the supervised task loss $L_{sup}(\hat{f})$ is bounded by the unsupervised loss, $L_{un}(f)$.

C - Skip-Connection Based Model. Suppose we use neural network with skip connection (f_2) to learn the contrastive embedding, according to Eq. (2), we have

$$L_{un}(f_2) = E[l(\{f_2(x)^T(f_2(x^+) - f_2(x^-)\}_{i=1}^k)] \tag{7}$$
$$= E[l(\{(f_I(x) + f_1(x))^T(f_I(x^+) + f_1(x^+) - f_I(x^-) - f_1(x^-)\}_{i=1}^k)] \tag{8}$$

where l is the logistic loss. Suppose the learned $f_1(x)$ is a trivial identity matrix I. As x is closer to x^+ than to x^-, $f_I(x)^T(f_I(x^+) - f_I(x^-)) = f_I(x)^T f_I(x^+) - f_I(x)^T f_I(x^-) >= 0$ holds. Accordingly,

$$L_{un}(f_2) = L_{un}(f_I + f_I) = E[l(4\{f_I(x)^T(f_I(x^+) - f_I(x^-)\}_{i=1}^k)]$$
$$\leq E[l(\{f_I(x)^T(f_I(x^+) - f_I(x^-)\}_{i=1}^k)] = L_{un}(f_I)$$

holds because l is monotonically decreasing. This means the upper bound of skip connection contrastive learning loss $L_{un}(f_2)$ is smaller than $L_{un}(f_I)$ which is the contrastive learning error of the original embedding. If SIMSKIP learns f_1 which is not an identity matrix through contrastive learning process, $L_{un}(f_2)$ is trivially less than $L_{un}(f_I + f_I)$, which induces that $L_{un}(f_2) \leq L_{un}(f_I)$ always holds. Therefore, the upper bound for using skip connections for contrastive learning should be lower.

We have observed that the proposed SIMSKIP exhibits several similar properties to Adapter [10] in that both employ skip connection as their fundamental components as shown in Fig. 2(d). However, Adapter is embedded within each layer of Transformer [33], while SIMSKIP is positioned outside the original model $h()$. Although

Adapter has been widely used in many Transformer [33] based models [4, 21], no theoretical proof has been given thus far. In this work, we present the first theoretical proof demonstrating why skip connection-based refinement does not degrade downstream tasks.

4 Experiments

Throughout the experiments, we want to show the effectiveness of SimSkip through downstream task metric improvement and its wide applicability to various pre-trained embeddings over different modalities including shallow knowledge graph embedding, deep graph neural network embedding, image embedding, and text embeddings. The datasets and benchmark methods used in the study are initially described, followed by the presentation of experimental results.

4.1 Experimental Setting

The study utilizes five datasets, as outlined below:

- The **movieQA** is a movie knowledge graph derived from the WikiMovies Dataset. It includes over 40,000 triples that provide information about movies.
- The **STL10** is an image dataset for image classification task. It has 10 classes and has 500 96 × 96 training images along with 800 test image per class.
- The **CIFAR10** is an image dataset for image classification tasks. It has 10 classes and has 6,000 32 × 32 color images per class. The dataset is split into 50000 training images and 10000 test images.
- The **Cora** is a graph dataset for node classification. It consists of 2,708 scientific publications as nodes with seven classes and 5,429 citations as edges.
- The **Pubmed** is a graph dataset for node classification. It consists of 19,717 scientific publication in Pubmed as nodes with three classes and 44,338 citations as edges.

The following methods are employed to learn the input embedding for contrastive learning:

- FedE [2] is a Federated Knowledge Graph embedding framework that focuses on learning knowledge graph embeddings by aggregating locally-computed updates. For the local Knowledge Graph embedding, we employed TransE [1]. This framework includes a client for each knowledge graph and a server for coordinating embedding aggregation.
- SimCLR [3] is a simple framework for contrastive learning of image representations. It first learns generic representations of images on an unlabeled dataset and then can be fine-tuned with a small number of labeled images to achieve good performance for a given classification task.
- GraphSAGE [8] is a general, inductive graph neural network (GNN) that leverages node feature information (e.g., text attributes). It samples and aggregates a nodes neighborhood's features to generate node embeddings.

– SimCSE [6] is a self-supervised text embedding that refines any pre-training transformer-based language models. Its main idea is to apply contrastive learning by treating two text embeddings obtained from the same input text with different dropout as positive pairs.

Throughout the experiment, we adhere to the original baseline experiment settings when running baselines. The embedding dimensions are 128 for FedE, 128 for SIM-CLR, 128 for GraphSage, and 768 for SimCSE. As for our proposed SIMSKIP, we explore various hyper-parameters over learning rate of 0.001, 0.0003, 000003, and 0.00001, and report its optimal performance. The encoder architecture of SIMSKIP is shown in Fig. 3. Layer 1 and layer 2 have the same structure, which contains a linear layer, a batch norm layer, a ReLU and a dropout layer. The projector is a two-layer feed forward network. When the dimension of the original embedding is d, the number of parameters for the encoder is $d \times d/2$ for layer 1, $d/2 \times d$ for layer 2, and $d \times d$ for the linear layer. The number of parameters for the project is $d \times d$ for both layer 1 and layer 2. For data augmentation, the masking augmentation randomly masked 20% of the vector, and Gaussian noise augmentation added noise sampled from a Gaussian distribution with mean 0 and variance 0.13.

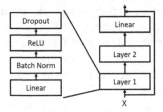

Fig. 3. The SimSkip Encoder. Layer 1 and Layer 2 have the same architecture.

4.2 SIMSKIP for Federated Knowledge Graph Embedding

This section evaluates SIMSKIP's performance of refining the embedding learned by federated learning. To have the federated knowledge graph learning setting, two knowledge graphs are randomly sampled from the movieQA knowledge graph. FedE [2] is used to learn the entity embeddings in a federated manner.

Table 1. Accuracy on different downstream tasks for movieQA knowledge graph

	KNN	Genre Classification	Movie Recommendation
FedE	58.5	84.8	7.47
SIMSKIP+ mask	62.8	**86.1**	**7.67**
SIMSKIP+ gaussian	**63.3**	86.0	7.01

In this experiment, we use three different downstream tasks - k-nearest neighbor same genre prediction (KNN), genre classification and movie recommendation. KNN is that, given a query movie, take 10-nearest movies to the query movie in the embedding space and count how many movies in those 10 movies have the same genre as the query movie. Genre classification downstream task predicts the genre of a movie according to its embedding and a 3 layered MLP was employed as a downstream classifier. Movie recommendation task recommends new movies to users according to the user's watching history. The user's watching history data is from Netflix dataset[1]. Given a user's watching history $V_1, ..., V_N$ in a chronological order, we treat $V_1, .., V_{N-10}$ as the training data, and predict 10 movies the user is most likely to watch. Then, we calculate how many movies in the 10 predicted movies belong to $V_{N-9}, ..., V_N$.

The results of different methods are shown in Table 1. SIMSKIP improved accuracy of all three downstream tasks. For KNN task, the improvement is about 4%. For genre classification and movie recommendation, the average improvement is 2% and 0.2%, respectively.

4.3 SIMSKIP for Image Embedding

In this experiment, we test SIMSKIP's performance for refining the self-supervised image embedding. We first use SimCLR [3] to learn the embedding, then we further refine the embedding with SIMSKIP. STL10 and CIFAR10 datasets were used for evaluation. The downstream task is the image classification task and a 3-layer MLP was employed as a downstream classifier. Table 2 presents the downstream image classification accuracy and shows that SIMSKIP refines the embedding space and improves the downstream task accuracy about 1% in average.

Table 2. Image classification accuracy on STL10 and CIFAR10

Image Classification	STL10	CIFAR10
SimCLR	76.09	66.88
SIMSKIP+ mask	75.84	65.93
SIMSKIP+ gaussian	**77.73**	**67.02**

4.4 SIMSKIP for Node embedding learned by Supervised Learning

In this section, we test SIMSKIP's performance of refining the embedding learned by supervised learning. We use GraphSAGE [8] as the supervised embedding learner. Core and Pubmed were used for evaluation which GraphSAGE used for its evaluation. In the experiment, we first train GraphSAGE in supervised setting and treat the output of the second to last layer as the node embedding, then we apply SIMSKIP. The downstream task is the node classification task and the same classification head of GraphSAGE was used as a downstream node classifier.

---

[1] https://www.kaggle.com/code/laowingkin/netflix-movie-recommendation/data.

Table 3 shows that node classification accuracy on Cora and Pubmed datasets. Because we originally thought that the embedding trained by supervised learning should fit the downstream task best, further refining it with SIMSKIP won't improve the downstream task performance. However, the experiment results show that SIMSKIP further improved the downstream task performance even with the embedding learned by a supervised task.

Table 3. Node classification accuracy on Cora and Pubmed

	Cora	Pubmed
GraphSAGE	82.60	81.6
SIMSKIP+ mask	82.60	81.8
SIMSKIP+ gaussian	**82.90**	**82.9**

4.5 SIMSKIP for Transformer-based Text Embedding

In this section, we test SIMSKIP's performance of refining the self-supervised transformer-based text embedding. Specifically, we further refine the pre-trained Sim-CSE [6] embedding with SIMSKIP and apply SIMSKIP text embedding to various NLP downstream tasks including CR [11], MPQA [36], MR [27], MRPC [5], SST-2 [32], SUBJ [26], and TREC [34]. These tasks are also used in [6]. We use accuracy as the metric, which means that a higher value indicates better performance. The results are presented in Table 4. Our findings suggest that stacking multiple embedding enhancing techniques (see SimCSE + SIMSKIP) keeps improving the downstream task performance.

Table 4. NLP task accuracy for self-supervised text embedding

Model	CR	MPQA	MR	MRPC	SST2	SUBJ	TREC
SimCSE	85.99	88.5	80.54	73.65	**86.47**	94.8	**82.19**
SimCSE + SIMSKIP	**86.36**	88.27	**80.82**	74.71	85.55	94.93	80.8
SimCSE + SIMSKIP (mask+noise)	85.44	**88.56**	78.25	**75.01**	82.96	**95.28**	80.1

4.6 Ablation Study

In section, we assess the effect of the skip connection in SIMSKIP. For comparison, we implemented SIMSKIP⁻ which is obtained by removing the skip connection from SIM-SKIP. For original embedding, we used embedding learned by SimCLR, STL10 and CIFAR10 as datasets, image classification as the downstream task. Table 5 shows that the accuracy of the downstream task with SIMSKIP⁻ is lower than SIMSKIP and even lower than that with the original SimCLR embedding (see Table 2). This aligns with our findings in Subsect. 3.1 which states that a wider and deeper network does not necessarily lead to a better approximation of the optimal function (see Fig. 2(b)). When the skip connection is omitted, the initial embedding obtained from the contrastive encoder becomes randomly dispersed across the entire embedding space. Consequently, subsequent updates have limited impact. However, with the inclusion of a skip connection, we ensure that the initial embedding from the contrastive encoder retains original useful information, facilitating the effectiveness of the subsequent updating process.

Table 5. Ablation study of SIMSKIP on image classification downstream task (unit: accuracy)

	STL10	CIFAR10
SIMSKIP + mask	75.84	65.93
SIMSKIP + gaussian	**77.73**	**67.02**
SIMSKIP⁻ + mask	47.1	56.5
SIMSKIP⁻ + gaussian	56.6	66.7

5 Related Work

5.1 Representation Learning

The goal of representation learning is to learn low dimensional vectors of the input data so that similar data points will be close to each other in the vector space, while dissimilar data points will be far from each other. It has been applied in many applications, such as dialogue system [17, 19], fact checking [15, 18–20], and question answering [16] and so on. Representation learning methods like TransE [1], RESCAL [25] and DistMult [37] embed entities in the knowledge graph as points in the low dimensional Euclidean space and model relations as linear or bilinear transformation in the space.

5.2 Contrastive Learning

Contrastive Learning focuses on minimizing the distance between the target embedding (anchor) vector and the matching (positive) embedding vector [14, 22, 35, 38], while maximizing the distance between the anchor vector and the non-matching (negative) embedding vectors. Recent work on contrastive learning have shown that discriminative or contrastive approaches can (i) produce transferable embeddings for visual objects through the use of data augmentation [3], and (ii) learn joint visual and language embedding space that can be used to perform zero-shot detection [12]. Given the sparseness and long-tailed property of scene graph datasets, application (i) of contrastive approach can help the model learn better visual appearance embeddings of (subject, object) pairs under limited resource settings. Moreover, in application (ii), contrastive learning gives a clearer separation of the visual embeddings and language embeddings compared to the traditional black-box neural fusion approaches [9], which allows more control over both the symbolic triples input and the final output embedding spaces.

One thing to note is that the direct comparison of SIMSKIP to other contrastive techniques is not the primary focus. The main claim of SIMSKIP is its ability to further enhance the quality of embeddings learned by other contrastive methods through a skip-connection based encoder-projector architecture with contrastive learning in terms of downstream task performance. Accordingly, SIMSKIP serves as a facilitator of other techniques rather than a direct competitor. Additionally, since the embedding enhancing capability of SIMSKIP originates from the architecture rather than a specific contrastive learning training technique, SIMSKIP benefits from integration with other state-of-the-art contrastive learning techniques in various dimensions such as loss function and data augmentation.

6   Conclusion

In this paper, we propose a skip connection based contrastive learning framework (SIMSKIP) which refine the input embedding space. We theoretically prove that the downstream task error upper bounds with using SIMSKIP embedding as its input will not be larger than that with the original embedding. The experiment results show the effectiveness of the proposed method. For future work, we intend to explore diverse data augmentation methods in embedding space and continue reducing the error bound in theoretical analysis. Besides, we plan to analyze how SIMSKIP and related architectures can address the issues raised in Fig. 1.

References

1. Bordes, A., N, U.: Translating embeddings for modeling multi-relational data. In: Advances in Neural Information Processing Systems, vol. 26 (2013)
2. Chen, M., Zhang, W., Yuan, Z., Jia, Y., Chen, H.: FedE: embedding knowledge graphs in federated setting. In: The 10th International Joint Conference on Knowledge Graphs, IJCKG 2021, pp. 80–88. Association for Computing Machinery, New York (2021). https://doi.org/10.1145/3502223.3502233
3. Chen, T., Kornblith, S., Norouzi, M., Hinton, G.: A simple framework for contrastive learning of visual representations. In: III, H.D., Singh, A. (eds.) Proceedings of the 37th International Conference on Machine Learning. Proceedings of Machine Learning Research, vol. 119, pp. 1597–1607. PMLR (2020). https://proceedings.mlr.press/v119/chen20j.html
4. Devlin, J., Chang, M.W., Lee, K., Toutanova, K.: BERT: pre-training of deep bidirectional transformers for language understanding (2019)
5. Dolan, W.B., Brockett, C.: Automatically constructing a corpus of sentential paraphrases. In: Proceedings of the Third International Workshop on Paraphrasing (IWP2005) (2005). https://aclanthology.org/I05-5002
6. Gao, T., Yao, X., Chen, D.: SimCSE: simple contrastive learning of sentence embeddings. In: Empirical Methods in Natural Language Processing (EMNLP) (2021)
7. Grover, A., Leskovec, J.: node2vec: scalable feature learning for networks (2016). https://doi.org/10.48550/ARXIV.1607.00653, https://arxiv.org/abs/1607.00653
8. Hamilton, W., Ying, Z., Leskovec, J.: Inductive representation learning on large graphs. In: Guyon, I., et al. (eds.) Advances in Neural Information Processing Systems, vol. 30. Curran Associates, Inc. (2017). https://proceedings.neurips.cc/paper/2017/file/5dd9db5e033da9c6fb5ba83c7a7ebea9-Paper.pdf

9. He, K., Zhang, X., Ren, S., Sun, J.: Deep residual learning for image recognition. In: 2016 IEEE Conference on Computer Vision and Pattern Recognition (CVPR), pp. 770–778 (2016). https://doi.org/10.1109/CVPR.2016.90
10. Houlsby, N., et al.: Parameter-efficient transfer learning for NLP (2019)
11. Hu, M., Liu, B.: Mining and summarizing customer reviews. In: Proceedings of the Tenth ACM SIGKDD International Conference on Knowledge Discovery and Data Mining, KDD 2004, pp. 168–177. Association for Computing Machinery, New York (2004). https://doi.org/10.1145/1014052.1014073
12. Jiang, H., Wang, R., Shan, S., Chen, X.: Transferable contrastive network for generalized zero-shot learning. CoRR abs/1908.05832 (2019). http://arxiv.org/abs/1908.05832
13. Khosla, P., et al.: Supervised contrastive learning. In: Larochelle, H., Ranzato, M., Hadsell, R., Balcan, M., Lin, H. (eds.) Advances in Neural Information Processing Systems, vol. 33, pp. 18661–18673. Curran Associates, Inc. (2020). https://proceedings.neurips.cc/paper_files/paper/2020/file/d89a66c7c80a29b1bdbab0f2a1a94af8-Paper.pdf
14. Khosla, P., et al.: Supervised contrastive learning. arXiv preprint arXiv:2004.11362 (2020)
15. Liu, L., Du, B., Fung, Y.R., Ji, H., Xu, J., Tong, H.: KompaRe: a knowledge graph comparative reasoning system. In: Proceedings of the 27th ACM SIGKDD Conference on Knowledge Discovery and Data Mining, KDD 2021, pp. 3308–3318. Association for Computing Machinery, New York (2021)
16. Liu, L., Du, B., Ji, H., Zhai, C., Tong, H.: Neural-answering logical queries on knowledge graphs. In: Proceedings of the 27th ACM SIGKDD Conference on Knowledge Discovery and Data Mining, KDD 2021, pp. 1087–1097. Association for Computing Machinery, New York (2021)
17. Liu, L., Hill, B., Du, B., Wang, F., Tong, H.: Conversational question answering with reformulations over knowledge graph. arXiv preprint arXiv:2312.17269 (2023)
18. Liu, L., Ji, H., Xu, J., Tong, H.: Comparative reasoning for knowledge graph fact checking. In: 2022 IEEE International Conference on Big Data (Big Data) (2022)
19. Liu, L., Tong, H.: Knowledge graph reasoning and its applications. In: Proceedings of the 29th ACM SIGKDD Conference on Knowledge Discovery and Data Mining, pp. 5813–5814 (2023)
20. Liu, L., et al.: Knowledge graph comparative reasoning for fact checking: problem definition and algorithms. Data Eng. **45**, 19–38 (2022)
21. Liu, Y., et al.: RoBERTa: a robustly optimized BERT pretraining approach (2019)
22. Luo, Z., Xu, W., Liu, W., Bian, J., Yin, J., Liu, T.Y.: KGE-CL: contrastive learning of tensor decomposition based knowledge graph embeddings. In: Proceedings of the 29th International Conference on Computational Linguistics, pp. 2598–2607. International Committee on Computational Linguistics, Gyeongju (2022). https://aclanthology.org/2022.coling-1.229
23. Mikolov, T., Sutskever, I., Chen, K., Corrado, G.S., Dean, J.: Distributed representations of words and phrases and their compositionality. In: Burges, C., Bottou, L., Welling, M., Ghahramani, Z., Weinberger, K. (eds.) Advances in Neural Information Processing Systems, vol. 26. Curran Associates, Inc. (2013). https://proceedings.neurips.cc/paper/2013/file/9aa42b31882ec039965f3c4923ce901b-Paper.pdf
24. Nickel, M., Kiela, D.: Poincaré embeddings for learning hierarchical representations. In: Guyon, I., et al. (eds.) Advances in Neural Information Processing Systems, vol. 30, pp. 6341–6350. Curran Associates, Inc. (2017). http://papers.nips.cc/paper/7213-poincare-embeddings-for-learning-hierarchical-representations.pdf
25. Nickel, M., Tresp, V., Kriegel, H.P.: A three-way model for collective learning on multi-relational data. In: Proceedings of the 28th International Conference on International Conference on Machine Learning, ICML 2011, pp. 809–816. Omnipress, Madison (2011)

26. Pang, B., Lee, L.: A sentimental education: sentiment analysis using subjectivity summarization based on minimum cuts (2004). https://doi.org/10.48550/ARXIV.CS/0409058, https://arxiv.org/abs/cs/0409058

27. Pang, B., Lee, L.: Seeing stars: exploiting class relationships for sentiment categorization with respect to rating scales. In: Proceedings of the 43rd Annual Meeting of the Association for Computational Linguistics (ACL 2005), pp. 115–124. Association for Computational Linguistics, Ann Arbor (2005). https://doi.org/10.3115/1219840.1219855, https://aclanthology.org/P05-1015

28. Perozzi, B., Al-Rfou, R., Skiena, S.: DeepWalk: online learning of social representations. In: Proceedings of the 20th ACM SIGKDD International Conference on Knowledge Discovery and Data Mining, KDD 2014, pp. 701–710. Association for Computing Machinery, New York (2014). https://doi.org/10.1145/2623330.2623732

29. Robinson, J., Chuang, C.Y., Sra, S., Jegelka, S.: Contrastive learning with hard negative samples (2021)

30. Saunshi, N., Ash, J., Goel, S.: Understanding contrastive learning requires incorporating inductive biases. arXiv (2022)

31. Saunshi, N., Plevrakis, O., Arora, S., Khodak, M., Khandeparkar, H.: A theoretical analysis of contrastive unsupervised representation learning. In: Chaudhuri, K., Salakhutdinov, R. (eds.) Proceedings of the 36th International Conference on Machine Learning. Proceedings of Machine Learning Research, vol. 97, pp. 5628–5637. PMLR (2019). https://proceedings.mlr.press/v97/saunshi19a.html

32. Socher, R., et al.: Recursive deep models for semantic compositionality over a sentiment treebank. In: Proceedings of the 2013 Conference on Empirical Methods in Natural Language Processing, pp. 1631–1642. Association for Computational Linguistics, Seattle (2013). https://aclanthology.org/D13-1170

33. Vaswani, A., et al.: Attention is all you need (2017)

34. Voorhees, E.M., Tice, D.M.: Building a question answering test collection. In: Proceedings of the 23rd Annual International ACM SIGIR Conference on Research and Development in Information Retrieval, SIGIR 2000, pp. 200–207. Association for Computing Machinery, New York (2000)

35. Wang, L., Zhao, W., Wei, Z., Liu, J.: SimKGC: simple contrastive knowledge graph completion with pre-trained language models. In: Proceedings of the 60th Annual Meeting of the Association for Computational Linguistics (Volume 1: Long Papers). Association for Computational Linguistics, Dublin (2022)

36. Wiebe, J., Wilson, T., Cardie, C.: Annotating expressions of opinions and emotions in language. Lang. Resour. Eval. **39**, 165–210 (2005)

37. Yang, B., tau Yih, W., He, X., Gao, J., Deng, L.: Embedding entities and relations for learning and inference in knowledge bases (2015)

38. You, Y., Chen, T., Sui, Y., Chen, T., Wang, Z., Shen, Y.: Graph contrastive learning with augmentations. In: Advances in neural information processing systems, vol. 33, pp. 5812–5823 (2020)

In-Use

Automation of Electronic Invoice Validation Using Knowledge Graph Technologies

Johannes Mäkelburg[1]([✉]) [ID], Christian John[2], and Maribel Acosta[1] [ID]

[1] TUM School of Computation, Information and Technology, Technical University of Munich, Heilbronn, Germany
{johannes.makelburg,maribel.acosta}@tum.de
[2] Einkaufsbüro Deutscher Eisenhändler GmbH, Wuppertal, Germany
christian.john@ede.de

Abstract. Invoicing is a crucial part of any business's financial and administrative activities. Nowadays, invoicing is handled in the form of Electronic Data Interchange (EDI), where invoices are managed in a standardized electronic or digital format rather than on paper. In this context, EDI increases the efficiency of creating, distributing, and processing invoices. The most used standard for representing electronic invoices is EDIFACT. Yet, the validation of EDIFACT invoices is not standardized. In this work, we tackle the problem of automatically validating electronic invoices in the EDIFACT format by leveraging KG technologies. The core of our proposed solution consists of representing EDIFACT invoices as RDF knowledge graphs (KGs). We developed an OWL ontology to model EDIFACT terms with semantic descriptions. The invoice KG can be validated using SHACL constraints acquired from domain experts. We evaluated our ontology and invoice validation process. The results show that our proposed solution is complete, correct, and efficient, and significantly undercuts the efforts of current human evaluation.

Keywords: Electronic Invoice · Ontology · EDIFACT · RDF · RML · SHACL

1 Introduction

In the current business landscape, efficient financial and administrative practices are paramount for organizations. Central to these operations is the invoicing process, a critical component that demands accuracy, timeliness, and standardization. Organizations increasingly rely on Electronic Data Interchange (EDI) to streamline invoicing, transitioning from traditional paper-based methods to electronic formats. Within EDI, EDIFACT is a widely adopted standard for representing electronic invoices, offering a uniform approach to their creation, distribution, and processing.

A. Meroño Peñuela et al. (Eds.): ESWC 2024, LNCS 14664, pp. 253–269, 2024.
https://doi.org/10.1007/978-3-031-60626-7_14

While EDIFACT has significantly enhanced the efficiency of invoicing procedures, a notable gap exists in the validation of electronic invoices. This gap affects business processes that rely on correct invoicing data in a timely manner. This is the case of the group purchasing organization Einkaufsbüro Deutscher Eisenhändler (E/D/E). As E/D/E is represented in 30 European countries, where many different processes and regulations exist for business documents, EDIFACT is used as the standard to handle invoices. Yet, as the validation of EDIFACT invoices is not standardized – i.e., there is no language for representing constraints over the invoices – in many cases, this process is carried out manually, which is time-consuming and prone to errors.

This work proposes a solution that leverages KG technologies to validate electronic invoices in the EDIFACT standard. The core of our approach is to model electronic invoices as RDF graphs. Shifting to the semantic web technology stack allows for applying existing open standards and solutions for managing machine-readable data. To achieve this, first, we present the EDIFACT Ontology. The ontology captures the terms to model the content of EDIFACT messages using RDF. Second, we propose the tool EDIFACT-VAL to validate the content of the original EDIFACT messages. The tool processes the invoices in the EDIFACT format and translates them into XML. From the XML files, EDIFACT-VAL creates the RDF graphs using the EDIFACT Ontology and the RDF Mapping Language (RML) [4]. EDIFACT-VAL validates the invoice RDF graph using constraints defined in the Shapes Constraint Language (SHACL) [13]. The constraints are created with input from domain experts based on the EDIFACT guidelines.

We evaluate the soundness of our proposed solution with an experimental evaluation using real-world EDIFACT invoices. The results show that EDIFACT-VAL produces complete RDF graphs in the order of seconds. The validation with SHACL shows that many real-world invoices do not fully comply with the EDIFACT standard. Lastly, we conducted an in-use evaluation with domain experts, who compared EDIFACT-VAL with the current manual process. The experts found the tool's performance remarkable, and are working on integrating it into their invoice workflows. This shows the potential impact of our proposed solution.

In summary, our contributions are:

- An OWL ontology to represent terms from the EDIFACT standard to model electronic invoices as RDF KGs.
- A tool dubbed EDIFACT-VAL to automatically validate invoices using SHACL.
- An experimental evaluation that shows the soundness of our proposed solutions.
- An in-use evaluation where domain experts at the E/D/E assessed the performance and applicability of EDIFACT-VAL to their business processes.

2   Preliminaries

First, we introduce the EDIFACT invoice concept, which forms the basis for the knowledge graphs. Then, we introduce the purchase-to-pay ontology, which builds the basis for an EDIFACT ontology.

Listing 1.1. Excerpt of an EDIFACT invoice

```
1 UNH+1+INVOIC:D:96A:UN:EANOO8'
2 BGM+380+4031541+43'
3 DTM+137:20220908:102'
4 NAD+IV+4317784000000::9++Einkaufsbuero DeutscherEisenhaendler:GmbH+EDE PLatz 1+Wuppertal
    ++42389+DE'
5 LIN+1++4016671029277:EN::9'
6 PRI+AAA:16.78:::1:PCE'
7 MOA+79:100.68'
8 MOA+124:19.13'
9 UNT+37+1'
```

EDIFACT Invoice. EDIFACT is the most commonly used and most comprehensive international standard for electronic data interchange. EDIFACT is used across almost all business sectors; the individual sectors are delimited in EDIFACT by so-called subsets. The maintenance of the standard lies under the responsibility of the United Nations and the Economic Commission for Europe.

Documents transmitted in EDIFACT are all types of messages of the business processes area. The structure of the messages is based on segments; these, in turn, consist of data elements and data element groups. These three components together are referred to as the EDIFACT elements.

Listing 1.1 shows an excerpt of a real-world EDIFACT message. The segments are split into three sections: header-, detail- and summary section. In all three sections, some segments are required, meaning all three sections are always represented in an EDIFACT message. However, there are some segments within the sections that are optional. For example, the header section contains eight mandatory segments and six conditional segments.

In the header, general information about the invoice is displayed, e.g., the invoice number (Line 2), the document date (Line 3), and information about the involved organizations (Line 4). Information about the sold items, including the net price (Line 6) or the article number (Line 5), is allocated in the detail section. The summary section contains the total amounts of the invoice, e.g., the total item amount (Line 7) or the total tax amount (Line 8). Above the header and below the summary section are segments allocated for the EDIFACT structure, e.g., the version and type of message (Line 1) or the end character (Line 9).

P2P-O: Purchase to Pay Ontology. We use the Purchase-to-Pay Ontology (P2P-O) [21] as a foundation for modeling concepts from the EDIFACT standard. P2P-O is an OWL ontology that models semantic representation of invoices based on the *core invoice model* of the European Standard EN 16931-1:2017 [6].

P2O-O is divided into seven modules: *item, price, documentline, organization, document, invoice, process*. The *item* module allows for describing products listed on the invoices. Not only sold items are mentioned with the term, but also working hours. The *price* module makes it possible to describe the prices of the items and the amounts of money in an invoice. The possibility of making statements about the positions of prices and items on documents is enabled by the model *documentline*. The participating organizations are described by the mod-

Table 1. Competency Questions for the EDIFACT Ontology

Name 0	Competency Question
CQ 0	What invoices are all listed in an EDIFACT message?
CQ 1	Which organizations are involved in the invoice?
CQ 2.1	What role does organization S play in the invoice?
CQ 2.2	Which organization is the buyer in the invoice?
CQ 3.1	What information is displayed about the involved organizations ?
CQ 3.2	What is the address of the buyer?
CQ 4	What items are sold in the invoice?
CQ 5.1	What information is displayed about the items sold?
CQ 5.2	What is the net price of the items sold in the invoice?
CQ 6.1	What are the invoice details of the invoice?
CQ 6.2	What is the invoice amount of the invoice?
CQ 6.3	What is the invoice number?
CQ 7	What information must be provided so that the file format is valid?
CQ 8	To which business process can the invoice be assigned?

ule *organization*. In the *document* module ontology resources which are essential for the purchase-to-pay process are provided. The document-type invoice is described more specifically in the module *invoice*. Therefore the two modules *document* and *documentline* are used. In the *invoice* module, all the mandatory constraints from EN 16931 are implemented. Lastly, the *process* module contains classes for a specific description of the kind of purchase-to-pay process (Table 1).

3   Our Approach

Given an electronic invoice in the EDIFACT format, the problem tackled in this paper is to validate the correctness of the invoice by representing the invoice as an RDF knowledge graph (KG). Our proposed solution comprises two parts: (1) a proposed ontology to represent EDIFACT concepts (§ 3.1), based on the concepts presented in Sect. 2, and (2) the EDIFACT-VAL tool to perform the validation of invoices using KG technologies (§ 3.2).

3.1   EDIFACT Ontology

Competency Questions. We use the NeOn method [25] as a systematic approach to structure and develop the EDIFACT ontology according to an established principle. The ontology development process involves the formulation of the requirements, framework, and competency questions of the ontology. In this work, these steps were carried out in collaboration with Electronic Data Interchange experts from E/D/E. Table 1 shows the competency questions have been drawn up based on the EDIFACT guideline. The competency questions (CQ) are grouped according to different aspects of the invoices. CQs in groups 1 and 2 (i.e., CQ 1, 2.1, and 2.2) are concerned with organizations and their role in the invoice. CQ 3.1 and 3.2 address specific information about the involved organizations. CQ 4, 5.1, and 5.2 request information about the items listed in the

Table 2. Overview of linked ontologies and vocabularies in the EDIFACT Ontology

Prefix Name	Prefix	Frequency of use
agentRole	https://archive.org/services/purl/domain/modular_ontology_design_library/agentrole#	5
dc	http://purl.org/dc/elements/1.1#	2
frapo	http://purl.org/cerif/frapo/	3
schema	http://schema.org/	3
org	http://www.w3.org/ns/org#	1
p2p-o-doc-line	https://purl.org/p2p-o/documentline#	2
p2p-o-doc	https://purl.org/p2p-o/document#	1
p2p-o-inv	https://purl.org/p2p-o/invoice#	3
p2p-o-item	https://purl.org/p2p-o/item#	2
p2p-o-org	https://purl.org/p2p-o/organization#	4
vcard	http://www.w3.org/2006/vcard/ns#	2

invoice. CQ6 handles the identifier or number of the invoice. Lastly, CQ 7 and 8 address aspects of the invoice metadata, i.e., the EDIFACT structure elements and related business processes.

Reuse of Ontology Design Patterns and Existing Vocabularies and Ontologies. Following ontology design best practices [17,25], we reuse existing resources. An overview of the reused resources and the number of reuses can be found in Table 2. We apply the agent role pattern from the Modular Ontology Design Library (MODL) [22] for modelling the participation of organizations in invoices. In particular, the same organization can have many roles (seller, supplier, etc.) in the same or different invoices, to which different property values can be associated depending on its role. The ontologies listed in Table 2 also provide adequate solutions for our purpose. Most of them are reused in the class *AgentRole*, for instance, the country code from the Funding, Research Administration, and Projects Ontology [23] or the address of the vCard Ontology [16]. Also, four of the seven different main classes we defined in our EDIFACT Ontology are linked to concepts of these vocabularies or ontology. For example, the class *FormalOrganization* is linked to the Core Organization Ontology (*org*), the *E-Invoice* to the P2P-O module *document*, and *Item* to the P2P-O module *item*.

Ontology Description. Based on the gaps identified in existing vocabularies and ontologies discussed in the previous section, we propose the EDIFACT Ontology tailored to represent the concepts and fields of the EDIFACT standard. The proposed OWL ontology comprises 28 classes, 10 OWL object properties, 233 OWL data properties, and 6 annotation properties. The ontology was developed using WebProtégé [11]. Figure 1 illustrates the main concept of the EDIFACT ontologies by showing the connections between the different classes. The EDIFACT Ontology can be found under https://purl.org/edifact/ontology. Additionally, it has been integrated into the LOV catalogue, available at https://lov.linkeddata.es/dataset/lov/vocabs/edifact-o. The classes,

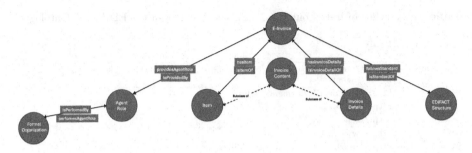

Fig. 1. Main concepts of the EDIFACT ontololgy. Source: WebVOWL [15]

their properties, and how they address the competency questions from Table 1 are explained in more detail in the following.

E-Invoice This is the main class of the EDIFACT ontology, and its individuals or entities represent electronic invoices. This class is connected to other classes through object properties, including, *EDIFACT-Structure* via the property *followsStandard*, *Item* via the property *hasItem*, *InvoiceDetails* via the property *hasInvoiceDetails*, and *AgentRoles* via the *isProvidedBy* property to capture the role of organizations in the invoice. The only data property used to describe this class is *belongsToProcess* which displays the business process of the invoice. This class allows for handling the competency questions CQ 0 and CQ 8.

EDIFACT-Structure This class contains all the information that ensures that the invoice file meets the requirements of the EDIFACT file format. This information appears at the beginning and end of a message and, therefore, has no correspondence with the content of the individual invoices. For this reason, this class and the *InvoiceContent* class are disjoint, specified with the *owl:disjointWith* predicate. Nevertheless, as this information belongs to an invoice, this class is connected to the class *E-Invoice* via the object property *followsStandard*. Among the datatype properties of the entities of this class, we have *creationDate*, *dataExchangeCounter*, *messageReferenceNumber*, and *senderIndicator*. This class is associated with the competency question CQ 7.

Invoice Details This class contains all the information that can only occur in the header- and summary sections of the invoices. Exemplary datatype properties of this class are the document date (*hasDocumentDate*), the document number (*hasDocumentNumber*), the delivery condition (*deliveryCondition*) and the invoice amount (*hasInvoiceAmount*). The connection of the *InvoiceDetails* class to the *E-Invoice* class is done via the object property *hasInvoiceDetails*. The class *Invoice Details* allows handling the competency questions CQ 6.1, 6.2, and 6.3.

Item This class allows for representing an item listed in the invoice, which is found in the detail section of the invoice. Examples of the datatype properties for an item are the name of the item (*p2p-o-item:Item*), the net price of the item (*hasNetPriceOfItem*), the net weight of an item (*hasNetWeight*), the number assigned to a manufacturer's product according to the International Article Numbering Association (*internationalArticleNumber*), etc. It is to mention that

Listing 1.2. Representation of organizations using *AgentRole* and *FormalOrganization*

```
1  @prefix invoice: <http://www.ede.com/edifact/invoice#>.
2  @prefix edifact-o: <https://purl.org/edifact/ontology#> .
3  @prefix agentRole: <https://archive.org/services/purl/domain/modular_ontology_design_
        library/agentrole#> .
4  @prefix frapo: <http://purl.org/cerif/frapo/> .
5  @prefix p2p-o-org: <https://purl.org/p2p-o/organization#> .
6  @prefix vcard: <http://www.w3.org/2006/vcard/ns#> .
7
8  invoice:r999995 a edifact-o:DeliveryPartyRole;
9    frapo:hasCountryCode "DE";
10   edifact-o:hasCity "Wuppertal";
11   edifact-o:hasCountry "Deutschland";
12   vcard:hasStreetAddress "In der Fleute 153";
13   vcard:postalCode "42389";
14   agentRole:isProvidedBy invoice:i999999;
15   p2p-o-org:formalName "E/D/E-GmbH Anlieferstelle L205" .
16
17 invoice:r999996 a edifact-o:InvoiceeRole;
18   frapo:hasCountryCode "DE";
19   edifact-o:hasCity "Wuppertal";
20   edifact-o:hasCountry "Deutschland";
21   vcard:hasStreetAddress "EDE Platz 1";
22   vcard:postalCode "42389";
23   agentRole:isProvidedBy invoice:i999999;
24   p2p-o-org:formalName "Einkaufsuero Deutscher Eisenhaendler".
25
26 invoice:4317784000000 a edifact-o:FormalOrganization;
27   agentRole:performsAgentRole invoice:r999995, invoice:r999996 ;
28   p2p-o-org:globalLocationNumber 4317784000000 .
```

the ratio between the classes *E-Invoice* and *Item* is 1:N. Therefore, here we were confronted with the design decision of modeling the relationship between these two classes as a multi-valued property or as an RDF collection to represent a close list. We decided on the former option by defining the object property *hasItem* (and its inverse *isItemOf*) between the *Item* and the *E-Invoice* classes, as this facilitates the validation and querying of the invoices. The *Item* concept allows for handling the competency questions CQ 4, 5.1, and 5.2.

InvoiceContent This class allows for modelling the information that can be found in all three sections of the invoices. As a result, the information of the *Invoice-Content* class is defined as the union of the classes *Item* and *InvoiceDetails*, i.e., *InvoiceContent owl:unionOf (Item InvoiceDetails)*. Creating the *InvoiceContent* class makes it possible to define properties that apply to all parts of the invoice by using this class as the domain or range of those properties. This ensures the consistency of the ontology.

AgentRole According to the guidelines, an organization can, or sometimes must, have many different roles. For example, in the warehousing business, one organization needs to have three roles: *Buyer Role, Invoicee Role*, and *Delivery Party Role*. We model these cases using the Role ontology pattern as defined by Shimizu et. al [22] and Grüninger & Fox [8]. In our ontology, the purpose of the class *AgentRole* is to represent the manifold roles an organization can have in an invoice. As a solution, several subclasses have been created, each representing one of these roles. The connection to the *E-Invoice* class exists through the object property

Fig. 2. Overview of EDIFACT-VAL

isProvidedBy with the *E-Invoice* class in the range. The *AgentRole* class allows for addressing the competency questions C 2.1, 2.2, 3.1, and 3.2.

FormalOrganization This class represents the organizations involved in the messages, which addresses the competency question CQ 1. The properties assigned to the *FormalOrganization* class are solely responsible for allocating the role that an organization plays in an invoice through *performsAgentRole*, and for providing a globally unique identifier for organizations, i.e., *globalLocationNumber*. In the EDIFACT ontology, the only connection of the *FormalOrganization* class with another class is with the *AgentRole* class via the object property *performsAgentRole* with the *AgentRole* class in the range.

We have opted against the intuitive option of assigning the information regarding the organization to the class of *FormalOrganization*. Instead, we assign them to the class *AgentRole*, more precisely to one of its subclasses. This is because some of the displayed information can change for the same organization depending on its role in an invoice. We illustrate this in the following example.

Listing 1.2 showcases the RDF triples of the different agent roles related to warehouse business invoices. The address of the organization changes depending on its role: the address of the warehouse *In der Fleute 153* (Line 12) is used when the organization has the DeliveryParty Role, while the main location *EDE Platz 1* (Line 21) is used for the Invoicee Role. However, all roles belong to the same organization, as evidenced by both agent roles being performed by the same *FormalOrganization* (Line 27 and 28).

3.2   The EDIFACT-VAL Tool

Now that we have modelled the terms of the EDIFACT standard in an OWL ontology (see Sect. 3.1), the EDIFACT-VAL tool processes the electronic invoices in the EDIFACT format to validate their content using KG technologies. Concretely, the tool carries out the following steps (cf. Figure 2). **(1) Invoice Preprocessing:** Translates the invoice files into XML files. **(2) RML Mapping:** Creates RDF knowledge graphs from the invoice XML files using generated RML mappings. **(3) RDF Validation:** Validates the invoice RDF graphs using constraints based on EDIFACT guidelines and reports from domain experts, which are formulated using the Shapes Constraints Language (SHACL) [13].

Invoice Pre-processing. This step aims to prepare the EDIFACT message in a way that meets the requirements for the creation of an RDF graph. Despite that EDIFACT is an open standard, the file format is not yet widely compatible with existing tools, especially the ones related to knowledge graphs. Hence, a transformation into a compatible format is required. Our choice of format is XML, justified by the ease of modification and the numerous processing possibilities. Then, EDIFACT-VAL extends the resulting XML files, to incorporate additional XML elements and attributes that capture the qualifiers in EDI files. EDI files use qualifiers to secure file compactness, i.e., several values are packed into a single field. An example of this is shown in Line 4 of Listing 1.1. There the NAD-segment with the qualifier *IV* is used to display information about the Invoicee of the invoice. To handle such qualifiers, the pre-processing separates this field into several XML elements, creating a one-to-many correspondence between EDIFACT elements and XML elements.

RDF Graph Creation for Invoices. EDIFACT invoices can be transformed into RDF representations using our proposed EDIFACT Ontology described in Sect. 3.1. To generate RDF knowledge graphs based on a (semi-structured) data source, we define mapping rules. For this, we use the RDF Mapping Language (RML) [4], which provides a way to transform heterogeneous data into the RDF data model. Yet, manually creating RML mapping is a complex task [12], especially when a large number of terms are involved. To ease the definition of the mapping rules, languages like YARRRML [26] and ShExML [7] provide a human-friendly serialization of RML. In this work, we use YARRRML [26] as it allows users to specify simpler mapping rules compared to RML;[1] these mappings are then automatically transformed into the RML mappings, later used to transform the invoice XML file into an RDF graph.

Our YARRRML mapping contains six mappings, each representing one different class of the EDIFACT ontology. Compared to the EDIFACT ontology, only six of the seven classes are displayed, as the class *InvoiceContent* only has the purpose of providing a domain or range for some resources. When creating each mapping, two types of rules must be defined: (i) rules for defining the data sources, i.e., the XML files of the EDIFACT invoices, and (ii) rules for generating the RDF triples. In total, there are 245 different data source rules, since a rule is generated for each EDIFACT element. Depending on which information is to be displayed in the mappings, the data sources are assigned to the mappings. The rules for generating the RDF triples are divided into two parts. Part one defines the rule for identifying the subject of the RDF triple. In our tool, this is done by the added attributes from the preprocessing 3.2. The second part defines the rules for the predicates and objects of the RDF triple. The predicate rule contains the ontology resources, and the object rule contains the identification of the value of the predicate. In our mapping, these values are either the values of XML elements or ontology resources. The RML mappings obtained with YARRRML and the invoice XML files are then fed into the RMLMapper[2] tool, to obtain the invoice RDF graph. Exemplary transformation of the *NAD*

---

[1] E.g., our YARRRML file has 2,924 lines vs. 366,709 lines in the generated RML file.

[2] https://github.com/RMLio/rmlmapper-java.

segment from an EDIFACT message to an RDF graph can be seen in Line 4 in Listing 1.1 to Lines 17–24 and Lines 26–28 in Listing 1.2.

Listing 1.3. SHACL constraint for mandatory and single property values

```
1  :ExistenceDocumentNumber
2     a sh:NodeShape;
3     sh:targetClass edifact-o:InvoiceDetails;
4     sh:property [
5        sh:path edifact-o:hasDocumentNumber;
6        sh:minCount 1;
7        sh:maxCount 1; ] .
```

Listing 1.4. SHACL constraint for formatting check (datatype and length)

```
1  :LengthDocumentNumber
2     a sh:NodeShape;
3     sh:targetClass edifact-o:InvoiceDetails;
4     sh:property [
5        sh:path edifact-o:hasDocumentNumber;
6        sh:datatype xsd:string;
7        sh:maxLength 12; ] .
```

Invoice Validation using SHACL. Once the invoice RDF graph has been created, the next step is validating the knowledge graphs. We use the W3C recommended language, Shapes Constraint Language (SHACL) [13], for the validation of RDF graphs. Constraints in SHACL can be specified over specific classes of the ontology using the *sh:targetClass* predicate, and over specific properties of the target class using the *sh:path* predicate. In our work, the shapes were created based on the input of the domain experts. For validating the correctness of the representation we create shapes based on the EDIFACT invoice guidelines. For example, shapes can express mandatory or conditional modules according to these guidelines. Listing 1.3 shows a SHACL constraint for specifying that the *documentNumber* property is mandatory (*sh:minCount* 1) and single valued (*sh:maxCount* 1) for entities of the class *InvoiceDetails*. SHACL can also be used to specify constraints about the formatting of the EDIFACT elements, including length, number, and permitted characters. Listing 1.4 shows a SHACL constraint for checking the datatype (*sh:datatype*) and the length (*sh:maxLength*) for the *documentNumber* property of the target class *InvoiceDetails*.

In addition to checking the structure and completeness of the invoices, EDIFACT-VAL also checks the logical correctness of the information with SHACL. This includes, for example, that the total net values of the goods in an invoice correspond to the sum of the net amounts of the items sold in the invoice. This type of constraint that involves aggregations over values of several RDF triples can be expressed in SHACL using SPARQL queries. Listing 1.5 shows how the aforementioned constraint is formulated with SHACL and SPARQL.

Listing 1.5. SHACL constraint to check that the total net value of the invoice is equal to the sum of the value of the items listed in the invoice

```
1  :SumNetPrice a sh:NodeShape ;
2     sh:targetClass edifact-o:InvoiceDetails ;
3     sh:sparql [
4        a          sh:SPARQLConstraint ;
5        sh:message "edifact-o:hasTotalLineItemAmount must equal the sum of all values of
                    edifact-o:hasLineItemAmount";
6        sh:prefixes [ sh:declare [
7                      sh:prefix   "edifact-o" ;
8                      sh:namespace "https://purl.org/edifact/ontology#"^^xsd:anyURI ; ] ] ;
9        sh:select
10       """SELECT $this (edifact-o:hasTotalLineItemAmount AS ?path) (?totalAmount AS ?value)
11          WHERE { $this a edifact-o:InvoiceDetails ; edifact-o:hasTotalLineItemAmount ?
                    totalAmount .
12             { select $this (sum(?itemAmount) as ?sum) {
```

```
13              ?item edifact-o:isItemOf ?invoice; edifact-o:hasLineItemAmount ?itemAmount
14              ?invoice edifact-o:hasInvoiceDetails $this . } group by $this }
15         FILTER (?sum != ?totalAmount) }""" ; ] .
```

4 Evaluation

In this section, we empirically evaluate the developed EDIFACT ontology and EDIFACT-VAL tool in terms of soundness and performance. Concretely, we focus on the following core questions:

Q1 Does the EDIFACT ontology meet the state-of-the-art standards? (§4.2)
Q2 Are RDF graphs created with EDIFACT-VAL sound? (§4.3)
Q3 Are the validation of EDIFACT-VAL results complete and correct? (§4.4)
Q4 How long does EDIFACT-VAL take to validate an EDIFACT invoice? (§4.5)
Q5 Is EDIFACT-VAL applicable to real-world business processes? (§4.6)

4.1 Experimental Set up

Dataset: We use 44 different real-world EDIFACT messages from 12 different suppliers and 6 different business cases to evaluate the performance of the EDIFACT-VAL tool. The selection of messages has been made together with the EDI experts from E/D/E to have a range of messages representing the day-to-day business workflow. Since each supplier may include different segments and data elements in the invoices, the selected messages represent a wide range of segments and segment combinations.

Tool Implementation: EDIFACT-VAL is implemented in Python 3. For SHACL validation, we use pySHACL [24], an open-source Python library. The EDIFACT-VAL is available online.[3] EDIFACT-VAL is equipped with 394 SHACL constraints provided by the experts following the EDFICAT standard. All experiments have been conducted on a machine with Intel Core i5 CPU and 8 GB of RAM.

4.2 Results of the Ontology Evaluation

Compliance with Ontology Best Practices. We assessed our EDIFACT Ontology concerning current standards and best practices for ontology publication using OOPS! [19]. OOPS! (OntOlogy Pitfall Scanner!) is a web service[4] that receives the URI of the ontology and performs checks in the structural, functional, and usability profiling. In total, OOPS! tests for 41 pitfalls concerning modelling decisions, ontology language, ontology clarity, ontology understanding, no inference, wrong inference, application context, common sense, and requirement completeness. The OOPS! results show that our EDIFACT ontology meets

---

[3] https://github.com/DE-TUM/EDIFACT-VAL.
[4] https://oops.linkeddata.es.

state-of-the-art ontology standards. Only one pitfall (P22) occurred during the evaluation regarding naming conventions in the ontology terms. Since we incorporate resources from eleven ontologies, some have different naming conventions. For example, in the vCard Ontology [16], the term 'has' is always in front of a data property, http://www.w3.org/2006/vcard/ns#hasStreetAddress, whereas the Dublin Core-Ontology [2] does not do it, http://purl.org/dc/elements/1.1/ date. Yet, this pitfall is marked as "minor level" in OOPS!, which means that it does not affect the overall quality of the EDIFACT Ontology.

Coverage of the Competency Questions. To ensure that the EDIFACT Ontology satisfies the competency questions CQ (cf. Listing 1), we have translated each CQ into a SPARQL query. The SPARQL queries are available in a Jupyter notebook.[5] Using the ontology and one EDIFACT message from the dataset, we execute the SPARQL queries using the Python library `rdflib` [14]. The results show that the CQ can be answered with our RDF representations.

4.3 Completeness of the Invoice KG Generated by EDIFACT-Val

In this work, we define completeness as follows: An invoice RDF graph is complete if every *relevant* data element in an EDIFACT message is represented in the graph. Since relevance is a domain-specific criterion, we consulted with domain experts to categorize elements in EDIFACT as relevant or negligible. Then, we analyze the RDF graph completeness considering the three cases of how the EDIFACT data corresponds with RDF representations:

CASE I (No Correspondence): The EDIFACT element is deemed negligible by experts and, therefore, is not encoded in the RDF graph.

CASE II (One-to-one Correspondence): The EDIFACT element is directly represented as one element in the EDIFACT Ontology.

CASE III (One-to-many Correspondence): The EDIFACT element is represented using several elements of the EDIFACT Ontology.

Distinguishing between these cases allows us to better understand and quantify the completeness of the generated RDF graphs. They ensure that no important information in the EDIFACT messages is omitted or incorrectly translated into the RDF representations.

CASE I indicates that the negligible EDIFACT elements should not affect the KG completeness. In our dataset, we identified 44 of 438 of these elements.

CASE II allows for a straightforward measurement of completeness by computing the ratio between the number of RDF triples – for which the terms have a one-to-one correspondence with EDIFACT – and the number of these EDIFACT data items in the messages. After this evaluation, we obtained that the RDF graphs produced by EDIFACT-VAL are complete (i.e., completeness 1.0).

---

[5] https://github.com/DE-TUM/edifact-ontology/blob/main/CompetencyQuestions/CQ-SPARQL.ipynb.

Table 3. Validation results of SHACL constraints for different business processes (BP). BP names are omitted in accordance with privacy regulations

Business Process	BP1	BP2	BP3	BP4	BP5	BP6	Total
EDIFACT Invoices	96	9	7	12	906	8	1,038
SHACL Constraints	86	81	48	36	57	86	394
Violations	692	41	45	58	859	38	1,733

CASE III entails a more sophisticated transformation (compared to the previous case) from the EDIFACT format into the RDF representation. The one-to-many correspondence in EDIFACT elements occurs in the representation of organizations, where EDIFACT combines the qualifier and identification of the organization in a single element. Yet, these are represented using several (more fine-grained) properties in the EDIFACT Ontology as they model different pieces of information about the organizations. The one-to-many correspondence case is handled by EDIFACT-VAL in the invoice pre-processing step (cf. Section 3.2). In new qualifiers can be defined during production by the users, EDIFACT-VAL implements a built-in mechanism to display unknown qualifiers for elements. In all tested 1,038 EDIFACT invoices, only 2 invoices yield 'unknown qualifier', i.e., the completeness of the RDF graph is 0.998 (out of 1.0). We inspected these invoices and found that these cases only occur due to non-standardized and self-created qualifiers in the original EDIFACT invoice. Based on these results, we can conclude that all admissible qualifiers are included in the resulting RDF graphs. I.e., EDIFACT-VAL achieves 1.0 completeness for admissible qualifiers.

4.4   Results of Invoice Validation with EDIFACT-Val

First, as a controlled evaluation, we manually introduced errors in a sample of EDIFACT invoices to test whether EDIFACT-VAL can detect them. These errors would cause violations of the SHACL constraints defined in our approach. These results show that EDIFACT-VAL successfully detected all the synthetic errors.

Next, we perform the validation over the entire dataset of EDIFACT messages. For this analysis, we present the results of the SHACL validations over the constraints defined for each business process in E/D/E. Table 3 presents the results of this evaluation. The results show the original EDIFACT invoices indeed contain errors that may affect the correctness of the invoice and compromise the integrity of related business processes.

4.5   Runtime Performance of EDIFACT-Val

We measure the efficiency of the EDIFACT-VAL when processing the invoices, i.e., the elapsed time between the tool receiving an EDIFACT message and producing the validation result. This time includes the generation of the RDF graph and its evaluation using the SHACL constraints. We selected one EDIFACT message

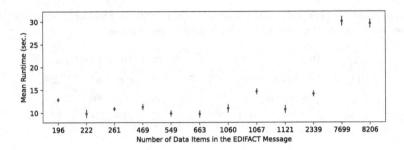

Fig. 3. EDIFACT-VAL runtime (sec.) for processing an EDIFACT message

per supplier, which may contain several invoices with varying numbers of EDI-FACT data items. We ran EDIFACT-VAL 10 times on each EDIFACT message using the `hyperfine` [18] command-line benchmarking tool. Figure 3 reports the average and variance of the runtime per message. These results indicate that the EDIFACT-VAL runtime is impacted by the number of elements in the messages, as expected. Overall, the mean time for validating an EDIFACT message is within an interval of 10 to 12 s. Only EDIFACT messages with more than 5,000 data items require more processing time, around 30 s.

4.6 In-Use Evaluation

Since the development was motivated by the group purchasing organization E/D/E, we also evaluated the applicability of our proposed solution approach to the E/D/E invoice processes. We asked the domain experts to validate the EDIFACT messages following the usual procedure, which is done manually. On average, it takes the experts around 30 min to validate one EDIFACT message. In comparison to EDIFACT-VAL (cf. Figure 3), the longest runtime is in the order of 30 s, which results in a time saving of around 98%.

We also asked the domain experts to analyse the violations reported by EDIFACT-VAL using the SHACL constraints. We learned that while the violations correctly capture the strictly formulated EDIFACT guidelines, certain violations are not critical in practice. This information can be included in the SHACL constraints implemented by EDIFACT-VAL using the *sh:severity* predicate of SHACL. This will be part of our future work.

5 Related Work

Ontologies for Electronic Invoices Schulze et al. [21] proposes an ontology, the Purchase-To-Pay Ontology (P2P-O) to represent electronic invoices. P2P-O relies on the European Standard EN 16931-1:2017 intending to provide ontology resources for all mandatory information in a purchase-to-pay process, which includes the invoicing process. In contrast, we aim not only to provide mandatory information for processing an electronic invoice, we also aim to validate the completeness and syntactical correctness of the invoices regarding their specific format. Furthermore, the terms used in the EDIFACT messages are not captured in the P2P-O ontology. This is why the creation of a standard-specific ontology, the EDIFACT ontology, is crucial in this case.

In the literature, we also found several ontologies and vocabularies that provide invoice-related definitions. The ones we reuse can be found in Table 2. However, they only present small parts of invoices. For example, schema.org [9] provides terms for the amount of money and currencies. Yet, these vocabularies are not specific enough to model all the EDIFACT terms.

Invoice Validation Several works have tackled the problem of invoice validation from different perspectives. Emmanuel and Thakur [5] present an approach that implements an EDI invoice validation framework; this work focuses on checking the completeness of invoices (defined as the number of fields) by comparing actual values to expected values as given in an XML file. The expected value is defined in a rule set. Similarly, our work validates invoices but uses KG technologies and more expressive SHACL constraints formulated by experts and the EDIFACT standard. Other approaches also perform invoice validation but not in the context of EDI. For example, Sál [20] presents a tool to check whether the invoice fields are provided correctly to a digital system w.r.t. to the original invoice in PDF. Other solutions [1,3,10] propose processing and classifying invoices in PDF using Machine Learning (ML) models. All these works check for the correctness of the translation of the invoices into machine-readable formats, and not the correctness of the original invoices. These approaches greatly differ from EDIFACT-VAL, as it processes invoices that are already in a machine-readable format and validates the correctness of the original EDIFACT invoice. Lastly, the work by Schulze et al. [21] uses SPARQL queries to perform analytical tasks on the invoices. Examples of these include finding items that are sold in large cities. In contrast, EDIFACT-VAL is tailored to check the conformance of the invoices to the EDIFACT standard and other constraints defined by experts.

6   Conclusion and Future Work

We presented a novel automatic approach EDIFACT-VAL that assists EDI experts in validating EDIFACT messages with the help of an invoice knowledge graph. The EDIFACT messages are transformed into RDF graphs using the proposed EDIFACT Ontology and RML mappings. The graphs are further validated using SHACL constraints specified with the input from domain experts.

Our experiments show that our proposed solutions enable the validation of EDIFACT invoices effectively. In particular, the evaluation with domain experts revealed that EDIFACT-VAL can considerably reduce the manual efforts. One crucial takeaway from this evaluation is that the benefit of automatic approaches like EDIFACT-VAL relies on the quality of the validation constraints. Creating too strict constraints can result in mistakenly flagging invoices as faulty even though the errors may not affect the invoice processing workflow. Therefore, assigning proper severity levels to the constraints is essential to develop usable solutions.

Future work can extend our solution for different business documents in different EDIFACT formats, i.e. purchase order message (ORDERS) or despatch advice message (DESADV). Furthermore, the obligatory implementation of electronic invoices in Germany by 2025 will make the application of open standards more prominent, especially for organizations that find the cost or time required for EDIFACT implementation too demanding. In this line, we hope that our work contributes to the implementation of electronic invoice processing using open knowledge graph and semantic web technologies.

References

1. Baviskar, D., Ahirrao, S., Kotecha, K.: Multi-layout unstructured invoice documents dataset: a dataset for template-free invoice processing and its evaluation using AI approaches. IEEE Access **9**, 101494–101512 (2021). https://doi.org/10.1109/ACCESS.2021.3096739
2. DCMI Usage Board: DCMI Metadata Terms. https://www.dublincore.org/specifications/dublin-core/dcmi-terms/ (2020)
3. Desai, D., Jain, A., Naik, D., Panchal, N., Sawant, D.: Invoice processing using RPA & AI. In: Proceedings of the International Conference on Smart Data Intelligence (ICSMDI 2021) (2021)
4. Dimou, A., et al.: RML: A generic language for integrated RDF mappings of heterogeneous data. Ldow **1184** (2014)
5. Emmanuel, M., Thakur, S.: An approach to develop invoice validation framework. Int. J. Eng. Res. Technol. (IJERT) **1** (2012)
6. EN16931-1:2017: Electronic invoicing-part 1 : Semantic data model of the core elements of an electronic invoice. Technical report, European Committee for Standardization (2017)
7. García-González, H., Boneva, I., Staworko, S., Labra-Gayo, J.E., Lovelle, J.M.C.: ShExML: improving the usability of heterogeneous data mapping languages for first-time users. PeerJ Comput. Sci. **6**, e318 (2020)
8. Grüninger, M., Fox, M.S.: The role of competency questions in enterprise engineering. In: Rolstadås, A. (ed.) Benchmarking — Theory and Practice. IAICT, pp. 22–31. Springer, Boston, MA (1995). https://doi.org/10.1007/978-0-387-34847-6_3
9. Guha, R.V., Brickley, D., Macbeth, S.: Schema.org: evolution of structured data on the web. Commun. ACM **59**(2), 44–51 (2016)
10. Gunaratne, H., Pappel, I.: Enhancement of the e-invoicing systems by increasing the efficiency of workflows via disruptive technologies. In: Chugunov, A., Khodachek, I., Misnikov, Y., Trutnev, D. (eds.) EGOSE 2020. CCIS, vol. 1349, pp. 60–74. Springer, Cham (2020). https://doi.org/10.1007/978-3-030-67238-6_5
11. Horridge, M., Gonçalves, R.S., Nyulas, C.I., Tudorache, T., Musen, M.A.: Webprotégé: a cloud-based ontology editor. In: Companion Proceedings of The 2019 World Wide Web Conference, pp. 686–689 (2019)
12. Iglesias-Molina, A., Chaves-Fraga, D., Dasoulas, I., Dimou, A.: Human-friendly RDF graph construction: which one do you chose? In: Garrigós, I., Murillo Rodríguez, J.M., Wimmer, M. (eds.) Web Engineering, ICWE 2023. Lecture Notes in Computer Science, vol. 13893, pp. 262–277. Springer, Cham (2023). https://doi.org/10.1007/978-3-031-34444-2_19
13. Knublauch, H., Kontokostas, D.: Shapes constraint language (SHACL). W3C recommendation, W3C (2017). https://www.w3.org/TR/2017/REC-shacl-20170720/

14. Krech, D., et al.: RDFLib (2023). https://doi.org/10.5281/zenodo.6845245, https://github.com/RDFLib/rdflib

15. Lohmann, S., Link, V., Marbach, E., Negru, S.: WebVOWL: web-based visualization of ontologies. In: Lambrix, P., et al. (eds.) EKAW 2014. LNCS (LNAI), vol. 8982, pp. 154–158. Springer, Cham (2015). https://doi.org/10.1007/978-3-319-17966-7_21

16. McKinney, J., Iannella, R.: vCard Ontology - for describing People and Organizations. W3C note, W3C (2014). https://www.w3.org/TR/2014/NOTE-vcard-rdf-20140522/

17. Noy, N.F., et al.: Ontology development 101: a guide to creating your first ontology (2001)

18. Peter, D.: hyperfine (2023). https://github.com/sharkdp/hyperfine

19. Poveda-Villalón, M., Gómez-Pérez, A., Suárez-Figueroa, M.C.: Oops!(ontology pitfall scanner!): an on-line tool for ontology evaluation. Int. J. Semantic Web Inf. Syst. (IJSWIS) **10**(2), 7–34 (2014)

20. Sál, J.: Data mining as tool for invoices validation. IT Pract. **2018**, 121 (2018)

21. Schulze, M., Schröder, M., Jilek, C., Albers, T., Maus, H., Dengel, A.: P2P-O: a purchase-to-pay ontology for enabling semantic invoices. In: Verborgh, R., et al. (eds.) ESWC 2021. LNCS, vol. 12731, pp. 647–663. Springer, Cham (2021). https://doi.org/10.1007/978-3-030-77385-4_39

22. Shimizu, C., Hirt, Q., Hitzler, P.: MODL: a modular ontology design library. arXiv preprint arXiv:1904.05405 (2019)

23. Shotton, D.: FRAPO, the Funding, Research Administration and Projects Ontology. Technical report (2017). http://purl.org/cerif/frapo

24. Sommer, A., Car, N.: pySHACL (2022). https://doi.org/10.5281/zenodo.4750840, https://github.com/RDFLib/pySHACL

25. Suárez-Figueroa, M.C., Gómez-Pérez, A., Fernández-López, M.: The NeOn methodology for ontology engineering. In: Suárez-Figueroa, M.C., Gómez-Pérez, A., Motta, E., Gangemi, A. (eds.) Ontology Engineering in a Networked World, pp. 9–34. Springer, Heidelberg (2012). https://doi.org/10.1007/978-3-642-24794-1_2

26. Van Assche, D., Delva, T., Heyvaert, P., De Meester, B., Dimou, A.: Towards a more human-friendly knowledge graph generation & publication. In: ISWC2021, The International Semantic Web Conference, vol. 2980. CEUR (2021)

Towards Cyber Mapping the German Financial System with Knowledge Graphs

Markus Schröder[1]([⊠])(iD), Jacqueline Krüger[2](iD), Neda Foroutan[1],
Philipp Horn[2], Christoph Fricke[2], Ezgi Delikanli[2], Heiko Maus[1](iD),
and Andreas Dengel[1](iD)

[1] Smart Data & Knowledge Services Department, DFKI GmbH,
Kaiserslautern, Germany
{markus.schroeder,neda.foroutan,heiko.maus,andreas.dengel}@dfki.de
[2] Deutsche Bundesbank, Frankfurt am Main, Germany
{jacqueline.krueger,philipp.horn,christoph.fricke,
ezgi.delikanli}@bundesbank.de

Abstract. The increasing outsourcing by financial intermediaries intensifies the interconnection of the financial system with third-party providers. Concentration risks can materialize and threaten financial stability if these third-party providers are affected by cyber incidents. With the goal of preserving financial stability, regulators are interested in tracing cyber incidents efficiently. One method to achieve this is cyber mapping, which allows them to analyze the connections between the financial network and the cyber network. In this paper, a provenance-aware knowledge graph is constructed to model this kind of mapping for investment funds which are part of the German financial system. As a first application, we provide a front-end for analyzing the funds' outsourcing behaviors. In a user study with ten experts, we evaluate and show the application's usability and usefulness. Time estimations for certain scenarios indicate our application's potential to reduce time and effort for supervisors. Especially for complex analysis tasks, our cyber mapping solution could provide benefits for cyber risk monitoring.

Keywords: Knowledge Graph Construction · RDF · Ontology ·
German Financial System · Cyber Mapping · Cyber Incidents · Fund
Prospectus

1 Introduction

Financial intermediaries are increasingly outsourcing processes and services to third parties, which manifests in growing interconnectedness of the financial system with entities outside this network [20]. The trend is further intensified by increasing digitization, as information and communication technology becomes the core infrastructure for all financial processes [13]. Concentration risks may arise if outsourcing activities rely on only a few large service providers (e.g., cloud service providers) [4,23]. With growing third-party dependencies and increasing concentration risks, the question arises as to possible transmission channels

A. Meroño Peñuela et al. (Eds.): ESWC 2024, LNCS 14664, pp. 270–288, 2024.
https://doi.org/10.1007/978-3-031-60626-7_15

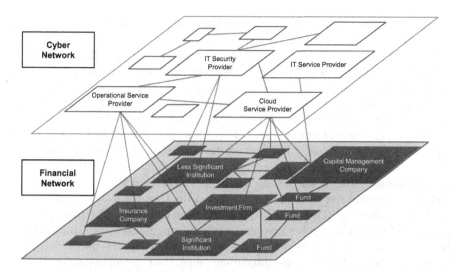

Fig. 1. Conceptual presentation of the cyber mapping methodology showing the cyber network (top) and financial network (bottom). Red arrows indicate the actual *mapping* between them. Source: Deutsche Bundesbank. (Color figure online)

between third-party providers (TPPs) and financial intermediaries. Cyber incidents entail large economic impacts, with average estimated direct costs per incident of €365,000 [2]. The number of cyber incidents is growing rapidly. The financial sector is particularly affected, with the average growth rate of cyber incidents between 2020 and 2022 twice as high as the growth rate of cyber incidents across all sectors [27]. With 22% of ransomware attacks targeting IT companies, possible TPPs of financial intermediaries are severely prone to cyber attacks [10]. Cyber incidents affecting both financial intermediaries and their supply chains pose a risk to financial stability if they significantly impair the provision of key economic functions by the financial system [36,39]. Recent cyber attacks have already been seen to affect not only but also to have wider impacts [14] and have also induced turmoil in financial markets [26]. Thus, an understanding of the supply chains is crucial in order to assess any risks to financial stability or intermediaries and address them properly.

Cyber mapping enables the network analysis of the financial system and its TPPs (see Fig. 1). Mapping, in this context, refers to the connection of nodes from two distinct networks - in our case, from the financial network and the cyber network. It thus allows for the identification of vulnerabilities related to outsourcing activities. Due to limited data availability [21], cyber mapping has mostly been more of a theoretical concept thus far [8]. However, recent regulatory changes [9,42] enlarge the data basis and enable cyber mapping to be realized.

Cyber mapping has been recognized as essential in cyber risk monitoring of the financial sector and also provides benefits for banking supervision and financial market infrastructure oversight [3,21,38]. Its key feature is the immediate provision of information on potentially affected financial entities in the event of a cyber incident, thus enabling decision-makers to determine necessary ad hoc measures. It also helps supervisors and overseers to quickly gain an overview of outsourcing risks and third-party dependencies of selected financial entities. By identifying relevant TPPs bearing concentration risks for the financial system, cyber mapping can also help to determine suitable regulatory measures or contribute to the development of the regulatory framework.

However, to the best of our knowledge, cyber mapping of the German financial system as we envision it has not yet been performed. This work takes the first steps in filling the gap. Our contribution is an outcome of the transfer lab "Cybermapping"[1] which was set up by the Deutsche Bundesbank[2] (the central bank of the Federal Republic of Germany) and the German Research Center for Artificial Intelligence[3] (DFKI) to foster research on this topic. The paper reports on the development of a cyber mapping method using Knowledge Graphs (KGs) [22] and the Resource Description Framework (RDF) [40]. In particular, KGs enable us to model the networks' interconnections, their mappings and metadata by integrating various data silos. The graph is built using a dedicated cyber mapping ontology and linking it to structured financial data as well as information extracted from unstructured texts. Traceability is ensured by including additional provenance information.

The remainder of the paper is organized as follows: Sect. 2 covers related work on building Knowledge Graphs in the cybersecurity and financial domain as well as relevant projects. Subsequently, our own construction approach is described in Sect. 3. Next, initial numbers on our KG and a first user application are provided (Sect. 4). By conducting a user study, discussed in Sect. 5, we present first results on our application's usefulness. Finally, Sect. 6 provides a conclusion and an outlook for further research.

2   Related Work

In the literature, Knowledge Graphs (KGs) have been built for various domains (for a survey see [1]). In the field of Information and Communication Technology (ICT), there are efforts to build cybersecurity KGs. To achieve this, the works identify relevant concepts and relationships in (un)structured sources using various methods such as Named Entity Recognition (NER) [33], Relation Extraction [37], word embeddings [15] and Extraction-Transformation-Loading (ETL) [34]. Similarly, there are approaches which aim to cover parts of the financial system with KGs. To construct these financial KGs, several sources are considered such as annual financial reports [44], financial news articles [17], financial research reports [43] and data on equity [31].

The referenced papers give a comprehensive overview of a variety of KG construction approaches. Although cybersecurity and financial domains are covered,

---

[1] https://www.dfki.uni-kl.de/cybermapping.

[2] https://www.bundesbank.de/en.

[3] https://www.dfki.de/en/web.

these works lack the actual mapping between them which is a fundamental aspect of cyber mapping. We have therefore investigated related research projects on this topic, too.

The Financial Supervisory Authority of Norway (Finanstilsynet[4]) together with the Central Bank of Norway (Norges Bank[5]) drafted a first solution on financial sector mapping [32]. Their mapping approach is based on a poll among ministries to identify fundamental national functions, e.g. the execution of financial transactions. In a second step, the relevant organisations and critical service providers related to these functions are identified and mapped to the fundamental national functions. Thus, their top-down concept of a financial sector map differs from our data-driven approach. Moreover, there is no publicly available report on the applied technology or its implementation.

For Germany, the Bundesanstalt für Finanzdienstleistungsaufsicht[6] (German Federal Financial Supervisory Authority, or BaFin) engaged a research team from Innsbruck University to explore possible scenarios regarding future developments in the financial industry [12,13]. Based on these scenarios, the research team recommended identifying the relevant ICT service provider by mapping the financial system. As a consequence, BaFin produced a first map of the financial system [11], but it is still restricted to structured data of one financial sector.

In conclusion, a cyber mapping approach with semantic technologies does not seem to be available yet. Therefore, we implemented an approach to construct a KG for cyber mapping the financial system.

3 Knowledge Graph Construction

Our Knowledge Graph (KG) uses the Resource Description Framework (RDF) [40] to make formal statements about cyber mapping. Figure 2 presents an overview of our KG construction. With the help of a dedicated cyber mapping ontology (covered in Sect. 3.1), structured data about the German financial system in Excel, CSV and XML format is mapped to RDF statements (Sect. 3.2). Similarly, unstructured data in fund prospectus PDFs is extracted using Natural Language Processing (NLP) techniques such as Named Entity Recognition (NER) and Relation Extraction (RE) which are outlined in Sect. 3.3. Either way, the origins of created RDF statements are recorded in a special provenance box for traceability reasons (Sect. 3.4).

3.1 Cyber Mapping Ontology

To express facts about cyber mapping, we are in need of an appropriate ontology [25]. In the literature, some ontologies for the domains of finance and cybersecurity can be found. A prominent one is the Financial Industry Business Ontology[7]

---

[4] https://www.finanstilsynet.no/en/.
[5] https://www.norges-bank.no/en/.
[6] https://www.bafin.de/EN/.
[7] https://spec.edmcouncil.org/fibo/.

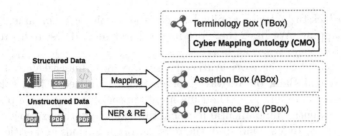

Fig. 2. Knowledge graph construction overview: a cyber mapping ontology (terminology box) provides necessary vocabulary to import (un)structured data into a KG (assertion box). The provenance box keeps track of each RDF statement's origin.

(FIBO) [7], which focuses on the business of finance. Another one is the Unified Cybersecurity Ontology (UCO) [41], covering the cybersecurity domain. However, none of them fully satisfy our requirements for cyber mapping. We therefore defined our own Cyber Mapping Ontology (CMO, prefixed cmo) which is still a work in progress. It has been defined with the well-known Protégé ontology editor[8] and is published with WIDOCO[9]. A first draft of the ontology's specification is available online[10] in English and German.

Since we are interested in mapping the German financial system, we consulted data provided by BaFin. Its database about companies lists 44 company types, which we reviewed and generalized to a class tree with 12 nodes under a branch node cmo:Company. Similarly, its investment funds database helped us to distil 7 fund classes from 15 types with a generalizing cmo:Fund class. All classes in our ontology are generalized to an upper class named cmo:Entity.

During the examination of sources, we came across various ways to identify entities, which is reflected in our ontology with individual properties. The BaFin uses internal references (cmo:baFinRef) and external IDs (cmo:baFinId), while the European Central Bank[11] (ECB) manages its own IDs (cmo:ecbId). Funds are usually identified by the International Security Identification Number[12] (cmo:isin), while companies are commonly recognized by the Legal Entity Identifier (cmo:lei). The Register of Institutions and Affiliates Database (RIAD) [19] proposes its own identifier (cmo:riadId).

To perform cyber mapping, we are interested in certain relationships between two companies. One such relation is the outsourcing of a service from one company to another (cmo:outsourcesTo). Since this typically involves a source entity (cmo:source), target entity (cmo:target) and the subject matter of the outsourcing (cmo:Outsourcing), we decided to model this statement as an n-ary relation[13]

---

[8] https://protege.stanford.edu/.

[9] https://github.com/dgarijo/Widoco.

[10] https://www.dfki.uni-kl.de/cybermapping/ontology.

[11] https://www.ecb.europa.eu/.

[12] https://www.isin.org/.

[13] https://www.w3.org/TR/swbp-n-aryRelations/.

(cmo:OutsourcingStatement). In order to keep track of where this statement was made, we make use of PROV-O's prov:wasDerivedFrom property [6] and the NLP Interchange Format[14] (NIF) [29]. This way, we are able to record, for instance, in which sentence of a PDF file an outsourcing statement occurred. Section 3.3 shows how these concepts are applied in practice, while more on the provenance topic is covered in Sect. 3.4.

To model addresses, concepts from DBpedia's ontology[15] and the ontology for vCard[16] are used: the vcard:hasAddress property with a vcard:Address blank node store dbo:address (street name and house number), dbo:postalCode, dbo:city and dbo:country. The property dbo:subsidiary models typical company structures.

With our defined ontology, we are able to map structured data about the financial system to our KG, which is described in the next section.

3.2 Structured Data: Financial Domain

The Deutsche Bundesbank and BaFin legally collect regulatory data from German financial entities (e.g., banks). However, at the beginning of our transfer lab useful data for our project was either still in the process of being collected or subject to strict confidentiality. We therefore decided to initially use public data to construct the KG.

Table 1 lists six publicly available data sources about financial intermediaries and funds provided by BaFin, the ECB and the European Securities and Markets Authority[17] (ESMA). It provides an overview of the number of records (#Rec.), columns (#Col.) and types (#Typ.) as well as the source's data format (XML, Excel or CSV). While entities are always named (mapped to skos:prefLabel), different sets of identifiers are provided by each data source, which is indicated by a check mark (✔). In some cases, data is less complete, which is indicated in the table by a tilde sign (~). Regarding the *Address* column, this means that only the country is mentioned (mapped to dbo:country). The selection of additional properties also greatly varies per dataset.

With the CMO (Sect. 3.1) and information from Table 1, structured data can be lifted to RDF using an appropriate technique, for instance, KG generation with the RDF Mapping Language[18] (RML) [16]. To give an illustrative example of a resource in our graph, Listing 1.1 depicts in Turtle syntax [5] a fictional German stock company with type, label, identifier, address and subsidiary information. The resource is identified with a Universally Unique Identifier (UUID) in our cyber mapping resource namespace (prefixed cmr). Since such entities can be named differently (e.g. abbreviations), skos:altLabel records alternative labels.

---

[14] https://persistence.uni-leipzig.org/nlp2rdf/ontologies/nif-core.

[15] https://dbpedia.org/ontology.

[16] https://www.w3.org/2006/vcard/ns#.

[17] https://www.esma.europa.eu/.

[18] https://rml.io/.

Table 1. Six publicly available data sources about companies and funds in the financial system. For each entry the number of records (#Rec.), columns (#Col.) and types (#Typ.) is given as well as its format. A check mark (✓) indicates that this property can be found in the dataset, while a tilde (~) indicates incomplete data. The last column lists additional properties available in the dataset, for instance, some funds' refer to their Capital Management Company (CMC).

Data Source	#Rec	#Col	#Typ	Format	skos:prefLabel	baFinId	ecbId	lei	isin	riadId	Address	Additional Properties
BaFin Company Database[a]	7,151	8	44	XML	✓						✓	arbitration board
BaFin Investment Funds Database[b]	14,427	10	15	XML	✓	✓			~		~	structure; name ref. to CMC
ECB Supervised Entities[c]	900	5	4	Excel	✓			✓			~	subsidiary; grounds for significance
ECB Investment Funds[d]	78,932	19	3	Excel	✓		✓	✓	✓		✓	capital variability; investment policy; net asset value size class; ref. to CMC
ECB Monetary Financial Institutions[e]	5,720	14	4	CSV	✓				✓	✓	✓	country of registration; metadata about head
ESMA Money Market Funds[f]	472	15	4	CSV	✓			✓			~	legal framework; ref. to CMC

[a] https://portal.mvp.bafin.de/database/InstInfo/?locale=en_US
[b] https://portal.mvp.bafin.de/database/FondsInfo/?locale=en_US
[c] https://www.bankingsupervision.europa.eu/banking/list/html/index.en.html
[d] https://www.ecb.europa.eu/stats/financial_corporations/list_of_financial_institutions/html/index.en.html#if
[e] https://www.ecb.europa.eu/stats/financial_corporations/list_of_financial_institutions/html/daily_list-MID.en.html
[f] https://registers.esma.europa.eu/publication/searchRegister?core=esma_registers_mmf04

Listing 1.1. A fictional example of a mapped resource in the cyber mapping KG expressed in Turtle syntax.

```
cmr:bc57a47d-d990-486f-9b7f-4af78aded30a
   rdf:type          cmo:SignificantInstitution ;
   skos:prefLabel    "Mercurtainment Bank Aktiengesellschaft" ;
   skos:altLabel     "Mercurtainment Bank AG" ;
   cmo:baFinRef      "303846" ;
   cmo:lei           "G9QIEQ1BITM5RF3YCDRQ" ;
   cmo:riadId        "DE70255" ;
   vcard:hasAddress  [
     dbo:address     "Musterstr. 42" ;
     dbo:postalCode  "60312" ;
     dbo:city        dbr:Frankfurt_am_Main ;
     dbo:country     dbr:Germany
   ] ;
   dbo:subsidiary cmr:c613bf97-07da-46cd-ab5c-eba0454679a9 .
```

While the above-mentioned data lists the majority of entities in the German financial system, it does not depict specific relationships for the purpose of performing cyber mapping. Helpfully, such information can be found, at least partly, in unstructured texts, which is covered in the following section.

3.3   Unstructured Data: Fund Prospectus

Section 164[19] of the German Capital Investment Code (Kapitalanlagegesetzbuch, or KAGB) governs the creation of sales prospectuses, or more specifically fund prospectuses. Here, Capital Management Companies (CMCs) are obligated to describe which activities are outsourced to specific companies (see §165, Sec. 2, No. 33 KAGB). By interpreting these texts, we are able to map funds to companies with our cmo:outsourcesTo property. Acquiring such outsourcing relations from financial intermediaries to TPPs makes it possible to perform an initial cyber mapping.

In prior work [24], we compiled a corpus of 1,054 fund prospectuses (PDFs). From these documents, 948 extracted sentences were manually annotated with 5,969 named entity annotations and 2,573 Outsourcing–Company relationship annotations. The resulting German dataset on Company Outsourcing in Fund prospectuses (CO-Fun) is used in this paper to acquire structured RDF statements. Our initial plan had been to fully automate the extraction from such unstructured data; however, there was no available NLP model that met our requirements. We therefore decided to manually annotate outsourcing relationships in documents to build a ground-truth dataset which is used to populate our KG. What such RDF statements look like is presented in Listing 1.2. While the cmo:outsourcesTo property simply relates a fund to a company, the corresponding statement (cmo:OutsourcingStatement) additionally states the outsourcing and provenance information (prov:wasDerivedFrom). This way, we are

---

[19] https://www.gesetze-im-internet.de/kagb/__164.html.

Listing 1.2. Illustration of an outsourcing statement lifted from a sentence in a fund prospectus. For readability, some UUIDs are shortened and some literals are formatted.

```
cmr:1c956834
  cmo:outsourcesTo cmr:2d56a950 .

cmr:8ea294fd-9f0e-4158-a2e3-c14f93c2b4b2
  rdf:type              cmo:OutsourcingStatement ;
  cmo:source            cmr:1c956834 ;
  cmo:target            cmr:2d56a950 ;
  cmo:outsourcing       cmr:DataCenterService ;
  prov:wasDerivedFrom   cmr:s96048cb .

cmr:s96048cb
  rdf:type              nif:String, nif:Sentence ;
  nif:anchorOf          "The company has outsourced data center
                        services to Mercurtainment & CO KGaA." ;
  nif:referenceContext  <file://fund-prospectus.pdf> .

<file://fund-prospectus.pdf>
  rdf:type              nfo:FileDataObject ;
  nfo:fileName          "fund-prospectus.pdf" ;
  dct:hasPart           cmr:s96048cb , cmr:s5d53b88 , cmr:sd4219e7;
  cmo:managementCompany cmr:cmcf78d3 ;
  cmo:fund              cmr:1c956834 .
```

able to reconstruct the sentence in a certain fund prospectus that leads to an outsourcing statement. Regarding the prospectus itself, we can record its CMC (cmo:managementCompany) and the fund (cmo:fund) it pertains to.

Providing additional information to trace the origins of statements has proven to be very useful. We therefore considered provenance for all statements in the ABox, which is discussed next.

3.4 Provenance Information

A special feature in our cyber mapping KG is the existence of a Provenance Box (PBox in analogy to TBox and ABox, see Fig. 2). Its purpose is the storage of additional statements for *every* statement asserted in the ABox to enable comprehensive traceability. To implement this, we make use of RDF-star[20] [28], which allows us to annotate statements in RDF with metadata. Listing 1.3 gives an example of how the asserted skos:altLabel statement from Listing 1.1 is annotated. Using RDF-star, the triple is quoted (<<...>>) on subject position. With the provenance ontology (PROV-O) [6], several aspects about the skos:altLabel-statement are recorded: one is the agent who is responsible for creating the statement by using the prov:wasAttributedTo property. Usually, this involves a certain activity, for instance, an importing procedure or interface usage, which is stated with a prov:wasGeneratedBy statement. To note the origin of the quoted statement, a prov:wasDerivedFrom property refers to the data source (e.g., a

---

[20] https://www.w3.org/2021/12/rdf-star.html.

Listing 1.3. An example of how provenance information is annotated with RDF-star and PROV-O. URIs and UUIDs are shortened for readability.

```
<< cmr:bc57a47d skos:altLabel "Mercurtainment Bank AG" >>
   rdf:type              prov:Entity ;
   prov:wasAttributedTo  <https://.../agent/smith> ;
   prov:wasGeneratedBy   <https://.../activity/1868ccf> ;
   prov:wasDerivedFrom   <https://.../entity/some.csv> ;
   dct:date              "2023-07-19T14:32:54.812Z"^^xsd:dateTime .
```

CSV file). Further RDF statements are made about agent, activity and source to provide useful metadata about them. Dublin Core's[21] dct:date attribute is used to be able to reconstruct a chronological order.

By performing all steps discussed in Sect. 3, our construction approach results in a first version of a cyber mapping KG. In the next section, we discuss the graph's content and a first utilization of it.

4   Knowledge Graph Application

Graph. An initial version of our cyber mapping KG contains 1,725,383 RDF statements about 108,030 entities, including 93,253 funds and 14,777 companies in the financial system. The latter are separated into 8,184 (financial) institutions, 5,307 capital management companies and 1,286 insurance companies. However, duplicates likely still exist, particularly in the case of cmo:Fund instances, as these were imported from several independently managed sources (Sect. 3.2).

Regarding documents, metadata about 917 fund prospectuses with 686 sentences are available, as indicated in Listing 1.2. We acquired 7,239 outsourcing statements, which refer to 375 outsourcing entities. 4,033 distinct cmo:outsourcesTo relationships between funds and companies could be identified. Such statements are essential in modeling the actual cyber mapping.

Our Provenance Box (PBox) graph contains 10,338,786 triples attributed to two users, one of whom was responsible for initiating the main data import steps. Statements were generated by 17 activities, primarily RDF mapping procedures and a few user interactions via interfaces. Resulting triples were derived from 15 sources: besides the main data sources (see Table 1), this also includes auxiliary data and users. The total number of provenance triples is higher than expected due to multiple iterations of import steps and kept provenance information.

Gathering such an RDF dataset about parts of the German financial system naturally raises concerns about potential misuse. We therefore abstain from making our KG publicly available and only provide fictional examples.

Application. Having successfully constructed such a KG enables us to provide useful applications for end users such as overseers and supervisors. One use case is the analysis of outsourcing relations in the context of cyber risks. With a focus on

---

[21] http://purl.org/dc/terms/.

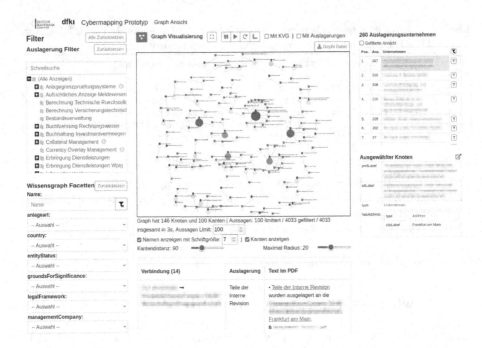

Fig. 3. Web application (in German) to interactively analyze outsourcing relationships between funds and companies. Left: Graph filter options; Center: Graph visualization and connection list; Right: Top outsourcing companies and selected node.

outsourcing companies, we would like to allow users to perform data exploration using an initial application, which is presented in Fig. 3. The application provides several features (F) for inspecting and filtering the KG.

Regarding inspection, a graph visualization (F1) derived from the KG is presented in the center view showing funds and companies as connected nodes. Larger red nodes suggest a high number of incoming edges, indicating where funds mostly outsource their services to. Below, a table (F2) lists outsourcing relationships together with the relevant text passage from the linked fund prospectus (F3). In the top right corner, a table shows for the current view outsourcing companies ordered by their incoming edges. Additional properties of a selected node (F4) are presented on the bottom right.

In the case of filters (left), a taxonomy of outsourcing categories (top left, F5) lets users restrict the graph to a certain outsourcing type such as Information Technology (IT). Outsourcing categories have expandable definitions attached for a better understanding (F6). Properties in our KG (bottom left, F7) can be used to further filter nodes, for instance, by name (skos:prefLabel), location (dbo:country) or other metadata (e.g., cmo:groundsForSignificance).

In the next section, a study is presented in which the application's usefulness for potential users is evaluated.

5 User Study

A user study was conducted in order to evaluate our application regarding its perceived user experience, features and potential time saving for given scenarios. Its setup is described in Sect. 5.1, followed by a description and interpretation of the results in Sect. 5.2.

5.1 Setup

We conducted a study with ten selected experts (E1–E10, 9 male, 1 female) from three different departments of the Deutsche Bundesbank with at least three years of work experience. In particular, the experts' average work experience is 16.1 years (s.d. 8.8, min. 4, max. 33). The distribution among age groups is almost balanced starting from 25 to older than 55 years (ten year spanned).

Sessions were conducted in one-on-one interviews or small group meetings with up to three experts. Each expert was provided with individual access to the application described in Sect. 4. At the beginning of each session, a short introduction was provided by a conductor (an author of this paper) which took about 15 min. The introduction consisted of the provision of basic information, such as the elaboration of the cyber mapping concept and its data basis. After that, the conductor provided a practical induction for the application by present- ing its key features. In a subsequent testing phase, the experts were asked to use our application for at least ten minutes in order to familiarize themselves with the application and to explore its features. Questions could be asked anytime, followed by further clarifications provided by the conductor. In the end, experts spent an average of 20.2 min testing our application (s.d. 4.35, min. 14, max. 25).

After the testing phase, the experts were provided with a structured ques- tionnaire, consisting of four parts. In the first part, 26 questions were asked by employing a standardized User Experience Questionnaire (UEQ) [35] to measure the experts' experience regarding the following six factors: attractiveness, per- spicuity, efficiency, dependability, stimulation and novelty. In the second part, questions were asked with respect to the perceived usefulness of the seven fea- tures of our application (F1–F7) using a 7-point Likert scale from useless (1) to useful (7). Furthermore, feedback from the experts was collected with regard to further desired features that should be added to the current application.

Last, participants had to estimate the time needed to fulfil tasks in given scenarios (S1–S3) with and without our application. For the definition of the scenarios, we considered essential cyber risk monitoring features of cyber map- ping.

Affected Intermediaries (S1) "You learn of a cyber incident at an IT com- pany that is no longer able to perform its tasks. Suppose you would like to find out which financial intermediaries relevant to you are potentially affected."

Outsourcing Relations (S2) "Suppose you would like to know to whom a specific financial intermediary has outsourced its processes to."

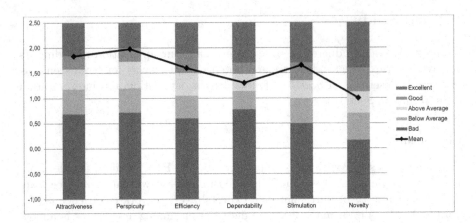

Fig. 4. Mean and distribution of the six factors from the User Experience Questionnaire (UEQ) [35] derived from all expert answers to the 26 questions.

Outsourcing Relevance (S3) "Suppose you would like to identify the top ten outsourcing companies for the financial system regarding outsourcing of accounting services."

Notably, the identification of possible transmission channels between a TPP and financial intermediaries in the event of a cyber incident were included in S1. S2 covers the understanding of the supply chain of a financial intermediary. S3 represents a first step in the analysis of potential concentration risks for a defined scope. Since our application was tested in different departments, the experts were asked to assume the application would already include their relevant data.

5.2   Results

The scaled UEQ results (see Fig. 4) show that all factors were rated higher than the values of the benchmark dataset based on 468 studies [30]. The ratings show that attractiveness and perspicuity achieved the highest ratings close to "excellent" results, while stimulation and efficiency are located in the "good" area, followed by dependability and novelty being "above average". A rather low value for the ergonomic quality aspect of dependability (i.e. predictable, secure) might reveal that further explanations of the way the application functions and simplifications towards a more intuitive interface could be helpful. The quality aspect of novelty (i.e. innovative, creative) might be low because the application's front-end mainly comprises data visualization and filtering options, which is rather standard and thus expected by the experts. Still, overall results show us that our initial application gave participants a good user experience with potential for improvements.

The results regarding the experts' ratings of the usefulness of the seven specific features (F1–F7) and the scenario results are depicted in Table 2. With an overall mean value of 6.1 close to the max. value of 7 (useful), we perceive that

Table 2. Questionnaire results stating for each Expert (E) their feature ratings and estimated times. F1 to F7 encompass the questions posed regarding the features' perceived usefulness. S1 to S3 cover the estimated times needed in minutes for solving the three scenarios with (w) and without (w/o) our application ('n' denotes unsolvable; minutes rounded for better readability). Difference (diff) between the estimations is provided. Below, mean and standard deviation (s.d.) values are calculated.

E	F1	F2	F3	F4	F5	F6	F7	S1			S2			S3		
								w	w/o	diff	w	w/o	diff	w	w/o	diff
E1	6	4	6	5	6	7	7	5	5	0	5	5	0	5	n	–
E2	5	6	6	5	7	7	5	5	120	115	5	60	55	5	240	235
E3	7	7	7	3	7	7	5	8	60	53	8	30	23	8	n	–
E4	7	7	7	5	5	6	5	10	n	–	5	n	–	15	n	–
E5	7	4	7	7	7	7	5	60	n	–	1	60	59	15	n	–
E6	5	7	7	6	6	7	7	5	120	115	1	60	59	10	180	170
E7	7	7	4	6	4	6	5	5	n	–	5	n	–	5	n	–
E8	7	7	7	7	6	6	5	1	1	0	1	n	–	60	n	–
E9	7	7	6	6	7	5	5	15	120	105	10	30	20	15	n	–
E10	7	6	7	5	7	7	7	1	60	59	0	15	15	45	n	–
mean	6.5	6.2	6.4	5.5	6.2	6.5	5.6	5.6	69.4	63.9	4.1	37.2	33	7.9	210	202.1
s.d	0.8	1.2	1	1.2	1	0.7	1	4.4	52.7	50.4	3.2	23.1	24.3	5.6	42.4	46

the features of our application are well received. The definition of outsourcing categories (F6) and the visualization of the graph (F1) are rated most useful (mean 6.5), followed closely by the linkage of fund prospectuses (F3 with 6.4). The filter option for outsourcing categories (F5) and the display of the relevant text passages from fund prospectuses (F2) both receive 6.2 on average. The lowest values are for the filter options for the knowledge graph facets (F7 with 5.6) and the selected company's information (F4 with 5.5). Interpreting the results, visualizations, explanations and links to further information are perceived as rather useful. However, in its current state, the KG's properties for filtering and inspection provide room for improvement.

Our application is perceived to save time in solving the given scenarios (S1–S3). Using our application, participants estimate that they could be completed in less than 10 min on average. Conversely, without our application experts report a completion time of about 30–60 min for the same tasks. Considering S1 and S2, the reduction in time expenditure is estimated to be a factor of approx. 10. Looking at absolute numbers, the estimated time saving ranges on average from around 30 min (S2) and one hour (S1) up to three hours (S3). Especially for S3, which encompasses a broader cross-sectoral scope, eight out of ten experts stated that they would be likely unable to solve this task without our cyber mapping application (indicated with 'n'). The remaining experts estimated the time needed to solve this task via a manual workaround would be 3–4 h.

The study also collected feedback on desired features in our application. Mostly, experts asked for new features regarding the integration of more data sources and functions. In particular, capabilities such as full-text search, auto completion and fine-grained filter options were suggested. Since our application currently covers only a part of the financial system, participants recommend extending the KG with further data relevant to their jobs.

In conclusion, results in our user study show that our initial application provides a good user experience, notably regarding perspicuity and attractiveness. Features relating to visualization, definitions and references were perceived as most useful, yet further improvements in the KG's content and filter operations need to be implemented. Time estimations indicate that our application has the potential to reduce the time needed to investigate cyber incidents in the financial system. Especially for complex analysis tasks (like S3), our application could provide benefits for cyber risk monitoring.

6   Conclusion and Outlook

After establishing the importance of cyber mapping for ensuring financial stability, we presented a first approach towards this goal by utilizing Knowledge Graphs (KGs). Although there were some endeavors in the past, no approach applying semantic technologies has been published so far. Therefore, a KG construction approach was proposed which consists of a dedicated cyber mapping ontology, the integration of (un)structured knowledge and its traceability. Having such a KG at hand, we implemented an application to let users analyze outsourcing relations of funds and TPPs. A user study with ten experts was conducted to collect feedback about the usability and usefulness and to estimate the possible time saving potential of our application. With room for improvements, the results have indicated a good user experience and the features' usefulness. In particular, time estimations have shown that our application has the potential to reduce time and effort for supervisors. In the case of complex tasks, our cyber mapping solution could provide benefits for cyber risk monitoring. With this work, we took a first step towards cyber mapping the German financial system with KGs.

Our collaborative research lab is still running. The main objective is to gather and interconnect data in such a way that indications for cyber incidents become visible. For a comprehensive cyber mapping, incorporating regulatory data sources would be essential. Also, we aspire to use the acquired feedback to improve our KG and provide further applications. Regarding potential measures for node impact in the financial network, we intend to implement criticality measures as specified by [18]. Moreover, our ambition is to improve our data integration pipeline: on the one hand by introducing virtual knowledge graphs to keep the data up-to-date. We envisage using RDF-star again to add *valid from/to* properties for temporal aspects. On the other hand, we intend to process unstructured financial data with state-of-the-art technology in the field of neural networks and large language models. Regarding our proposed ontology, our plan is to perform ontology matching to enrich its terminology with existing ones.

References

1. Abu-Salih, B.: Domain-specific knowledge graphs: a survey. J. Netw. Comput. Appl. **185**, 103076 (2021). https://doi.org/10.1016/J.JNCA.2021.103076
2. Adamcyk, M., Drougkas, A., Philippou, E., Abel, P., Gratiolet, F., Maaskant, E.: NIS investments - cybersecurity policy assessment, November 2023. Technical report, European Union Agency for Cybersecurity (ENISA) (2023). https://www.enisa.europa.eu/publications/nis-investments-2023
3. Adelmann, F., et al.: Cyber Risk and Financial Stability: It's a Small World After All. IMF Staff Discussion Notes (2020). https://www.imf.org/en/Publications/Staff-Discussion-Notes/Issues/2020/12/04/Cyber-Risk-and-Financial-Stability-Its-a-Small-World-After-All-48622
4. Bank of England: Operational resilience: Critical third parties to the UK financial sector. PRA Discussion Paper 3/22 and FCA Discussion Paper 22/3 (2022). https://www.bankofengland.co.uk/prudential-regulation/publication/2022/july/operational-resilience-critical-third-parties-uk-financial-sector
5. Beckett, D., Berners-Lee, T., Prud'hommeaux, E., Carothers, G.: RDF 1.1 Turtle (2014). https://www.w3.org/TR/turtle/
6. Belhajjame, K., et al.: PROV-O: The PROV Ontology (2012). http://www.w3.org/TR/prov-o/
7. Bennett, M.: The financial industry business ontology: best practice for big data. J. Bank. Regul. **14**(3), 255–268 (2013). https://doi.org/10.1057/jbr.2013.13
8. Brauchle, J.P., Göbel, M., Seiler, J., von Busekist, C.: Cyber mapping the financial system. Technical report, Carnegie Endowment for International Peace (2020). http://www.jstor.org/stable/resrep24291
9. Bundesamt für Justiz [German Federal Office of Justice]: Gesetz zur Stärkung der Finanzmarktintegrität (Finanzmarktintegritätsstärkungsgesetz – FISG) [Act to Strengthen Financial Market Integrity (Financial Market Integrity Strengthening Act)] (2021). https://www.bgbl.de/xaver/bgbl/start.xav#__bgbl__%2F%2F*%5B%40attr_id%3D%27bgbl121s1534.pdf%27%5D__1699873137312
10. Bundesamt für Sicherheit in der Informationstechnik [Federal Office for Information Security]: Die Lage der IT-Sicherheit in Deutschland 2023 [The state of IT security in Germany in 2023]. Technical report, Bundesamt für Sicherheit in der Informationstechnik [Federal Office for Information Security] (2023). https://www.bsi.bund.de/SharedDocs/Downloads/DE/BSI/Publikationen/Lageberichte/Lagebericht2023.html
11. Bundesanstalt für Finanzdienstleitsungsaufsicht [German Federal Financial Supervisory Authority] (BaFin): Auslagerungen: Landkarten bieten Orientierung [Outsourcing: Maps Provide Orientation] (2022). https://www.bafin.de/SharedDocs/Veroeffentlichungen/DE/Fachartikel/2022/fa_bj_2208_Auslagerungen_Landkarten.html
12. Bundesanstalt für Finanzdienstleitsungsaufsicht [German Federal Financial Supervisory Authority] (BaFin): Wertschöpfungsketten im Finanzsektor: Empfehlungen zur IT-Aufsichtspraxis [Value chains in the financial sector: recommendations for IT supervisory practice] (2022). https://www.bafin.de/SharedDocs/Veroeffentlichungen/DE/Fachartikel/2022/fa_bj_2207_uni_innsbruck_wertschoepfungsketten.html
13. Böhme, R., Pesch, P.J., Fritz, V.: Auswirkungen sich verändernder Wertschöpfungsketten im Finanzsektor auf die IT-Sicherheit [Effects of changing value chains in the financial sector on IT security] (2022). https://www.bafin.de/

SharedDocs/Downloads/DE/Bericht/dl_abschlussbericht_forschungsprojekt_uni_innsbruck.pdf?_blob=publicationFile

14. Clancy, L., Mourselas, C.: Ion cyber outage continues as banks rely on workarounds (2023). https://www.risk.net/derivatives/7955967/ion-cyber-outage-continues-as-banks-rely-on-workarounds

15. Deng, Y., Lu, D., Huang, D., Chung, C., Lin, F.: Knowledge graph based learning guidance for cybersecurity hands-on labs. In: Proceedings of the ACM Conference on Global Computing Education, CompEd 2019, Chengdu, Sichuan, China, 17–19 May 2019, pp. 194–200. ACM (2019).https://doi.org/10.1145/3300115.3309531

16. Dimou, A., Sande, M.V., Colpaert, P., Verborgh, R., Mannens, E., de Walle, R.V.: RML: a generic language for integrated RDF mappings of heterogeneous data. In: Proceedings of the Workshop on Linked Data on the Web co-located with the 23rd International World Wide Web Conference (WWW 2014), Seoul, Korea, 8 April 2014. CEUR Workshop Proceedings, vol. 1184. CEUR-WS.org (2014). http://ceur-ws.org/Vol-1184/ldow2014_paper_01.pdf

17. Elhammadi, S., et al.: A high precision pipeline for financial knowledge graph construction. In: Proceedings of the 28th International Conference on Computational Linguistics, COLING 2020, Barcelona, Spain (Online), 8–13 December 2020, pp. 967–977. International Committee on Computational Linguistics (2020). https://doi.org/10.18653/V1/2020.COLING-MAIN.84

18. European Banking Authority (EBA), European Insurance and Occupational Pensions Authority (EIOPA), European Securities and Markets Authority (ESMA): Joint European Supervisory Authorities' Technical Advice (ESA 2023 23). Technical report, European Banking Authority (EBA) and European Insurance and Occupational Pensions Authority (EIOPA) and European Securities and Markets Authority (ESMA) (2023). https://www.eba.europa.eu/sites/default/files/document_library/Publications/Other%20publications/2023/JC%20technical%20advice%20on%20DORA/1062226/Joint-ESAs%E2%80%99%20response%20to%20the%20Call%20for%20advice%20on%20the%20designation%20criteria%20and%20fees%20for%20the%20DORA%20oversight%20framework_final.pdf

19. European Central Bank (ECB): Guideline (EU) 2018/876 of the European Central Bank of 1 June 2018 on the Register of Institutions and Affiliates Data (ECB/2018/16). Official Journal of the European Union, pp. 3–21 (2018). https://eur-lex.europa.eu/eli/guideline/2018/876

20. European Systemic Risk Board (ESRB): Systemic cyber risk. Technical report, European System of Financial Supervision (ESFS) (2020). https://www.esrb.europa.eu/pub/pdf/reports/esrb.report200219_systemiccyberrisk~101a09685e.en.pdf

21. European Systemic Risk Board (ESRB): Mitigating systemic cyber risk. Technical report, European System of Financial Supervision (ESFS) (2022). https://www.esrb.europa.eu/pub/pdf/reports/esrb.SystemiCyberRisk.220127~b6655fa027.en.pdf

22. Fensel, D., et al.: Knowledge Graphs - Methodology, Tools and Selected Use Cases. Springer, Cham (2020). https://doi.org/10.1007/978-3-030-37439-6

23. Financial Stability Board (FSB): Third-party dependencies in cloud services - Considerations on financial stability implications. Technical report, FSB (2019). https://www.fsb.org/wp-content/uploads/P091219-2.pdf

24. Foroutan, N., Schröder, M., Dengel, A.: CO-fun: a German dataset on company outsourcing in fund prospectuses for named entity recognition and relation extraction. CoRR abs/2403.15322 (2024). https://arxiv.org/abs/2403.15322

25. Gruber, T.: A translation approach to portable ontology specifications. Knowl. Acquis. **5**(2), 199–220 (1993)
26. Handelsblatt: Nach Cyberangriff: Evotec verlässt MDax wegen Fristverletzung [After cyber attack: Evotec leaves MDax due to deadline violation] (2023). https://www.handelsblatt.com/finanzen/maerkte/aktien/chart-des-tages-nach-cyberangriff-evotec-verlaesst-mdax-wegen-fristverletzung/29133970.html
27. Harry, C., Gallagher, N.: Classifying Cyber Events: A Proposed Taxonomy. Center for International and Security Studies at Maryland (CISSM), Cyber Attacks Database (2018). https://cissm.liquifiedapps.com/#about
28. Hartig, O.: Foundations of RDF⋆ and SPARQL⋆ (an alternative approach to statement-level metadata in RDF). In: Proceedings of the 11th Alberto Mendelzon International Workshop on Foundations of Data Management and the Web, Montevideo, Uruguay, 7–9 June 2017. CEUR Workshop Proceedings, vol. 1912. CEUR-WS.org (2017). https://ceur-ws.org/Vol-1912/paper12.pdf
29. Hellmann, S., Lehmann, J., Auer, S., Brümmer, M.: Integrating NLP using linked data. In: Alani, H., et al. (eds.) ISWC 2013. LNCS, vol. 8219, pp. 98–113. Springer, Heidelberg (2013). https://doi.org/10.1007/978-3-642-41338-4_7
30. Hinderks, A., Schrepp, M., Thomaschewski, J.: User Experience Questionnaire, Data Analysis Tools. Website (2023). https://www.ueq-online.org/
31. Huakui, L., Liang, H., Feicheng, M.: Constructing knowledge graph for financial equities. Data Anal. Knowl. Discov. **4**(5), 27–37 (2020)
32. (IMF), I.M.F.: Norway: Financial Sector Assessment Program. Technical Note - Cybersecurity Risk Supervision and Oversight. IMF Staff Country Report 2020/262. Technical report, IMF (2020). https://www.imf.org/~/media/Files/Publications/CR/2020/English/1NOREA2020004.ash
33. Jia, Y., Qi, Y., Shang, H., Jiang, R., Li, A.: A practical approach to constructing a knowledge graph for cybersecurity. Engineering **4**(1), 53–60 (2018). https://doi.org/10.1016/j.eng.2018.01.004
34. Kiesling, E., Ekelhart, A., Kurniawan, K., Ekaputra, F.: The SEPSES knowledge graph: an integrated resource for cybersecurity. In: Ghidini, C., et al. (eds.) ISWC 2019. LNCS, vol. 11779, pp. 198–214. Springer, Cham (2019). https://doi.org/10.1007/978-3-030-30796-7_13
35. Laugwitz, B., Held, T., Schrepp, M.: Construction and evaluation of a user experience questionnaire. In: Holzinger, A. (ed.) USAB 2008. LNCS, vol. 5298, pp. 63–76. Springer, Heidelberg (2008). https://doi.org/10.1007/978-3-540-89350-9_6
36. Panetta, F.: The Quick and the Dead: building up cyber resilience in the financial sector. Technical report, European Central Bank (ECB) (2023). https://www.ecb.europa.eu/press/key/date/2023/html/ecb.sp230308~92211cd1f5.en.html
37. Pingle, A., Piplai, A., Mittal, S., Joshi, A., Holt, J., Zak, R.: Relext: relation extraction using deep learning approaches for cybersecurity knowledge graph improvement. In: ASONAM 2019: International Conference on Advances in Social Networks Analysis and Mining, Vancouver, British Columbia, Canada, 27–30 August 2019, pp. 879–886. ACM (2019). https://doi.org/10.1145/3341161.3343519
38. Resano, J.R.M.: Digital resilience and financial stability. The quest for policy tools in the financial sector. Technical report, Banco de España (2022). https://dx.doi.org/10.2139/ssrn.4336381
39. Ros, G.: The Making of a Cyber Crash: A Conceptual Model for Systemic Risk in the Financial Sector. ESRB: Occasional Paper Series No. 2020/16 (2020). https://dx.doi.org/10.2139/ssrn.3723346
40. Schreiber, G., Raimond, Y.: RDF 1.1 Primer (2014). https://www.w3.org/TR/rdf11-primer/

41. Syed, Z., Padia, A., Finin, T., Mathews, M.L., Joshi, A.: UCO: a unified cyber-security ontology. In: Martinez, D.R., Streilein, W.W., Carter, K.M., Sinha, A. (eds.) Artificial Intelligence for Cyber Security, Papers from the 2016 AAAI Workshop, Phoenix, Arizona, USA, 12 February 2016. AAAI Technical Report, vol. WS-16-03. AAAI Press (2016). http://www.aaai.org/ocs/index.php/WS/AAAIW16/paper/view/12574

42. The European Parliament and the Council of the European Union: Regulation (EU) 2022/2554 of the European Parliament and of the Council of 14 December 2022 (2022). https://eur-lex.europa.eu/legal-content/EN/TXT/PDF/?uri=CELEX:32022R2554&from=FR

43. Wang, W., Xu, Y., Du, C., Chen, Y., Wang, Y., Wen, H.: Data set and evaluation of automated construction of financial knowledge graph. Data Intell. 3(3), 418–443 (2021). https://doi.org/10.1162/DINT_A_00108

44. Zehra, S., Mohsin, S.F.M., Wasi, S., Jami, S.I., Siddiqui, M.S., Raazi, S.M.K.: Financial knowledge graph based financial report query system. IEEE Access 9, 69766–69782 (2021). https://doi.org/10.1109/ACCESS.2021.3077916

Integrating Domain Knowledge for Enhanced Concept Model Explainability in Plant Disease Classification

Jihen Amara[1,2]([✉]) [iD], Sheeba Samuel[1,2] [iD], and Birgitta König-Ries[1,2] [iD]

[1] Heinz Nixdorf Chair for Distributed Information Systems, Friedrich Schiller University Jena, Jena, Germany
{jihene.amara,sheeba.samuel,birgitta.koenig-ries}@uni-jena.de
[2] Michael Stifel Center Jena, Jena, Germany

Abstract. Deep learning-based plant disease detection has seen promising advancements, particularly in its remarkable ability to identify diseases through digital images. Nevertheless, these systems' opacity and lack of transparency, which often offer no human-interpretable explanations for their predictions, raise concerns with respect to their robustness and reliability. While many methods have attempted post-hoc model explainability, few have specifically targeted the integration and impact of domain knowledge. In this study, we propose a novel framework that combines a tomato disease ontology with the concept explainability method Testing with Concept Activation Vectors (TCAV). Unlike the original TCAV method, which required users to gather diverse image concepts manually, our approach automates the creation of images based on relevant concepts used by domain experts in plant disease identification. This not only simplifies the concept collection and labelling process but also reduces the burden on users with limited domain knowledge, ultimately mitigating potential biases in concept selection. Besides automating the concept image generation for the TCAV method, our framework gives insights into the significance of disease-related concepts identified through the ontology in the deep learning model decision-making process. Consequently, our approach enhances the efficiency and interpretability of the model's diagnostic capabilities, promising a more trustworthy and reliable disease detection model.

Keywords: Explainable Artificial Intelligence · Plant Disease Classification · Tomato Disease Ontology · Deep Neural Networks

1 Introduction

Addressing global hunger for a projected 9 billion people by 2050 is a crucial challenge [31]. However, obstacles like limited crop productivity, environmental concerns, and rising plant diseases hinder progress in agriculture. Hence, innovative solutions are needed. Artificial intelligence (AI) technologies promise to provide such solutions. It offers unprecedented opportunities to enhance various facets of the field from precision agriculture to advanced automatized crop management [32]. One particularly promising path is the integration of deep learning, a subset of AI, in the identification of plant diseases through image data analysis. This not only enables fast and early disease detection but also paves the way for more effective and targeted intervention to minimize crop losses.

In recent years, we noticed a surge in the number of works successfully applying deep learning for plant disease image classification [2,3]. However, the reception of such models by plant scientists and farmers remains mixed due to their black-box nature. This uncertainty comes from the limited understanding of the internal process by which such models learn and encode plant disease traits and features. The absence of transparency throughout the decision-making process is a crucial concern in numerous critical application domains including plant disease diagnosis. Hence, explainability of deep learning models becomes a necessity for the swift realisation of AI practical applications in agriculture.

Different explainability methods have emerged to generate saliency heatmaps [26,30,33]. They rely mostly on the backpropagation of gradients to assess the impact of individual pixel changes on the model decision. However,compared to other fields of application of deep neural networks (DNN), plant disease classification carries an additional challenge. Plant diseases can have different symptoms such as discoloration, lesions, or abnormal growth patterns. These symptoms can be subtle and may vary depending on the disease stage making it hard to grasp without expert knowledge. Also, different plant diseases may exhibit similar symptoms. Therefore, common explanation methods such as saliency maps visualisation could not provide pertinent explanation on how much such visual concepts (i.e., color or symptom abnormalities) influence the model decision.

Hence different concept explanation methods [20,34] giving the attribution of concepts rather than pixels have been proposed. One of these methods is Testing with Concept Activation Vectors (TCAV) [4,20]. A concept represents an abstraction which could range from a simple color to an object or a complex idea [22]. Given any user-defined concept, TCAV detects if that concept is embedded within the latent feature space learned by the network [22]. Hence, in the original TCAV method [20], users were required to gather diverse image concepts manually. This posed a potential challenge, particularly for machine learning engineers lacking specialised knowledge of the specific domain in study. We propose to leverage semantic web methods to tackle this issue effectively. An ontology can provide relevant concepts experts use in identifying plant diseases and aids in automating the creation of images based on these concepts. This not only simplifies the concept collection and labelling process but also alleviates the burden on users with limited domain knowledge. It can also help avoid human bias, which may influence the choice of concepts to test since it could reflect the community's understanding. Moreover, the ontology could define abstract concepts that might not have direct visual representations but can be inferred from related concepts.

In essence, our proposed framework not only automates concept image generation for the TCAV method but also offers insights into how important these disease-related concepts identified by using the ontology are for the deep learning model's decision-making process. This approach enhances both the efficiency and interpretability of the diagnostic capabilities of the model. Hence, the aim of this paper is to provide a semantic aware concept explainability method for plant diseases based deep learning classification. We choose tomato diseases as a use case to understand what semantic concepts DNN learns. The tomato disease image dataset was extracted from the PlantVillage dataset [16].

In summary, our contributions are:

- A new ontology to represent symptoms and abnormalities associated with tomato diseases.
- Mapping of concepts learnt by DNN within its latent space for plant disease classification to semantic concept descriptions of plant diseases within the ontology using CAVs.
- Automated concept labelling and generation such as color and symptoms for TCAV within the context of plant diseases.
- Analysis of contribution of various disease-related characteristics to the predictions made by a deep neural network. This provides valuable insights on the significance of different features in the decision-making process which could help in improving the accuracy and interpretability of plant disease predictions.

The remainder of the paper is organised as follows. Section 2 introduces the possible use cases. Section 3 discusses related work on the use of ontologies in explainability. Section 4 explains the methods and proposed approach and Sect. 5 provides details about the experiment and results. Finally, Sect. 6 provides the conclusion.

2   Use Cases

Our proposed framework of combining ontology and concept explainability for tomato disease classification with deep learning can offer several benefits and use cases. Some of those potential applications could be:

- Explanation of predictions: Our framework can provide explanations for the predictions made by the deep learning model. Users such as plant experts, agriculture policy makers, regulators and stakeholders can understand why a specific classification was made, which is crucial for building trust in the model.
- Domain-Specific understanding: By incorporating the ontology, the framework can leverage domain-specific knowledge about tomato diseases. This helps in transferring insights from experts to the model developers which enhances their understanding of the context and improves model accuracy.

- Identification of relevant features: The framework can highlight the specific concepts or features within the input data that contributed the most to a particular classification. This can be valuable for researchers and practitioners to identify key indicators of tomato diseases and improve their collected dataset.
- Error analysis and improvement: The framework can help in understanding semantic errors made by the model, indicating which parts of the input data might have led to a misclassification. This information can guide further model refinement and training.

3   Related Work

In recent years, there has been an increasing interest in understanding and explaining the prediction behavior of deep neural networks. One of the most popular methods is the saliency and attribution approaches [26, 27, 30, 33], where the explanation for the DNN is given as an importance map highlighting the contribution of each feature in its decision. Even though these methods increase the explainability of the DNN, they are limited in their understandability, leaving it to the user to interpret such maps. For instance, the importance of a single pixel in the classification does not bring a meaningful explanation, and it is also contrived by the number of features [22].

Hence, methods such as TCAV [20] present the use of "human-interpretable concepts" for explaining DNN networks. Still, no information is provided concerning how these concepts are relevant to the output of the DNN. The user also needs to collect these concepts as images, making interpreting abstract concepts hard. Consequently, a lot of researchers argue that an effective explainability of deep learning models cannot be achieved without the use of domain knowledge through the integration of semantic web technologies [12].

In [25] the authors suggest employing ontologies as background knowledge framework to facilitate obtaining formulae that interpret the functioning of deep models. In their work, the network is trained to classify scene objects. Based on the classification output, they run a DL-Learner on the Suggested Upper Merged Ontology [24] to generate class expressions that act as explanations. However, their approach is constrained specifically in its need for labeled data with the required different concepts.

Similarly, in [12], the authors proposed a neuro-symbolic framework where the semantics in the knowledge base are aligned with the annotations in the dataset. The model to explain is a DNN model trained for multi-label image classification, and the explanation was generated in a logical language. The specific focus of their study is the classification of food recipes. A different approach was proposed in [7], where the authors introduced explainable classifiers using domain knowledge. Their approach involved creating synthetic images of Pizza for training the DNN based on the specifications outlined in the pizza ontology. Then, they proposed a method that integrates a DL model with a graph of tensors automatically generated from description logic assertions extracted from the relevant ontology.

Furthermore, in [28] the authors tried to provide better explanations by mapping the internal state of neural network (neuron activations) to the concepts of an ontology to find symbolic justification for the output of DNN. This mapping is achieved by training small neural networks to predict a single concept from the DNN neuron's activation. However, this method assumes the presence of concepts in the images. That's why they used synthetic image datasets of trains modeled accordingly to present the needed concepts.

In [8], the authors proposed combining ontology with deep learning cassava disease classification. However, the ontology was only used to infer diseases based on simulated sensor observations such as temperature and soil moisture and provide extra domain knowledge about the classified disease without explaining the trained model behavior.

While current methods integrating explainability and web semantics have shown potential, many rely on deep learning models trained on synthetic images featuring a predefined set of concepts, limiting their real-world applicability. Additionally, a real-world application is burdened with the need for multi-class annotation, such as the example in [25]. In our work, we address this limitation by focusing on explaining a deep neural network trained on real-world images of plant disease. Our approach involves automatically mapping semantic concepts to activations learned within the network. The proposed system integrates an ontology applicable to any images of tomato diseases, enhancing interpretability. This contributes to a more flexible and practical model. As far as we know, this is the first approach to automatically explore associating semantic concepts with visual concepts for plant disease classification.

4   Methods

The framework illustrated in Fig. 1 represents our workflow for enhancing and automatizing concept explainability using knowledge in the form of an ontology. In the preparatory phase (pre-runtime) we train the deep neural network on a set of tomato disease images and we create the ontology. Using only the image annotations (Target Classes), the ontology can provide necessary concepts, properties and axioms related to visual tomato disease identification. For example, a bacterial spot tomato disease can be described by its appearance on the leaf with symptoms such as black coloration and spots spreading. Hence, the bacterial spot target class can be defined using the ontology axiom as ($\exists$hasSymptom.BacterialSpotsOnLeaf $\sqcap$ $\exists$hasColor.Brown). Section 4.1 will give a more detailed explanation of the used ontology. In the following part, we briefly describe the steps in our framework. First, as described above, a disease concept ontology is employed to get concepts related to different tomato disease classes in our image dataset.

Based on the target class label, the ontology provides all important properties linked to the specified disease class that could be visible in the image. For example, some of these properties (concepts) could be color or symptom texture. The generated concept labels (i.e., color brown) are then used to automatically generate corresponding images (i.e., different shades of brown images). More details

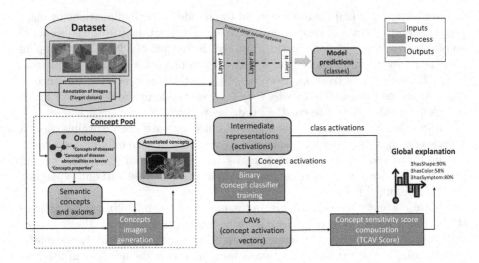

Fig. 1. The framework of the proposed method. Our modeled ontology is used to provide concepts related to the target class (disease label). These semantic concepts are then used for automatic image generation, and a list of annotated concepts such as (color, shape, and symptom) are created. Concept classifiers are then trained, and CAVs and TCAV scores are computed.

on the generation process will be described in Sect. 4.2. These generated concepts in the form of images with target class images and random images are subsequently employed to derive Concept Activation vectors (CAVs) [20] and compute the sensitivity score (TCAV score) (see Sect. 4.3). This helps us understand how sensitive the trained deep learning model is to these domain-specific concepts. For instance, we can quantify the influence of the concept ∃hasColor.Brown on the 'BacterialSpot' prediction as a single score. In the upcoming sections, we will describe various parts of our framework in detail. First, we will introduce our ontology and the modelling process (Sect. 4.1), then we will explain the process for generating images related to the concepts (Sect. 4.2). Finally, we will provide an overview of the TCAV algorithm (Sect. 4.3).

4.1 Ontology Development (Ontology Based Explanation)

This section describes the steps for developing the Tomato Disease Concepts (TomatoDCO) ontology. It uses OWL for modelling knowledge about classes, properties and axioms related to phenotype of various tomato diseases. We follow best practice recommendations on ontology engineering [23] to develop this ontology.

Ontology Requirements. The first step to developing any ontology is to define its scope, specifying the aspects it aims to model. In our work, the ontology is designed to include the tomato plant diseases domain. The ontology should also provide the different appearances (visual concepts) related to tomato disease. We integrate the TCAV explainability method with an ontology to provide a more comprehensive understanding of the model's decision-making process. The ontology will serve as a structured knowledge base, describing and modelling each disease class's specific symptoms and abnormalities. These properties will be then used to generate associated concept images. Following the scope, the ontology should be able to answer the following competency questions (CQs):

- What are the diseases that tomato plants could have?
- What are the possible symptoms and appearances of tomato disease X?
- What are the diseases that are caused by bacteria, fungi, viruses, or insect damage?
- What are the diseases if a tomato plant has symptoms/appearance A,B,..?

Ontologies Reuse. To model our ontology, we follow the recommendation to start by checking existing ontologies focused on plant diseases. We reuse relevant elements to help achieve our goal of creating explanations of our trained neural network, particularly concerning the diseases existing in our image dataset. To develop our TomatoDCO ontology, we reused and followed the disease hierarchy from [18,19], presenting a rice disease ontology (RiceDO) that helps identify rice diseases from existing symptoms in the plant. It was evaluated and assessed by ontology experts and senior agronomists, where important criteria such as appropriateness, consistency, and ontology satisfaction were considered [19]. The other most pertinent ontologies for our case are Plant Protection Ontology (PPO) [5], Plant Disease Ontology (PDO) [17], and Phenotype and Trait Ontology (PATO) [15]. PATO defines various phenotypic traits across different species. These traits include characteristics like color (e.g., brown, black), temperatures (e.g., high, low), and symptoms (e.g., swelling) [18]. PDO defines diseases in maize, wheat, and rice, categorized into bacteria, fungi, and viruses. PPO classifies barley disorders into abiotic and biotic (with further subcategories for bacteria, fungi, and viruses). RiceDO used and extended PDO, PPO, and PATO ontologies under the domain of rice diseases. It classifies diseases into bacteria, fungi, and viruses. Even though these ontologies serve as a valuable reference for comprehending and categorizing plant diseases and disorders, they are developed to integrate them with a decision expert system, which differs from our goal. Hence, we reused concepts that help our aim of providing properties associated with each disease visual manifestation that could be exploited later as concepts for our explainability algorithm. These existing ontologies (i.e., RiceDO and PDO) also don't include information on tomato diseases. Therefore, we adopt their approach of classifying diseases in defining specific classes relevant to tomato disease.

Concepts Identification. Since our image dataset is extracted from the PlantVillage dataset [16] available under an open licence [1], we use it as our primary knowledge source along with [6] to collect information about signs and symptoms associated with the mentioned diseases for our ontology. The most important concepts we identified in this step were different types of diseases such as bacterial (i.e., bacterial spot), fungal (i.e., early blight, late blight, leaf mold, septoria leaf spot, and target spot), viral (i.e., mosaic virus and yellow leaf curl virus) and diseases due to insect damage such as two-spotted spider mites disease. Diseases symptoms could be visual abnormalities such as changes in color, for example, black, brown, and yellow, and changes in leaf shape. Also, the emergence of textural changes on the leaf, such as blight or spots.

Classes Definition and Classes Hierarchy. The structure of our ontology, TomatoDCO, is shown in Fig. 2 . We have two top-level classes: 'TomatoDisease' and 'Abnormality' and three object properties (hasColor, hasShape and hasSymptom).

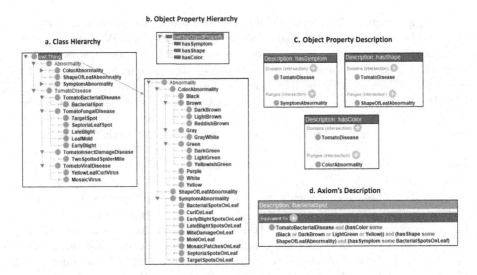

Fig. 2. The structure of TomatoDCO ontology is divided into three parts: (a) the class hierarchy of TomatoDCO; (b) the object properties of TomatoDCO; (c) the object properties descriptions (range and domain); and (d) an example of axioms representing concepts of abnormalities of the tomato bacterial spot disease.

A detailed description of these components is given in the following:

– Abnormality: We reuse this class from RiceDO ontology and PPO and extend it to meet our requirements. This class presents the kind of abnormalities visually noticed on a plant when it is affected by the disease. These abnormalities concepts are important to quantify how sensitive they are for our plant disease-trained model and to know to which extent our model is learning the true semantic representation of a disease. As shown in Fig. 2, they include:

- ColorAbnormality: The different color changes that could emerge because of the disease (e.g., Brown). The terms of colors are mapped to the existing ones in RiceDO ontology by using owl:equivalentClass, which is also mapped to the existing ones in PATO.
- ShapeOfLeafAbnormality: the abnormalities that happen to the shape of the leaf because of the disease.
- SymptomAbnormality: the symptoms of abnormalities of a leaf affected by a disease can vary according to the specific pathogen causing the problem, such as the emergence of spots or patches on the leaf (e.g., 'having a bacterial spot symptom on leaves' can be defined by BacterialSpotsOnLeaf)

- TomatoDisease: This class classifies the tomato diseases into bacterial, fungal, and viral, like the PDO and RiceDO ontologies, and also adds the class for diseases caused by insect damage since this could occur in the real world. It is worth noting also that PDO also lacks information regarding tomato diseases.

Object Properties. In our use case, we define three object properties that are necessary to describe the appearance of each tomato disease and are useful for extracting the required semantic concepts for the TCAV method.

- hasColor: This property defines a relation from TomatoDisease to ColorAbnormality.
- hasShape: This property defines a relation from TomatoDisease to ShapeOfLeafAbnormality.
- hasSymptom: This property defines a relation from TomatoDisease to SymptomAbnormality.

These properties will be then used to axiomatize the various visual abnormalities that occur on a leaf when affected by a certain disease.

Concept Definition. The appearance of each tomato disease is described in class description by using equivalent-to relation. For example, a tomato bacterial spot disease can cause the emergence of bacterial spot lesions that develop randomly on the leaflets, and they turn brown or black and sometimes have a yellow hallo. In some cases, entire leaves can turn yellow and wilt [6]. Since in our dataset images of the disease come from different stages, we make sure to integrate all the possible colours of abnormality. Hence as shown in Fig. 2.d, these could be described as:

BacterialSpot ≡TomatoBacterialDisease⊓

  (∃hasSymptom.BacterialSpotsOnLeaf)⊓

  (∃hasColor.(Black ⊔ Yellow ⊔ DarkBrown ⊔ LightGreen))⊓

  (∃hasShape.ShapeOfLeafAbnormality)

Hence, the TomatoDCO ontology is used to help the mapping between the visual level (target class image, i.e., bacterial spot) and the semantic level (what is the disease corresponding concepts (i.e., color, symptom, and shape)). In the following section, we describe how these concepts extracted thanks to the ontology could be defined visually as images.

4.2 Synthetic Concepts Images Generation

The texture of a leaf can provide valuable insights into the health of a plant, as changes in texture are frequently associated with specific diseases. Symptoms such as wilting, discoloration, or lesions may manifest, affecting the uniformity of the leaf surface. Hence, we propose visually representing the disease symptom (hasSymptom) by getting the texture details from the leaf image while excluding shape and color information. We design three different visual concept generation methods for texture (hasSymptom), shape (hasShape), and color (hasColor) separately.

Texture Generation Method. For texture generation, we follow the method described by Ge et al. [13]. The method is based on initially segmenting the leaf images from the background. Then, in order to eliminate color information, the segmented leaf is converted into a grayscale image. Subsequently, the grayscale image is divided into multiple square patches using an adaptive strategy where the patch size and location adjust according to the leaf size to include a broader range of texture information. If the overlap ratio between a patch and the original leaf segment exceeds a specified threshold τ (set to 0.99 in our experiments, indicating that over 99% of the patch area belongs to the object) the patch will be included in the patch pool. Four patches are randomly selected from the pool and then concatenated into a new texture image to capture both local (individual patch) and global (entire image) texture characteristics. This generated texture image corresponds to the target class disease symptom defined in our ontology. The segmentation step is omitted since we already have a segmented version of our image dataset. Figure 3 visualises the used method.

Fig. 3. The process for extracting texture. (a) Crop images and compute the overlap ratio between the 2D mask and patches. Patches with overlap > 0.99 are shown in a green shade, (b) add the valid patches to a patch pool. (c) is the final texture feature, the concatenation of k randomly selected patches from the patches pool. (Color figure online)

Color Generation Method. We extract the color concepts associated with the target class (e.g., bacterial spot) using the TomatoDCO ontology that specifies distinct colors linked to each tomato disease class. Then, we automate the generation of different images of the same color with varying intensities through the random perturbation of RGB values within the predefined color spectrum. These images will be used later with the TCAV method to quantify the model's sensitivity to the corresponding color when classifying a particular class.

Shape Generation Method. When a leaf is affected by a disease, its shape witnesses different changes. Some diseases cause damage along the edges of the leaves, resulting in distortion and curling. Hence, we visualize the shape abnormality concept by extracting only the shape edge using binary segmentation. An example is presented in Fig. 4.

Fig. 4. Example of extracted shape contours

4.3   Testing with Concept Activation Vectors (TCAV)

In this section, we provide a brief overview of CAVs and outline the approach employed in this study to compute TCAV scores. These scores measure the influence of a semantic concept on the predictions made by DNN. The TCAV method was proposed by kim et al. [20] to explain deep neural models without any retraining. A key component within TCAV is the concept activation vector (CAV) v_c^l, a vector representation of a concept within a specified convolutional layer l of DNN. To identify CAV in layer l, a set of positive and negative examples representing concept and non-concept (i.e., random) instances is needed. These examples are represented in the form of images and a binary linear classifier is trained to distinguish between them. The vector orthogonal to the decision boundary separating the two classes, i.e., the vector pointing in the direction of the representations of the concept images, is the CAV. To assess the impact of a CAV on a class of input images, the authors proposed the TCAV score metric. It uses directional derivatives, denoted as $S_{C,k,l}(x)$, , to gauge the contextual sensitivity of a concept across an entire input class, offering comprehensive explanations. The formula for calculating the TCAV score is as follows:

$$TCAV_{Q_{C,k,l}} = \frac{|x \in X_k; S_{C,k,l}(x) > 0|}{|X_k|} \quad (1)$$

where k denotes the class labels, X_k represents all inputs, and $S_{C,k,l}(x)$ is the directional derivative of a sample's activation x from layer l concerning class k and concept C. The TCAV score calculates the ratio of the class k's inputs positively influenced by concept C. To make sure that only meaningful CAVs are taken into account, a statistical significance two-sided t-test is performed [20].

5   Experiments and Results

5.1   Dataset and Trained Model

All experiments were performed using the Inception-V3 [4,29] model fine tuned on the tomato images from the PlantVillage dataset [16]. The model was created and loaded with pretrained weights on the ImageNet dataset [11] and top new layers were added. They consist of three dense layers with corresponding dropout layers. For training and optimizing the weights on the tomato disease dataset, we froze the first 51 convolutional layers and made the rest trainable for Inception V3. Training optimization was carried out via a stochastic gradient descent optimizer with a learning rate of 0.0001 and momentum of 0.9. We used a batch size of 20 and 20 epochs for training. We use data augmentation techniques to increase the dataset size in training and solve the class imbalance while including different variations. These variations consist of transformations such as random rotations, zooms, translations, shears, and flips to the training data as we train. The performance of the trained models is evaluated using recall, precision, and accuracy metrics [10].

Fig. 5. Sample images from the PlantVillage Dataset. (a) Bacterial Spot, (b) Early Blight, (c) Healthy, (d) Late Blight, (e) Leaf Mold, (f) Septoria Leaf Spot, (g) Two-spotted Spider Mites, (h) Target Spot, (i) Mosaic Virus, and (j) Yellow Leaf Curl Virus. (Color figure online)

The total number of images is 18,160, divided into ten classes (nine diseases and a healthy class). The data was separated into three sets, containing 80% of the data in the training set; the remaining 20% were divided between the testing and validation sets. Figure 5 presents one example of each disease class. The

Table 1. Per class precision, recall, F1-Score for the test set Class.

Class Label	Sample Count	Precision	Recall	F1-score
Bacterial Spot	191	1	0.64	0.78
Early Blight	119	0.99	0.62	0.76
Late Blight	178	0.89	0.99	0.94
Leaf Mold	77	0.96	0.92	0.94
Septoria Leaf Spot	198	0.9	0.96	0.93
Two-spotted Spider Mite	177	0.87	0.99	0.93
Target Spot	142	0.7	0.96	0.81
Yellow Leaf Curl Virus	534	0.99	1	0.99
Mosaic Virus	37	0.97	0.89	0.93
Healthy	163	0.98	0.99	0.98

trained model achieved the following training, validation and testing accuracies, respectively: 0.98, 0.92 and 0.92.

Table 1 shows the precision, recall, and F1-score for each class. The model was implemented using Keras [9], and was saved for subsequent interpretability analysis. We experimented on a server with a GPU that consists of two NVIDIA Tesla V100 with 128 GB of RAM.

5.2 Experimental Setup

In this work, our aim is to study the correlation between concepts derived from a domain ontology modelling the knowledge about diseases and those learned within the activation of the neural networks. For example, if a neural network is trained to identify late blight disease, then ontological concepts representing the disease like ∃hasColor.Black and ∃hasSymptom.Blight should be important for the decision. It is worth noting that none of these concepts were part of the predefined class labels of the network; rather they were all derived through ontology reasoning. Hence, the first step of our approach is exploiting the ontology to identify and automatically generate concepts specific to each disease class, as described in Sect. 4.2.

The subsequent step involves utilising the generated images that represent each concept for training the concept activation vectors (CAVs). To train CAVs, a set of 30 images per concept was generated. The selection of this number aligns with the recommendation in the original TCAV paper [20], where it is asserted that such a number suffices to learn CAVs. For the target classes, we randomly chose 30 images for each from the training set. Images for creating ∃hasSymptom concepts were selected randomly from the test set which the model was not trained on. We used the "mixed_8" bottleneck layer of the InceptionV3 model for these experiments. As demonstrated in [14,20], initial layers are better at capturing textures and colors while later ones are better at recognizing objects;

the choice of the "mixed_8" layer balances between these considerations. We used images of healthy tomatoes without any presence of disease as random (i.e., non concept) images. We believe this choice allows training CAVs with a better fine-grained recognition. The TCAV score is used to evaluate the concept's importance to a specific target class. To check statistical significance of learned concepts, we trained an additional 70 random CAVs. The distribution of random concept TCAV scores and actual concept TCAV scores was then compared by conducting a two-sided t-test [21] with ($\alpha = 0.05$) to assure significance of the found CAVs. In the results section, statistical insignificance is represented by stars. Our code and ontology can be accessed on GitHub[1].

5.3   Findings and Analyses

To evaluate our approach, we will concentrate on the quantitative evaluation of TCAV scores. Figure 6 presents the different sensitivity score (TCAV) highlighting the contribution of the semantic concepts to their relevant corresponding neural classification. Essentially, TCAV quantifies the impact of a given concept on a specific target class. For instance, in the case of Tomato Mosaic Virus, the results show a high TCAV score for all the semantic concepts such as ∃hasColor.Yellow (0.97) and ∃hasSymptom.MosaicPatchesOnLeaf (0.9). Contrarily, for the disease Target Spot the model did not learn the concept ∃hasSymptom.TargetSpotsOnLeaf, which may explain the low precision of the class as shown in Table 1. This suggests that the model may not be optimal for robustly detecting the TargetSpot disease and that gathering more training data with clear symptom texture existence could enhance the results. In contrast, for Septoria Leaf Spot, even though the symptom concept was not important, the abnormalities in concepts like hasColor and hasShape were sufficient for the model to identify the class. Additionally, for the class 'Two-spotted spider mite', none of the disease's semantic concepts made a significant contribution. This suggests that the model may not have effectively learned these important semantic concepts associated with this disease class. However, the class achieved a precision of 0.8, indicating that the model is learning to identify this class through another bias in the dataset. This insight highlights the need for a closer examination of the class and the model.

The results further highlight the significance of color concepts for different disease classes, supporting the findings of [16] where they observed an accuracy drop when the model was trained on grayscale images. In summary, these findings provide insights into the contribution of different semantic concepts in the decision of the model which shows to which extent the model is consistent with domain knowledge. Our approach not only relates the classified diseases to their symptoms and signs but also tries to quantify the contribution of these symptoms to the model decision.

---

[1] https://github.com/fusion-jena/XAI_TCAV_ONTO.

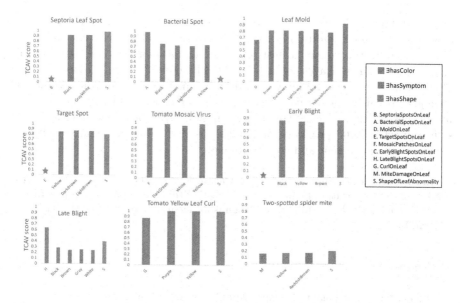

Fig. 6. Conceptual importance (TCAV scores) for the different disease semantic concepts for each class

6 Conclusion

With this work, we contribute to improving the explainability, dependability and trustworthiness of deep learning models by adding expert knowledge through an ontology. For our implementation, we focus on the identification of plant diseases as a use case. Our novel approach automatically generates concepts related to observable disease features using the ontology. This lets users peek into the model and see how its results depend on these concepts, all without needing to manually collect concepts. Through experimental evaluation, we showed the sensitivity of the model to these concepts. By formalizing expert knowledge in an ontology, we can enhance our comprehension of the relationships between various concepts within a model and also make the examination and correction of misclassifications and biases easier. We believe that our approach could be easily extended to other domains due to different points. First, our ontology is built upon a conceptual framework that involves color, symptom (texture), and shape abnormalities. This framework is not specific to tomato diseases and can be adapted to cover characteristics in other plant diseases or domains. Second, our ontology design is modular and flexible. Separating disease characteristics and types into distinct modules made the inclusion of new diseases and their corresponding specific concepts easier. Third, the most important features when describing images such as leaves or other objects are shape, color, and texture. Our proposed approach for generating images for such concepts is domain-independent, which shows its adaptability behind the current use case. Despite our findings, we acknowledge some challenges, like the difficulty of capturing all expert knowledge in an ontology. Also, further detailed tests with a high-quality dataset are needed for more comprehensive interpretation of the TCAV scores for this particular use case.

In future work, we aim to test our approach on more challenging plant diseases datasets where leaves could be infected by more than one disease. We plan also to consider how combining neural and semantic representation via knowledge graphs can be generalised to other problems such as object detection and image classification. The explainability framework can be integrated into a decision support system, providing actionable insights to farmers and stakeholders for disease management and crop protection.

Acknowledgement. Supported by the Carl Zeiss Foundation (project 'A Virtual Werkstatt for Digitization in the Sciences (K3)' within the scope of the programline 'Breakthroughs: Exploring Intelligent Systems for Digitization - explore the basics, use applications').

References

1. Plantvillage. www.plantvillage.psu.edu. Accessed 13 Nov 2023
2. Ahmad, A., Saraswat, D., El Gamal, A.: A survey on using deep learning techniques for plant disease diagnosis and recommendations for development of appropriate tools. Smart Agric. Technol. **3**, 100083 (2023)
3. Amara, J., Bouaziz, B., Algergawy, A.: A deep learning-based approach for banana leaf diseases classification. Datenbanksysteme für Business, Technologie und Web (BTW 2017)-Workshopband (2017)
4. Amara, J., König-Ries, B., Samuel, S.: Concept explainability for plant diseases classification. In: Proceedings of the 18th International Joint Conference on Computer Vision, Imaging and Computer Graphics Theory and Applications (VISIGRAPP 2023) - Volume 4: VISAPP, pp. 246–253 (2023)
5. Ammar, H.: Ontology for plant protection. https://sites.google.com/site/ppontology/home (2009)
6. Blancard, D.: Tomato Diseases: Identification, Biology and Control: A Colour Handbook. CRC Press, Boca Raton (2012)
7. Bourguin, G., Lewandowski, A., Bouneffa, M., Ahmad, A.: Towards ontologically explainable classifiers. In: Farkaš, I., Masulli, P., Otte, S., Wermter, S. (eds.) ICANN 2021. LNCS, vol. 12892, pp. 472–484. Springer, Cham (2021). https://doi.org/10.1007/978-3-030-86340-1_38
8. Chhetri, T.R., Hohenegger, A., Fensel, A., Kasali, M.A., Adekunle, A.A.: Towards improving prediction accuracy and user-level explainability using deep learning and knowledge graphs: a study on cassava disease. Expert Syst. Appl. **233**, 120955 (2023)
9. Chollet, F.: Deep learning with Python. Simon and Schuster (2021)
10. Davis, J., Goadrich, M.: The relationship between precision-recall and ROC curves. In: Proceedings of the 23rd International Conference on Machine Learning, pp. 233–240 (2006)
11. Deng, J., Dong, W., Socher, R., Li, L.J., Li, K., Fei-Fei, L.: ImageNet: a large-scale hierarchical image database. In: 2009 IEEE Conference on Computer Vision and Pattern Recognition, pp. 248–255. IEEE (2009)

12. Donadello, I., Dragoni, M.: SeXAI: introducing concepts into black boxes for explainable artificial intelligence. In: Proceedings of the Italian Workshop on Explainable Artificial Intelligence co-located with 19th International Conference of the Italian Association for Artificial Intelligence, XAI. it@ AIxIA 2020, Online Event, 25–26 November 2020, vol. 2742, pp. 41–54. CEUR-WS (2020)
13. Ge, Y., Xiao, Y., Xu, Z., Wang, X., Itti, L.: Contributions of shape, texture, and color in visual recognition. In: Avidan, S., Brostow, G., Cissé, M., Farinella, G.M., Hassner, T. (eds.) ECCV 2022. LNCS, vol. 13672, pp. 369–386. Springer, Cham (2022). https://doi.org/10.1007/978-3-031-19775-8_22
14. Ghorbani, A., Wexler, J., Zou, J.Y., Kim, B.: Towards automatic concept-based explanations. In: Advances in Neural Information Processing Systems, vol. 32 (2019)
15. Gkoutos, G.V., Green, E.C., Mallon, A.M., Hancock, J.M., Davidson, D.: Using ontologies to describe mouse phenotypes. Genome Biol. 6, 1–10 (2005)
16. Hughes, D., et al.: An open access repository of images on plant health to enable the development of mobile disease diagnostics. arXiv preprint arXiv:1511.08060 (2015)
17. Jaiswal, P., et al.: Planteome: a resource for common reference ontologies and applications for plant biology (2017)
18. Jearanaiwongkul, W., Anutariya, C., Andres, F.: An ontology-based approach to plant disease identification system. In: Proceedings of the 10th International Conference on Advances in Information Technology, pp. 1–8 (2018)
19. Jearanaiwongkul, W., Anutariya, C., Racharak, T., Andres, F.: An ontology-based expert system for rice disease identification and control recommendation. Appl. Sci. 11(21), 10450 (2021)
20. Kim, B., et al.: Interpretability beyond feature attribution: quantitative testing with concept activation vectors (TCAV). In: International Conference on Machine Learning, pp. 2668–2677. PMLR (2018)
21. Koch, G.G.: One-sided and two-sided tests and ρ values. J. Biopharm. Stat. 1(1), 161–170 (1991)
22. Molnar, C.: Interpretable machine learning. Lulu.com (2020)
23. Noy, N.F., et al.: Ontology development 101: a guide to creating your first ontology (2001)
24. Pease, A., Niles, I., Li, J.: The suggested upper merged ontology: a large ontology for the semantic web and its applications. In: Working Notes of the AAAI-2002 Workshop on Ontologies and the Semantic Web, vol. 28, pp. 7–10 (2002)
25. Sarker, M.K., Xie, N., Doran, D., Raymer, M., Hitzler, P.: Explaining trained neural networks with semantic web technologies: first steps. arXiv preprint arXiv:1710.04324 (2017)
26. Simonyan, K., Vedaldi, A., Zisserman, A.: Deep inside convolutional networks: visualising image classification models and saliency maps. arXiv preprint arXiv:1312.6034 (2013)
27. Smilkov, D., Thorat, N., Kim, B., Viégas, F., Wattenberg, M.: SmoothGrad: removing noise by adding noise. arXiv preprint arXiv:1706.03825 (2017)
28. de Sousa Ribeiro, M., Leite, J.: Aligning artificial neural networks and ontologies towards explainable AI. In: Proceedings of the AAAI Conference on Artificial Intelligence, vol. 35, pp. 4932–4940 (2021)
29. Szegedy, C., Vanhoucke, V., Ioffe, S., Shlens, J., Wojna, Z.: Rethinking the inception architecture for computer vision. In: Proceedings of the IEEE Conference on Computer Vision and Pattern Recognition, pp. 2818–2826 (2016)

30. Tjoa, E., Khok, H.J., Chouhan, T., Guan, C.: Enhancing the confidence of deep learning classifiers via interpretable saliency maps. Neurocomputing **562**, 126825 (2023)
31. Van Dijk, M., Morley, T., Rau, M.L., Saghai, Y.: A meta-analysis of projected global food demand and population at risk of hunger for the period 2010–2050. Nat. Food **2**(7), 494–501 (2021)
32. Wakchaure, M., Patle, B., Mahindrakar, A.: Application of AI techniques and robotics in agriculture: a review. Artif. Intell. Life Sci. 100057 (2023)
33. Zeiler, M.D., Fergus, R.: Visualizing and understanding convolutional networks. In: Fleet, D., Pajdla, T., Schiele, B., Tuytelaars, T. (eds.) ECCV 2014. LNCS, vol. 8689, pp. 818–833. Springer, Cham (2014). https://doi.org/10.1007/978-3-319-10590-1_53
34. Zhou, B., Sun, Y., Bau, D., Torralba, A.: Interpretable basis decomposition for visual explanation. In: Proceedings of the European Conference on Computer Vision (ECCV), pp. 119–134 (2018)

Generative Expression Constrained Knowledge-Based Decoding for Open Data

Lucas Lageweg[1]([✉]) and Benno Kruit[2]

[1] Statistics Netherlands, Henri Faasdreef 312, Den Haag, Netherlands
l.lageweg@cbs.nl
[2] Vrije Universiteit Amsterdam,, De Boelelaan 1105 Amsterdam, Netherlands
b.b.kruit@vu.nl

Abstract. In this paper, we present GECKO, a knowledge graph question answering (KGQA) system for data from Statistics Netherlands (Centraal Bureau voor de Statistiek). QA poses great challenges in means of generating relevant answers, as well as preventing hallucinations. This is a phenomenon found in language models and creates issues when attempting factual QA with these models alone. To overcome these limitations, the Statistics Netherlands' publicly available OData4 data was used to create a knowledge graph, in which the answer generation decoding process is grounded, ensuring faithful answers. When processing a question, GECKO performs entity and schema retrieval, does schema-constrained expression decoding, makes assumptions where needed and executes the generated expression as an OData4 query to retrieve information. A novel method was implemented to perform the constrained knowledge-based expression decoding using an encoder-decoder model. Both a sparse and dense entity retrieval method were evaluated. While the encoder-decoder model did not achieve production-ready performance, experiments show promising results for a rule-based baseline using a sparse entity retriever. Additionally, the results of qualitative user testing were positive. We therefore formulate recommendations for deployment help guide users of Statistics Netherlands data to their answers more quickly.

1 Introduction

Statistics Netherlands (Centraal Bureau voor de Statistiek; CBS) is an independent administrative body of the Dutch government tasked with the creation of statistics over a broad spectrum of social topics, and the responsibility to make them accessible to the general public. However, we have observed that non-experts currently struggle to find the correct tables for their needs in the vast amount of data available. The current research aims to develop a Question Answering (QA) system to provide specific statistical observations from this data as responses to natural-language user questions.

© The Author(s), under exclusive license to Springer Nature Switzerland AG 2024
A. Meroño Peñuela et al. (Eds.): ESWC 2024, LNCS 14664, pp. 307–325, 2024.
https://doi.org/10.1007/978-3-031-60626-7_17

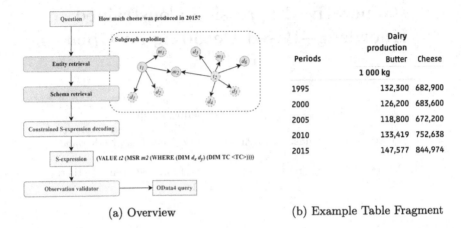

		Dairy production	
Periods		Butter	Cheese
	1 000 kg		
1995		132,300	682,900
2000		126,200	683,600
2005		118,800	672,200
2010		133,419	752,638
2015		147,577	844,974

(a) Overview   (b) Example Table Fragment

Fig. 1. (a) Overview of the GECKO pipeline from query to answer. Candidate nodes (green) are retrieved for the query and used for doing subgraph exploding (blue), after which it is used as input for the constrained S-expression decoding by either the baseline method or trained model. (b) Example CBS table fragment (from **7425eng**), showing one dimension (time periods) and two measures. (Color figure online)

QA systems can take several forms, with most recently free-form generative large language models (LLMs) like ChatGPT [1] and GPT4 [2] getting much attention. Due to the nature of these models, they are able to generalize very well on a large range of topics, but have shown to be prone to 'hallucinating', where plausible but incorrect or even nonsensical answers are being generated [3]. Especially for official data like governmental statistics, this is highly undesirable behaviour. Therefore, the main design goal for this system is the interpretability of the provenance of answers.

Knowledge graph question answering (KGQA) is a field where knowledge bases (KGs) containing real-world facts and relations in structured form are used as a basis for QA systems. Answers of such systems should always adhere to the KG. Therefore, assuming it contains correct information, answering by returning parts of the KG, or reasoning over it, cannot lead to nonsensical answers. In this paper, we introduce **GECKO** (**G**enerative **E**xpression **C**onstrained **K**nowledge-based decoding for **O**pen data), a proof of concept for a generation-based KGQA system for CBS data [1]. It will generate a response to a question in two parts: the answer itself following from the KG and a justification that should be able to explain why and what exactly is returned. The justification is crucial, as the KGQA system could return facts that, albeit correct, are not relevant for the question.

We focus on the retrieval of a single table cell from a relevant table based on a given natural language input question. In practise, this means that the system

---

[1] The source code, graph and models are made available at https://github.com/lagewel001/GECKO.

will be able to answer questions that have an answer in a single cell of a table present in the KG. For our investigation, we assume only Dutch input questions, and focus on all of the Dutch 'key figure' tables in the CBS data catalogue. These 60 tables, ranging over all different topics, contain the highest aggregated form of statistical data available at the CBS. This makes these tables suited for our specific proof of concept as this high-level data has little ambiguity.

As this project is still in the early stages of preparation for deployment, we present preliminary results of the application of these semantic technologies to this problem, and their fitness for purpose. In short, our contributions are as follows:

- we describe our Knowledge Graph Question Answering application on statistical observation data from Statistics Netherlands (Sect. 3),
- we evaluate the application (Sect. 4), both quantitatively on in-house annotated data (Sect. 4.1) and qualitatively with user tests (Sect. 4.2),
- and we discuss the impact and deployment of the application (Sect. 5).

2   Related Work

Systems mostly related to our approach are semantic parsing based QA systems, in particular text-to-SQL. In their survey, Qin et al. [4] explain several approaches on learning input and table schema representations (encoding), in order to later generate and parse SQL statements (decoding). Also, in earlier attempts at CBS, research was done on creating a text-to-SQL model for retrieval of micro-data (non-aggregated) statistics [5]. While similar to our approach in their query expression design, our system is specifically designed to tackle the case where a large number of heterogeneous tables and metadata concepts must be retrieved and used in queries. This quality is shared by KGQA systems, which aim to retrieve one or more small-scale facts (i.e. RDF triples) from expansive KBs, often using analogous text-to-SPARQL approaches. Damljanovic et al. [6] and Unger et al. [7] construct SPARQL templates and parse the questions using those templates. The downside of using templates is its limited flexibility, as only question types for which templates are created can be answered and thus no true free-form QA can be achieved. However, free-form generation of SPARQL (e.g. Ochieng [8]) is hard, as the syntax is quite abstract and intricate in the representation of data triples, and queries can become quite expansive. As the surveys by Lan et al. [9] and Gu et al. [10] show, more recent approaches use either ranking-based methods or generation-based approaches for creating alternative logical forms for querying the KG. Similarly, we generate simplified logical S-expressions, which are translated later into more complicated query language constructs.

Large Language Models (LLMs) have recently gained a lot of attention as free-form text generators that can be used as QA systems [1]. One major drawback of these systems is their tendency to generate false statements. Fact checking using KBs has been proposed as one solution [11]. Generative models are also considered for the task of generating logical forms to retrieve and reason over KBs themselves. However, adapting these models still provides a problem when it comes to producing queries/expressions that are faithful to the KG (i.e. to prevent querying non-existing triples). Current state-of-the-art methods for KGQA use KG grounding for constraining queries that adhere to the KG, as is shown by Gu et al. [12], Yu et al. [13] and Shu et al. [14]. All three examples take roughly the same overall approach, and serve as the main source of inspiration for this current research.

3   Application

Data. For this research, to narrow the scope, we will use CBS data that is publicly available and, more specifically, will use the *key figure* tables, which contain the most aggregated form of statistics. A complete overview of all different statistical research done by CBS is published yearly [15]. All data is made available via the ISO/IEC approved Open Data Version 4.01 protocol (OData4) [16,17], which is the API that will be queried to return observations to user questions. Each table observation consists of a single measure value (i.e. a statistic being measured) and values for all dimensions available in that table (i.e. filtering characteristics or properties for said measures). CBS maintains a public vocabulary of concepts, in which every unique measure and dimension has an identifier[2]. In the editorial process of publishing statistics, all measures and dimensions are standardised as much as possible in an attempt to maintain consistency between tables. However, this makes it possible for a single identifier to have multiple nuanced definitions based on the tables it is part of. For example, the standardised terms *Total imports of goods and services* and *Perception of (un)safety* have multiple definitions depending on how the goods/services are exactly defined and what specific question is asked to an respondent, which can vary in different surveys. This can also happen when a new but very near identical definition of a term is inadvertently added in the redaction process. In this work, we make use of a subset of this vocabulary encoded as RDF as our KG, corresponding to the schema descriptions of the tables in our target sample.

Knowledge Graph. Using RDF to represent the table structures creates a simple and intuitive system to work with for referencing observations later. Due to the tabular nature of the data at hand, and by only indirectly storing the references to observations using the tables' metadata, the KG remains relatively simple. Currently CBS makes its data available using the OData4 format, while the Statistical Data and Metadata eXchange (SDMX) format is currently the industry standard [18]. When transitioning to SDMX in the future, or when using SDMX for a different use case, our vocabulary can be used for setting up a KG that can be used by the subsequent methods that will be discussed in the following sections. This implies that our method can be applied by anyone using one of these standards.

---

[2] https://vocabs.cbs.nl/en/.

During the creation of the KG, whenever applicable we attempted to harmonize as many measures and dimensions as possible. This improves the system's metadata overall but will also help simplify the generation process for querying the KG as this results in less nodes overall that can cause ambiguity when matching to a query. This is an iterative process as new tables, measures and dimensions are constantly being created and new instances are found where a harmonization of two nodes can be beneficial. For example, at the time of writing there are two separate nodes/identifiers for the term *labour costs*, with one used for the labour cost survey and the other for production statistics. As said, nodes can contain more than one definition based on the nuances of the table and as this research puts its focus on the key figure statistics, undoubtedly a lot of harmonization can be done over the entirety of the CBS' data catalogue. Furthermore, this harmonization will also help in the editorial process as this enables the statisticians and editors to select standardised predefined options and only need to create new measures or dimension when no standardized code is available. One example of this in practise can be found in the standardization of CBS municipality codes (among other geographical related entities), which are also publicly shared in Wikidata[3], to be used by third parties to refer to related CBS data. Other examples include the use of the European-wide Classification of Products by Activity (CPA) [19] and Statistical classification of economic activities (NACE) [20].

Queries. In order to retrieve and return information from the KG, several solutions can be considered. For example, a single prefix tree (trie) can be returned as plain text, or a part of the KG can be returned in the form of the aforementioned RDF-format or bindings of a SPARQL query. Several other solutions to provide semantically meaningful representations for KGQA have been proposed, like graph query [21] or λ-calculus [22]. In this work we follow ArcaneQA's solution based on GrailQA's S-expression format [12] to provide a syntactic sugar for queries that can be executed over the KG.

S-expressions are a notation for logical expressions containing atoms and expressions (which are always S-expressions themselves) in a tree-like structure as nested lists. For our purpose, GrailQA proposes a format where the expression comprises of functions (AND, COUNT, JOIN, etc.) and entities (i.e. the atoms). The S-expressions always denote operations over the KG. ArcaneQA extends their definition with functions for general and temporal constraints but follows the same principle. The benefit of using S-expressions is that they are compact and concise, human-readable and machine-interpretable and, most importantly, easily converted to other types of querying formats like SPARQL or OData4. As our use case is quite different from ArcaneQA's, we use a modified version of S-expressions more suited to our KG. The S-expressions should be able to represent one or more OData4 observations. In practice, this means that the expression must be able to notate at least a table, measure and one or more dimensions to represent a single observation. Appendix A gives a supplementary

---

[3] https://www.wikidata.org/wiki/Property:P382 (Accessed: 14-07-2023).

table with the set of functions created for this purpose. An expression will always start with an aggregation/operator function to indicate the operation needed to be done over the values denoted in the expression that follows. Here, only a VALUE function is needed to retrieve a raw observation, but more functions for counting, averaging, etc. can be considered later. Two special atom placeholders exist for the time and geographical constraint functions (TC and GC).

Pipeline. The GECKO pipeline consists of four parts, described in Fig. 1. This pipeline is mostly inspired by the similar format shown by Shu et al. [14]. Due to its large size, it is infeasible to consider the entire graph when doing KGQA. We restrict the querying space by performing entity retrieval based on the query to determine relevant graph nodes. Two methods of entity retrieval were implemented: (1) Sparse retrieval, using BM25+ [23] to rank candidate nodes as a baseline. It takes the top 25 best ranking table, measure or dimension nodes based on their textual metadata (excluding time or geographical dimensions) and uses them for schema retrieval (see next step). The documents here are the table, measure and dimension entities and use the concatenated metadata of skos:prefLabel, skos:altLabel, skos:definition and dct:description as the search body. Dimension entities of type TimeDimension and GeoDimension are not considered in the search to significantly reduce the search space in the following steps. This method of sparse retrieval has shown to still be very competitive compared to more complex embedding-based dense search methods [24]. (2) Dense vector search, using context embedding vectors of the KG nodes in an inner product (IP) index and embed the given query to compare in the vector space to return the 75 closest nodes. The context embeddings are created using a sentence transformer [25] pre-trained language model (PLM) on the same textual metadata fields as mentioned above.

After obtaining the closest matching entities based on the query, we use schema retrieval to explode a subgraph using the entities given (Fig. 2). The result of the subgraph exploding is a graph containing all table nodes and their related measures and dimensions having nodes intersecting with the retrieved entities. This is visualized in Fig. 1a. The green nodes are those obtained using the entity retrieval step. The resulting graph from the schema retrieval are all depicted nodes, where the blue nodes are obtained trough the subgraph exploding. This results in a subgraph containing all relevant nodes connected to one or more tables. Relevant (hierarchical) relations and useful metadata for the different entities are also retrieved. Both the question and subgraph are input for the constrained S-expression generation. This step results in an expression as described above.

Observation validation. The observation validator translates the generated S-expressions to an OData4 query. As a single OData4 observation requires the query to explicitly state all dimension group filters, dimensions not included in the expression are assumed if possible using the table subgraph. For missing dimensions of a time or geographical type, the latest period available up until the current year or the biggest aggregate of the measure available (usually the

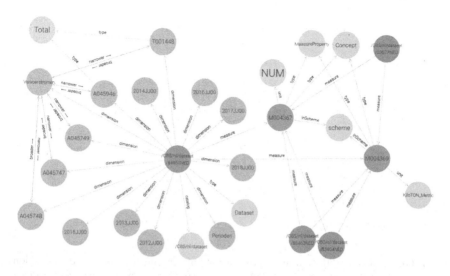

Fig. 2. Exploded subgraph for a small CBS table (`84957NED`: Pijpleidingenvervoer; kerncijfers) as visualized in GraphDB, containing the table nodes (red), dimensions and their hierarchical relations (blue), measures (purple) and several properties (yellow) like units.

Netherlands) are assumed respectively. In any other case, the validator checks if the group contains a dimension denoting a total. If no assumption can be made, the user should be asked to refine the given query and specify the missing dimension group. All assumptions made are included in the justification output to the user. S-expressions can contain a `<TC>` or `<GC>` placeholder. As there is rarely ambiguity between dates or geographical locations, these can be substituted by rules with the relevant dimension entities based on the subgraph and the given query. As ambiguity is still possible however, the largest dimension aggregation is chosen similar to the default assumptions if not specified in the question. This system also enables relative notations like 'last year' to be substituted correctly based on the current date. The final step in the pipeline is the execution of the OData4 query by translating the final expression to an OData4 query. The justification consists of the question, table, measures and dimensions selected and the default assumptions made.

Example. Following the example from Fig. 1, the following expression can be generated from the question "how much cheese was produced in 2015?":

    `(VALUE (7425zuiv (MSR D001544 (WHERE (DIM TC <TC>)))))`

Which will be rule-based substituted by the obervation validator into:

    `(VALUE (7425zuiv (MSR D001544 (WHERE (DIM TC 2015JJ00)))))`

And finally translated to its OData4 query counterpart:

    `https://odata4.cbs.nl/CBS/7425zuiv/Observations/?$filter=Measure in ('D001544') and Perioden in ('2015JJ00')`

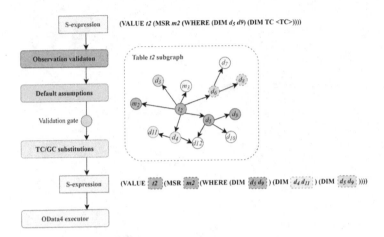

Fig. 3. Observation validation and substitutions of missing measures and dimensions in the generated S-expression. From initial nodes (teal), missing dimensions for observations (purple) are assumed if possible, and time and geographical constraints are filled in using rules (orange). If a required dimension is missing (red), validation fails. (Color figure online)

PLMs. As one of the preliminaries for creating our model we employ pretrained language models (PLMs) in two instances. The first is in creating the context embedding vectors for the dense vector search in entity retrieval as described above. These embeddings are created using a pretrained Dutch BERTje-based sentence transformer [26]. We will refer to this PLM as the GroNLP model. The second case is as a starting point for finetuning the constrained decoder model. For our model training, we finetuned the Dutch RoBERTa model, RobBERT [27], on a cleaned CBS dataset containing all written publications, articles and table descriptions. This was done using Masked Language Modeling (MLM). We will refer to our finetuned model as SNERT (Fig. 3).

3.1   Generative Models

A rule-based baseline was created for generating S-expressions to compare our model against. The S-expressions can be created token-by-token such that, given the subgraph, admissible tokens can be returned at every step. The baseline uses the BM25+ scores from the entity retrieval step to greedily determine what token from the admissible tokens to select. This deterministic approach implies that only the best scoring table and its measures and dimensions can be selected for the expression generation. Only scored nodes are considered for the expression output. Therefore, after the schema retrieval step, the BM25+ scores are calculated for all nodes in the subgraph to increase the recall. Scores of measures and dimensions denoting a total are boosted. All dimensions in the set of admissible tokens are greedily added to the expression. Therefore, a minimum threshold score per node is implemented to prevent dimensions with low scores to be considered in the output.

For the model solution, generation of the expressions is performed by a transformer-based seq2seq model. Utilising the encoder-decoder framework, the model will output the target S-expression by generating sequences of admissible tokens until a valid expression is constructed. We employ the core idea of dynamic contextualised encodings from Gu et al. [28] for encoding representative contextual embeddings for the code tokens. First, we extend the model's tokenizer vocabulary with all code nodes present in the KG. We will denote this extension to the SNERT tokenizer as SNERTe. Consequently, when initialising the model, the embedding weight matrix for these tokens are initialised randomly. Secondly, an index map is created, mapping all token identifiers to their respective position in the pre-computed IP-index, which contains all embeddings for the different code nodes based on their textual descriptions. Next, we update the embedding weight matrix for all code nodes to utilise the weights from the IP-index. In all following steps and during training, we fix the gradients for the embedding layer. This way, the decoder can learn the relations between the input question and given vectors from the schema retrieval step. The individual context vectors of nodes thus remain constant.

As the embedding layer, as part of the encoder, is not tasked with learning and creating the different embeddings for the KG nodes, new or altered codes only need to be added to the IP-index and the vocabulary. Our hypothesis is that this will help the model generalize, and:

(H₁) reduce the amount of training data needed for successfully training the model to recognize relations between questions and entities with the limited number of training samples available (learnability);

(H₂) solve the infeasibility of learning all input-output relations between questions and all possible codes in the KG (generalizability);

(H₃) and solve the infeasibility to retrain the model every time new nodes are added to the KG (maintenance).

For an incoming question a dynamic prompt is generated. We employ entity and schema retrieval as before using the dense vector search. For performance benefits during the subgraph exploding and to prevent overflowing the input prompt, we consider only the top 5 candidate table nodes, and for each table the top 10 scored measures and dimensions to add to our prompt. The question and KG tokens are then concatenated into the following sequence:

$$[CLS], q_1, ..., q_{|k|}, [SEP], t_1, |MSR|, m_1, ..., m_{|p|}, |DIM|, d_1, ..., d_{|q|}, [SEP], t_{|n|}, ...$$

where $\{q_i\} \subset Q_t$ denotes all wordpiece tokens following from the tokenizer and $(\{t_i\} \cup \{m_i\} \cup \{d_i\}) \subset W_t$ denotes respectively all table, measure and dimension tokens ordered by their graph relations and vector distances compared to the embedded query in the IP-index. Dimensions regarding time and geographical constraints are omitted from the prompt and are substituted with <TC> and <GC> in the target expressions for the model. In fact, these tokens are not added

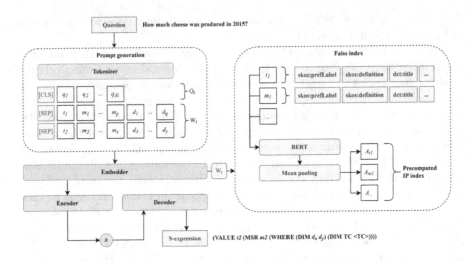

Fig. 4. Overview of the GECKO constrained S-expression encoder-decoder model. The question is tokenized into question tokens Q_t and candidate nodes W_t. At the embedding step, for all tokens in W_t the contextualised embedding matrix is obtained from the pre-computed Faiss index.

to the tokenizer's vocabulary and model's embedding matrix to avoid the model becoming too large, as there is an incredibly large number of these nodes in the KG (Fig. 4).

Using the different PLMs, the model is trained with the embedded prompt as input, and the S-expression as target sequence, using weighted cross-entropy with label smoothing as the objective function. As the opening and closing parentheses are the most predominant and constant tokens in the target expressions, the loss function is weighted for these specific tokens. Considering an extended vocabulary K, we define a vector ω and a weight parameter β to be

$$\omega_k = \begin{cases} \beta, & \text{if } k \in \{[\text{CLS}], [\text{SEP}], [\text{PAD}], (,) \} \\ 1.0, & \text{otherwise} \end{cases}, \quad \text{with} \quad \beta \in [0,1].$$

We also attempt to reduce model overconfidence in the beam search inference by applying label smoothing regularization [29]. Combined, this yields the following objective loss function:

$$\mathcal{L} = -\sum_{i=1}^{n} \sum_{k \in K} \omega_k q'(k|y_i) \log p(k|y_i),$$

$$\text{with} \quad q'(k|y_i) = (1 - \epsilon)q(k|y_i) + \frac{\epsilon}{K}$$

$$\text{and} \quad p(k|y_i) = \frac{\exp(z_k)}{\sum_{j=1}^{K} \exp(z_j)}$$

where y denotes the generated target expression, n the length of this sequence, q the ground truth (i.e. one-hot vector for every y_i) and p the softmax function for a given target token, denoting z as the logit for a given token. ϵ denotes the smoothing parameter in the regularization function q'.

Constrained Inference. As the decoder is not strictly constrained during the training phase, in contrast to ArcaneQA's approach, we perform constrained decoding during inference to prevent generating faulty sequences. This ensures that the generated S-expressions are syntactically correct and true to the KG. We employ the beam search algorithm [30] to generate the top-ranking sequences, i.e. expressions, over n beams. Using the constrained beam search method from De Cao et al. [31] we force the decoder to generate only admissible tokens following from the current S-expression and the exploded subgraph. In practice, this means that when a specific table identifier is generated for a specific beam, only measures and dimensions that have a relation with that table node can be passed as admissible tokens at a specific timestep t.

4   Evaluation

For determining the model performance we test the hypotheses $\mathbf{H}_1$, $\mathbf{H}_2$ and $\mathbf{H}_3$ stated above. The evaluation dataset consists of 120 unseen samples from our annotated key figures dataset. For the generalization/scalability test we let several annotators familiar with the data create ± 1.250 questions for the top-visited non key figure tables. These tables are unknown to the model and contain measures and dimensions not trained on. For the annotation task, we provided an interface serving a pivot table containing a single randomized OData4 observation and asked the annotators to come up with one or more corresponding questions.

To evaluate the entity retrieval step we will use the accuracy (Acc.), precision (P), recall (R) and F1-scores of the tables, measures and dimensions. These are calculated based on the generated prompts for a given query compared to the corresponding S-expression, using BM25+ and dense vector search. Mainly, a high accuracy or recall is desired for a correct outcome of the expressions in the generation step, as the model should learn to select the correct codes from a prompt into the generated expression. The same applies for the BM25+ baseline, as the code selection is performed using a greedy approach. For table and measure nodes, the accuracy scores denote the proportion of target nodes occurring in the input prompts, regardless of the length of the prompt. The precision scores also take into account how many table, measure or dimension nodes there are in the prompt (i.e. codes that are in the prompt but might not occur in the target expression). For tables, the mean reciprocal rank (MRR) is also calculated to indicate to what degree the correct tables and their selected measure and dimension codes occur at the start of the prompt, as the highest scoring tables from the entity retrieval step are placed at the beginning of a prompt.

To determine the performance of the baseline and trained model we use the ROUGE [32] and BLEU [33,34] metrics, often used in machine translation. This will determine the expressions' recall and precision respectively. To also syntactically evaluate matching expressions (i.e. correct relative placement of parentheses and functions), but disregard dimension filter ordering, the bi-gram version of ROUGE will be used alongside four-gram BLEU. For the individual table and measure codes in a generated S-expression, the exact match (EM) score will be used, as there can be only one of each in a given S-expression using the current expression format. The EM score therefore denotes the proportion of identical table or measure nodes between all target-prediction expression pairs. For the dimension tokens, of which there can be multiple in a single expression, we determine the F1 score.

Additionally, we use a human-judgement relevancy score (RS), manually annotating every model answer retrieved from OData4 following the generated expressions with either 1 (fully correct), 0.5 (minor issues) or 0 (wrong). All models were trained on 947 training samples from the key figures dataset, using 120 samples for evaluation. For our last configuration, we trained the SNERTe PLM on all samples available (key figures plus annotated top-visited non key figures), counting 2.069 training and 230 evaluation samples.

An important part of the evaluation is to test the generalizability of the model. As the KG and number of unique codes is vast, it is undesirable and unfeasible to annotate samples for all possibilities by hand. Next to that, frequent updates, additions and changes occur to the KG which can make the trained model outdated. To test our generalizability and maintenance hypotheses (see H_2 and H_3) we evaluate the model's generalizability and scalability on our metrics using controlled samples of S-expressions referencing never before seen tables. These tables and corresponding nodes were omitted completely from the KG during training and are added to the vocabulary before inference using the method described above.

4.1   Quantitative Results

Table 2 shows the evaluation scores of the table, measure (MSR) and dimension (DIM) nodes from the entity retrieval and prompt generation steps. Both the sparse (BM25+) and dense retrieval methods perform similarly, showing slight deviations in matching accuracy and precision per node type. Neither method significantly outperforms the other based on these scoring metrics. The overall results of the model performances can be seen in Table 1. Of the model-based solutions, the RobBERT-based model scores best on our evaluation set on both the ROUGE-2, BLEU, and matching criteria save from MSR EM, which is scored best by the SNERT-based model. The SNERT-based model trained on all data samples scores similarly to the smaller model during training.

The evaluation of the baseline model shows the highest scores on all metrics, significantly outperforming the model-based approaches. Both the sparse and dense vector search methods were tested for the entity retrieval (ER) step, with BM25+ scoring slightly higher on all measures except for table matching.

Table 1. Evaluation metrics and S-expression inference evaluation for the different models. The target-prediction expression similarity is expressed by ROUGE-2, BLEU, F1 and exact match (EM) scores. The relevancy score (RS) is the average of manually annotated relevant answers following from the questions and generated expressions by the model (i.e. disregarding exact target matches).

Model	ER		ROUGE-2	BLEU	RS	Table EM	MSR EM	DIM F1
Baseline	BM25+		**0.437**	**62.198**	**0.378**	0.347	**0.198**	**0.621**
	Dense		0.374	53.025	0.349	**0.396**	0.158	0.496
GroNLP	Dense		0.294	48.039	0.107	0.181	0.029	0.455
RobBERT	Dense		0.377	55.042	0.110	0.267	0.038	0.555
SNERTe	Dense		0.193	40.278	0.031	0.200	0.048	0.214
SNERTe (all samples)	Dense		0.318	46.182	0.167	0.188	0.100	0.398

Regarding the RS-score, for 80 samples, the BM25+ baseline test resulted in 27 relevant (=1) and 3 semi-relevant (=0.5) answers. The dense retrieval scored 21 and 11 in these categories respectively. This difference is also reflected in the similar RS and EM scores, but significantly different dimension F1.

Table 2. Evaluation results for entity retrieval performance using individual metrics for table, measure and dimension nodes by BM25+ and dense vector search.

		BM25+	Dense
TABLE	Acc.	**0.530**	0.496
	P	0.114	**0.184**
	MRR	**0.262**	0.239
MSR	Acc.	**0.448**	0.435
	P	0.018	**0.074**
DIM	P	0.023	**0.074**
	R	**0.592**	0.555
	F1	0.040	**0.054**

Table 3. Inference evaluation for non-key figure (unseen) tables for the different models. Target-prediction similarity is expressed by the exact matches (EM) of tables and measures nodes and the F1 score for dimensions. RS denotes the relevancy scores of answers generated. All models are significantly outperformed by the rule-based baseline.

Model	ER		RS	Table EM	MSR EM	DIM F1
Baseline	BM25+		**0.357**	**0.409**	**0.278**	**0.564**
	Dense		0.182	0.178	0.105	0.358
GroNLP	Dense		0.081	0.126	0.039	0.176
RobBERT	Dense		0.066	0.114	0.027	0.223
SNERTe	Dense		0.055	0.076	0.013	0.170

Noteworthy is the low precision and relatively high recall on the entity retrieval for dimensions, indicating the input prompts span a high number of dimensions (lowering the precision), containing for more than half of the samples the correct target dimension(s). This can also be seen in the resulting F1 scores for the dimensions in the S-expression evaluation, where for the baseline, GroNLP and RobBERT models the F1 scores are close to 0.5 (Table 3).

4.2 Qualitative User Testing

The test results from the previous section only show the performance on exact matches between the target and generated expressions, and the relevance of answers generated were determined based on an annotation guideline. In doing this, there remains a reality-gap between the model scores of the system and the actual usefulness to a user. To test this, there are a few questions that need to be answered. First of all, how well does a user parse the information that is presented to them when a question is answered? Is the pivot table shown as an answer an intuitive and readable way of presenting the statistical data? Secondly, how well do the users understand the justification given with each answer, and do they convey the right information when an assumption or ambiguity is present? This ties in with the question whether this approach helps to steer the users' querying of the system when unrelated or no answers are returned. Finally, as the main question, we would want to know if GECKO helps the users to find their statistical answers more quickly than without using our system.

Our user study comprised of 7 unique users. Example scenarios were constructed asking the users to find information on specific cases (e.g. finding different statistics and trends on solar power and number of solar panels in the Netherlands), as well as asking the users to bring scenarios of their own. The users have full control over what search queries to use and no external help influencing the results is provided. The overall consensus was that a system like GECKO is a very suitable option for finding statistics more quickly, especially for non-frequent users of our data. The users did indicate that the biggest drawback at the moment is the result comprising of only a single table cell, which does not help in showing what more information can be found in the table shown. This makes it impossible to do any associative searching by the user. Furthermore, it was shown that the way the answer justification is presented is vital to our use case. Presenting a textual prompt only containing the assumptions made and word matches between the query and measures/dimensions did not convince the users and raised more questions than it answered. Lastly, users would like an option to alter the assumptions made, in case a wrong default value is given, alongside the possibility to give feedback to the system in case a completely irrelevant answer is given.

5   Impact and Deployment

As a proof of concept, GECKO shows that it is possible to create a question answering system that is faithful to the CBS data and will not hallucinate, regardless of the discussed expression decoding methods used. Incorporating this system as part of a search engine can help a user get to a desired answer significantly faster.

In order to create a production-ready system that could be integrated as a search page functionality, a few steps need to be taken. First of all, using GECKO in its current form, an answer is always attempted based on a best effort (i.e. closest match), regardless of the input question. Albeit KG-faithful, it would still be undesirable to return a nonsensical answer to a question. Therefore, the system must be finetuned and incorporate a confidence threshold that can determine whether it is appropriate to return a generated answer.

Secondly, we would recommend the baseline as a viable option to continue optimizing for the current available S-expression functions. Compared to machine learning models, which can be considered as 'black boxes', this also improves the explainability of the system. Looking forward to the upcoming registry for algorithms[4] and the act for algorithm transparency at Dutch governmental institutions, as announced by the secretary of state for digitization [35], the importance of this aspect cannot be understated. A first step in this would be in investigating the possibilities of a reranking algorithm and combining the sparse and dense entity retrieval methods. When considering more complex S-expression functions, the current greedy baseline would need to be altered such that multiple aggregation functions can be considered by the model. As of the current state, the model-based approach does not yield reliable results for a production environment. To determine its viability, significantly more training data is needed. Synthetic training data can be considered alongside the annotated samples. We recommend conducting a cost-benefit analysis for this scenario and investigate if there is a functional requirement for being able to return more complex answers.

6   Conclusions and Future Work

In this paper we present GECKO as a question answering system to help guide users of CBS data to relevant answers for their questions. The system uses a knowledge graph containing table metadata and generates expressions that can be used for querying and retrieving observations from OData4. The results show that there is not a significant difference between the performance of the sparse and dense search methods for the entity retrieval step when it comes to exact matches of table, measure and dimension nodes compared to the questions' target expressions. When expanding the KG with more nodes however, the BM25+ sparse retrieval method outperforms the dense approach. When looking at the decoding performances of the different models compared to the baseline, the learnability hypothesis H_1 is disproven, as the models did not yield competitive results by training on the limited number of training samples available. We cannot conclude our generalizability hypothesis H_2. None of the models were able to generate more relevant answers from the candidate nodes in the prompt, and thus we cannot surely state that by fixing the embedding matrix the need of learning all input-output relations is omitted. This might be due to the limited number of training samples that were available and should thus be revisited in the future. Looking at the maintenance hypothesis H_3, we see that there is a slight drop in performance for the three different models when looking at the non-key figure results for unseen tables. As the general performances for both evaluation sets are too low however, this can neither support nor reject the claim that using the fixed embedding layer helps with generalizing over all possible (even unseen) nodes and remains to be tested when further improvements are made.

---

[4] https://algoritmes.overheid.nl/en (Accessed: 06-12-2023).

Currently, plans are being made to expand the system to be able to work on different datasets. In its current form, a KG can be generated from any OData4-based system. In the near future, we will expand the proof of concept to accommodate SDMX-based datasets, with SDMX being the industry standard that will also be utilised at CBS in the near future.

Future work can look at the possibility for more complex S-expressions in order to allow more complex and diverse questions to be answered. Next to more complex answers, future research could also look into the possibility for determining unanswerability of questions, as the current system will always attempt a best effort to answer the question given using the closest matching KG entities. Finally, combining the benefits of both sparse and dense entity retrieval methods might increase the relevance significantly for generating S-expressions. A combined distance metric for the entity retriever or a reranking solution can both be investigated.

Appendices

A   S-expression functions considered in GECKO

(See Table 4)

Table 4. Set of different S-expression functions for our system.

Function	Arguments	Description
VALUE	table entity	Returns all the raw OData4 observations (i.e. cell values) matching the table entity and corresponding selection following the expression.
MSR	(measure entity, WHERE-expression)	Denotes a selection filter for a table measure to retrieve.
WHERE	set of DIM-expressions	Function containing all sub-expressions for filtering dimensions on a specific measure.
DIM	(dimension group entity or TC/GC atom, dimension entity)	Denotes a selection filter for the table dimensions to retrieve.
TC / GC	dimension entity	Special temporal or geographical constraint function denoting a dimension filter for specific dimensions of type TimeDimension and GeoDimension respectively.
OR	set of dimension entities	Function for defining a selection filter on multiple dimension entities in the same dimension group

References

1. OpenAI. Introducing ChatGPT (2022). https://openai.com/blog/chatgpt. Accessed 14 June 2023
2. OpenAI. GPT-4 Technical Report (2023)
3. Zhang, M., Press, O., Merrill, W., Liu, A., Smith, N.A.: How Language Model Hallucinations Can Snowball (2023)
4. Qin, B., et al.: A Survey on Text-to-SQL Parsing: Concepts, Methods, And Future Directions (2022)
5. Gelsema, T., Heuvel, G.V.D.: Towards demand-driven on-the-fly statistics. https://www.researchgate.net/publication/349768489_Towards_demand-driven_on-the-fly_statistics (2021)
6. Damljanovic, D., Agatonovic, M., Cunningham, H.: Natural language interfaces to ontologies: combining syntactic analysis and ontology-based lookup through the user interaction. In: Aroyo, L., et al. (eds.) ESWC 2010. LNCS, vol. 6088, pp. 106–120. Springer, Heidelberg (2010). https://doi.org/10.1007/978-3-642-13486-9_8
7. Unger, C., Bühmann, L., Lehmann, J., Ngonga Ngomo, A.C., Gerber, D., Cimiano, P.: Template-based question answering over RDF data. In: Proceedings of the 21st International Conference on World Wide Web, WWW 2012, pp. 639-648, New York, NY, USA (2012). Association for Computing Machinery
8. Ochieng, P.: PAROT: translating natural language to SPARQL. Expert Syst. Appl. **176**(C) (2021)
9. Lan, Y., He, G., Jiang, J., Jiang, J., Zhao, W.X., Wen, J.R.: A Survey on Complex Knowledge Base Question Answering Methods, Challenges and Solutions (2021)
10. Gu, Y., Pahuja, V., Cheng, G., Su, Y.: Knowledge Base Question Answering: A Semantic Parsing Perspective (2022)
11. Li, X., et al.: Chain of Knowledge: A Framework for Grounding Large Language Models with Structured Knowledge Bases (2023)
12. Gu, Y., et al.: Three levels of generalization for question answering on knowledge bases. In: Proceedings of the Web Conference 2021. ACM (2021)
13. Yu, D., et al.: DecAF: Joint Decoding of Answers and Logical Forms for Question Answering over Knowledge Bases (2023)
14. Shu, Y., et al.: TIARA: multi-grained retrieval for robust question answering over large knowledge bases. In: The 2022 Conference on Empirical Methods in Natural Language Processing. ACL (2022)
15. Centraal Bureau voor de Statistiek. Wij maken en verspreiden statistieken - Overzicht statistieken (2022). https://www.cbs.nl/nl-nl/over-ons/wij-maken-en-verspreiden-statistieken. Accessed 13 July 2023
16. OData Version 4.01. Part 1: Protocol. Standard, OASIS international open standards consortium (2020)
17. Information technology - Open data protocol (OData) v4.0 - Part 1: Core (ISO/IEC 20802-1:2016). Standard, International Organization for Standardization, Geneva, CH (2016)
18. Statistical data and metadata exchange (SDMX). Standard, International Organization for Standardization, Geneva, CH (2013)
19. Commission Regulation (EU) No 1209/2014 of 29 October 2014 amending Regulation (EC) No 451/2008 of the European Parliament and of the Council establishing a new statistical classification of products by activity (CPA) and repealing Council Regulation (EEC) No 3696/93 (2014). http://data.europa.eu/eli/reg/2014/1209/oj

20. Regulation (EC) No 1893/2006 of the European Parliament and of the Council of 20 December 2006 establishing the statistical classification of economic activities NACE Revision 2 and amending Council Regulation (EEC) No 3037/90 as well as certain EC Regulations on specific statistical domains (2006). http://data.europa.eu/eli/reg/2006/1893/oj

21. Yih, W.T., Chang, M.W., He, X., Gao, J.: Semantic parsing via staged query graph generation: question answering with knowledge base. In: Proceedings of the 53rd Annual Meeting of the Association for Computational Linguistics and the 7th International Joint Conference on Natural Language Processing (Volume 1: Long Papers), pp. 1321–1331, Beijing, China (2015). Association for Computational Linguistics

22. Cai, Q., Yates, A.: Semantic parsing freebase: towards open-domain semantic parsing. In: Second Joint Conference on Lexical and Computational Semantics (*SEM), Volume 1: Proceedings of the Main Conference and the Shared Task: Semantic Textual Similarity, pp. 328–338, Atlanta, Georgia, USA (2013). Association for Computational Linguistics

23. Lv, Y., Zhai, C.: Lower-bounding term frequency normalization. In: Proceedings of the 20th ACM International Conference on Information and Knowledge Management, CIKM '11, pp. 7–16, New York, NY, USA (2011). Association for Computing Machinery

24. Sciavolino, C., Zhong, Z., Lee, J., Chen, D.: Simple Entity-Centric Questions Challenge Dense Retrievers (2022)

25. Reimers, N., Gurevych, I.: Sentence-BERT: sentence embeddings using siamese BERT-networks. In: Proceedings of the 2019 Conference on Empirical Methods in Natural Language Processing. Association for Computational Linguistics (2019)

26. De Vries, W., van Cranenburgh, A., Bisazza, A., Caselli, T., van Noord, G., Nissim, M.: BERTje: A Dutch BERT Model. arXiv:1912.09582 (2019)

27. Delobelle, P., Winters, T., Berendt, B.: RobBERT: a Dutch RoBERTa-based Language Model (2020)

28. Gu, Y., Su, Y.: ArcaneQA: dynamic program induction and contextualized encoding for knowledge base question answering. In: Proceedings of the 29th International Conference on Computational Linguistics, pp. 1718–1731, Gyeongju, Republic of Korea (2022). International Committee on Computational Linguistics

29. Szegedy, C., Vanhoucke, V., Ioffe, S., Shlens, J., Wojna, Z.: Rethinking the Inception Architecture for Computer Vision (2015)

30. Sutskever, I., Vinyals, O., Le, Q.V.: Sequence to Sequence Learning with Neural Networks (2014)

31. De Cao, N., Izacard, G., Riedel, S., Petroni, F.: Autoregressive Entity Retrieval (2021)

32. Lin, C.Y.: ROUGE: a package for automatic evaluation of summaries. In: Text Summarization Branches Out, pp. 74–81. Barcelona, Spain (2004). Association for Computational Linguistics

33. Papineni, K., Roukos, S., Ward, T., Zhu, W.J.: Bleu: a method for automatic evaluation of machine translation. In: Proceedings of the 40th Annual Meeting of the Association for Computational Linguistics, pp. 311–318, Philadelphia, Pennsylvania, USA (2002). Association for Computational Linguistics

34. Post, M.: A call for clarity in reporting BLEU scores. In: Proceedings of the Third Conference on Machine Translation: Research Papers, pp. 186–191, Belgium, Brussels (2018). Association for Computational Linguistics
35. Secretary of state for digitization. Stand van zaken Algoritmeregister (letter no. 2022-0000693912) (2022). https://www.rijksoverheid.nl/documenten/kamerstukken/2022/12/21/kamerbrief-over-het-algoritmeregister. Accessed 16 July 2023

OntoEditor: Real-Time Collaboration via Distributed Version Control for Ontology Development

Ahmad Hemid[1(✉)], Waleed Shabbir[2], Abderrahmane Khiat[2],
Christoph Lange[1,4], Christoph Quix[1,3], and Stefan Decker[1,4]

[1] Fraunhofer FIT, Data Science and Artificial Intelligence, Sankt Augustin, Germany
{ahmad.hemid,christoph.lange-bever,christoph.quix,
stefan.decker}@fit.fraunhofer.de
[2] Fraunhofer IAIS, Enterprise Information Systems, Sankt Augustin, Germany
{waleed.shabbir,abderrahmane.khiat}@iais.fraunhofer.de
[3] Hochschule Niederrhein, Krefeld, Germany
christoph.quix@hs-niederrhein.de
[4] RWTH Aachen University, Aachen, Germany
lange@cs.rwth-aachen.de, decker@dbis.rwth-aachen.de

Abstract. In today's remote work environment, the demand for real-time collaborative tools has surged. Our research targets efficient collaboration among knowledge engineers and domain experts in Ontology development. We developed a web-based tool for real-time collaboration, compatible with GitLab, GitHub, and Bitbucket. To tackle the challenge of concurrent modifications leading to potential inconsistencies, we integrated an Operational Transformation-based real-time database. This integration enables multiple users to concurrently collaborate to build and edit their ontologies, ensuring both consistency and atomicity. Furthermore, our tool enhances user experience by providing meaningful syntax error messages for ontologies expressed in various RDF serialization formats. This streamlined the manual correction process. Additionally, we established a reliable synchronization channel for users to allow pulling and committing changes to distributed repositories for their developed ontologies. Yielding promising results, our evaluation focused on two key aspects: first, assessing the tool's collaborative editing consistency via an automated typing script; second, conducting a comprehensive user study to evaluate its features and compare its functionalities with similar tools.

Keywords: Real-time collaboration · RDF serialization · Version Control Systems · Git integration · Error detection · Syntax validation · Ontology development

1 Introduction

The vision of the Semantic Web, fostered by the World Wide Web Consortium (W3C), aims to transform the web into a machine-interpretable platform, akin

A. Meroño Peñuela et al. (Eds.): ESWC 2024, LNCS 14664, pp. 326–341, 2024.
https://doi.org/10.1007/978-3-031-60626-7_18

to a global database [1]. However, even with all research efforts conducted to achieve this goal, the current landscape of the web lacks an interconnected data framework, leading to fragmented data confined within individual applications.

At the heart of the Semantic Web lies the Resource Description Framework (RDF) [2], pivotal for modeling data and its relationships. Yet, contemporary applications grapple with challenges such as limited real-time collaboration and inadequate syntax validation within RDF-based systems, demanding an integrated and comprehensive solution.

This paper introduces OntoEditor, an innovative Online Collaborative Ontology Editor designed to revolutionize real-time collaboration across various RDF serialization formats. By harnessing Version Control Systems (VCS) such as GitHub, GitLab, and Bitbucket, OntoEditor offers users live syntax validation and collaborative editing features, effectively tackling inherent limitations in current systems, particularly in conflict resolution.

OntoEditor fills a critical gap in the domain of collaborative ontology development, offering a robust platform that seamlessly integrates real-time collaboration capabilities with thorough syntax validation across multiple RDF serialization formats. Through this endeavor, we aim to alleviate the persistent challenges faced by users in effectively collaborating on ontology projects.

This paper is structured as follows: Sect. 2 presents a motivational example, while Sect. 3 explores related works, offering a comparative analysis of existing solutions concerning OntoEditor. In Sect. 4, an overview of OntoEditor, including its workflow and features, is provided. Section 5 details its implementation and the technologies employed. The evaluation, conducted through experimental tasks and a user study, is presented in Sect. 6. Section 7 discusses the current limitations of OntoEditor, while Sect. 8 explores strategies for its sustainable adoption. Finally, the paper concludes in Sect. 9, with potential avenues for future work highlighted in Sect. 10.

2   Motivation

Suppose John, along with his colleagues Robert and Lisa, aims to develop an ontology together, illustrating the need for real-time collaboration as depicted in Fig. 1. Their expertise in ontology engineering leads them to prefer plain text editors and VCS such as Git[1] for collaborative ontology development. However, existing tools lack efficient collaboration, communication, and real-time syntax error detection. This results in a cumbersome and error-prone process where users must separately write, check syntax, communicate changes, and repeat this cycle, causing inefficiency and errors.

Research from Queens University of Charlotte highlights that about 75% of employers highly value teamwork and collaboration, emphasizing the need for streamlined collaboration tools[2]. This inspired us to devise a solution that

---

[1] https://git-scm.com/.

[2] https://blog.bit.ai/collaboration-statistics/.

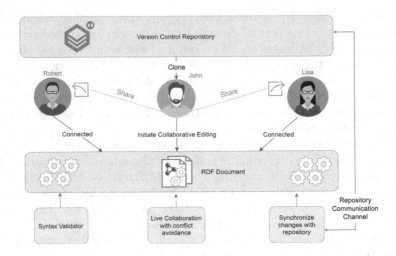

Fig. 1. Collaboration Scenario. An example motivating collaborative ontology development with a built-in syntax validator. Both Robert and Lisa seek to work seamlessly with John without requiring software downloads or installations.

enables multiple users to collaborate seamlessly during ontology development without constant syntax-checking interruptions.

Digging deeper into the scenario presented in Fig. 1, John, Robert, and Lisa aim to collaborate on an ontology. They can create a new ontology or import an existing one from VCS such as GitHub. John initiates collaborative editing, sharing the link with his colleagues, enabling real-time simultaneous document viewing and tracking of each user's edits.

With the syntax checker enabled, any modifications trigger instant error detection visible to all users, fostering discussions and corrections in real-time via a shared chat. This approach ensures everyone's awareness of errors and each other's editing progress.

The final step involves users synchronizing their work with the remote repository. Only authorized users can commit changes, ensuring controlled access and ownership permissions within the repository.

3   Related Work

This section delves into collaborative ontology development, ontology synchronization with VCS, and tools supporting various RDF formats and validation.

3.1   Parsing RDF and Syntax Checking

Numerous online and desktop tools specialize in validating RDF data and supporting syntax checking. The W3C RDF validation tool [11] primarily focuses on parsing and validating RDF/XML[3] exclusively. Meanwhile, isSemantic RDF Tools [7] offer syntax checking and format conversions among other visualization and generation services but lack comprehensive support for collaborative editing or VCS integration. Our objective revolves around supporting ontology development within distributed VCS, a feature noticeably absent in existing tools.

Table 1. Feature Comparison. OntoEditor, TurtleEditor, and WebProtégé for Collaborative Ontology Development.

Feature	Tool		
	OntoEditor	TurtleEditor	WebProtégé
Real-time Collaboration	✓	✗	✓
Textual RDF Editor	✓	✓	✗
RDF Serializations	Turtle, RDF/XML, JSON-LD	Turtle	Turtle, RDF/XML
Integration with Git	✓ (GitHub, GitLab, Bitbucket)	✓ (GitLab only)	✗ (Own Protégé Server)
Conflict Resolution	✓	✗	✓
Export/Download Option	✓	✗	✓

3.2 Ontology Editors

Our exploration identified tools with collaborative capabilities but notable limitations. Existing tools often bound ontology development to specific formats or lack crucial collaboration features. Table 1 compares OntoEditor with similar tools such as TurtleEditor and WebProtégé. TurtleEditor [10] specializes in syntax checks for the Turtle RDF format[4] but confines users to a single format and a specific repository service, limiting real-time collaboration. WebProtégé [13] enables collaboration through a server-based approach but mandates hosting on its servers, necessitating user accounts for collaboration, and it does not support text editing. VocBench [12] supports collaborative SKOS thesaurus editing but involves a complex deployment setup.

Taking from platforms such as Overleaf[5], OntoEditor stores content in a database, generating a unique shareable link. This link fosters collaboration by inviting contributors to work within the same document, offering users the flexibility to choose between individual or collaborative work based on their preferences.

OntoEditor fulfills the ongoing demand for a versatile tool enabling ontology development in any RDF format, syntax parsing, real-time collaboration, and seamless integration with VCS.

---

[3] https://www.w3.org/TR/rdf-syntax-grammar.
[4] https://www.w3.org/TR/turtle.
[5] https://www.overleaf.com/learn.

4 OntoEditor: A Collaborative Ontology Editor

OntoEditor is an Online Collaborative Ontology Editor, built on Distributed VCS. It aims to support collaborative ontology development across different RDF serialization formats: Turtle, JSON-LD[6], and RDF/XML. The following discusses its processing workflow as well as the integrated components for empowering collaborative editing.

4.1 OntoEditor Workflow

Figure 2 provides an overarching view of the fundamental steps within OntoEditor. The process commences with user authentication in Git, where credentials are provided. Upon successful authentication, users gain access to remote repository data, including repository names, branches, and file details. Subsequently, upon file and RDF serialization format selection, the editing phase commences.

The unique project link allows new users to view the names and cursor positions of those already connected to the document. This collaborative environment enables multiple users to concurrently edit RDF content, with every change being visible to all collaborators. Additionally, users have the option to individually enable or disable the syntax checker. When activated, the system parses the RDF document, identifying and displaying any syntax errors present. Upon completion of changes, users can commit and push their updates and changes to the distributed repository.

4.2 Collaborative Editing Components

OntoEditor encourages collaborative ontology development among multiple users, driven by essential components that facilitate seamless interaction:

Customizable Editor: CodeMirror[7] as a JavaScript-based Editor was selected for its robust programmable API and advanced editing capabilities such as auto-indentation, auto-completion, syntax highlighting, and search functionalities. As an open-source editor widely used in various projects, CodeMirror inherently supports syntax highlighting for over a hundred programming languages, including Turtle, XML, and JSON-LD. Crucially, it enables collaboration by detecting changes through *onChange* events.

Real-Time Communication Channel: Real-time communication is vital for collaborative editing. WebSocket [5] technology enables immediate and bidirectional data exchange between web browsers (clients) and servers, facilitating seamless interactions. Upon initiating document editing, a WebSocket connection is established. This channel relays all modifications to the server, managing live chat, user details, cursor positions, and notifications among connected users.

ShareDB: To enable seamless collaboration, a real-time database was imperative. After careful research, ShareDB[8] emerged as the optimal choice. ShareDB,

---

[6] https://www.w3.org/TR/json-ld11.

[7] https://codemirror.net/.

[8] https://share.github.io/sharedb/.

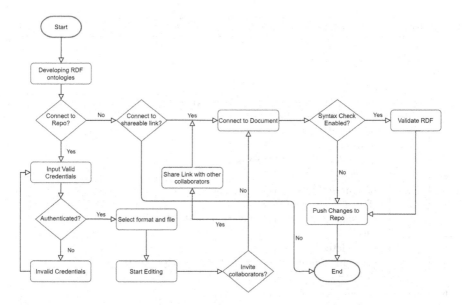

Fig. 2. OntoEditor's Workflow. The diagram illustrates the authentication process in Git, file selection, and the initiation of RDF editing. It highlights the collaborative nature facilitated by a unique shareable link, enabling simultaneous editing and syntax checking. Completed changes can be committed and pushed to the remote repository.

built on Operational Transformation (OT) [4], operates as a real-time in-memory database. It stores JavaScript objects on the server and facilitates their sharing among multiple clients through WebSockets. Documents in ShareDB include properties such as Version (incrementing from 0), Type (e.g., OT-text, OT-json1), and Data which is the intended content for storage within the database.

Algorithm 1 was designed to operationalize this approach. Upon a user's initiation of RDF document editing, a new document with an initial version of 0 and the RDF data to be inserted is created in ShareDB. If the document already exists in ShareDB, its existing path is returned to the user. For Operational Transformation, we leverage Plain text Operational Transformation[9]. This Operational Transformation type is utilized for editing plain text documents and supports operations including skipping forward N characters, inserting *str* at the current position, and deleting N characters at the current position.

Clients subscribe to ShareDB documents, updating the document's state with insertions and deletions by modifying the index position and content. Each change increments the version number and is stored in ShareDB. These operations are transmitted over WebSockets, updating local document states.

RDF Validator: OntoEditor validates RDF serialization formats-Turtle, RDF/ XML, and JSON-LD-in real-time, allowing immediate syntax error detection while typing. Users can toggle validation on or off as needed, with the tool

---

[9] https://github.com/ottypes/text.

Algorithm 1: The pseudo-code of Collaboration

  Data: inputFile, repoDetails
  Result: docChanges,users
 1 users = [ ]
 2 **if** *editingMode* **then**
 3 | path = startEditing(inputFile,repoDetails)
 4 | users.push(name, cursorPosition)
 5 | doc = webSocket(path)
 6 | **if** *doc.subscribe* **then**
 7 | | Initialize CodeMirror(doc.value)
 8 | | **if** *doc.change* **then**
 9 | | | user.updatePosition(userCursor)
10 | | | **if** *addedText* **then**
11 | | | | sharedb.submitOp([position, addedText])
12 | | | **else**
13 | | | | deletedText
14 | | | sharedb.submitOp([position, length.deletedText])
15 | | | doc.version += 1
16 | | **if** *sharedb.receivedOp* **then**
17 | | | **if** *op == sender* **then**
18 | | | | return
19 | | | **else**
20 | | | | **if** *op == insertion* **then**
21 | | | | | codeMirror.replaceRange(newData,position)
22 | | | | **if** *op == deletion* **then**
23 | | | | | codeMirror.removeRange(' ', startPosition, endPosition)
24 | **Function** *startEditing(inputFile, repoDetails)*
25 | | **if** *File exists in shareDB* **then**
26 | | | shareLink = (repoDetails) + sharedb.fetchDoc()
27 | | | return shareLink
28 | | **else**
29 | | | shareLink = (repoDetails) + sharedb.createDoc(inputFile)
30 | | | return shareLink
31 | **end**

---

providing coherent error messages. Even with syntax errors, users retain the ability to push file changes to the remote repository.

5   Implementation

Considering Fig. 3, OntoEditor comprises three key modules, each serving a distinct role within the system. The initial module manages communication with remote repositories, while the second module is designed to enable real-time collaboration among users. Finally, the third module is dedicated to comprehensive syntax validation.

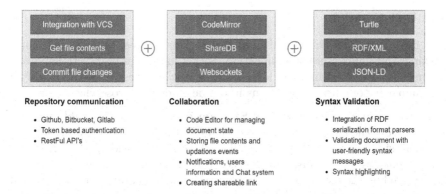

Repository communication

- Github, Bitbucket, Gitlab
- Token based authentication
- RestFul API's

Collaboration

- Code Editor for managing document state
- Storing file contents and updations events
- Notifications, users information and Chat system
- Creating shareable link

Syntax Validation

- Integration of RDF serialization format parsers
- Validating document with user-friendly syntax messages
- Syntax highlighting

Fig. 3. OntoEditor's Modules. OntoEditor comprises three modules: Repository Communication, Collaboration, and Syntax Validation.

5.1 Repository Communication

We utilize GitHub, GitLab, and Bitbucket's APIs for web-based RDF editing, bypassing the need for local Git installation. Authentication requires a username and access token for GitHub and GitLab, while Bitbucket needs an access token with an empty username. Users access repositories and branches and filter files by formats such as *ttl*, *rdfxml*, etc. Available actions include file operations, commits (requiring authentication), and link sharing.

To manage conflicts or concurrent edits, we monitor file history for new commits every 60 s. Conflict resolution leverages the Mergely JavaScript library[10], offering users a side-by-side comparison, shown in Fig. 4.

Fig. 4. Git Conflict Resolution Merge View. A user interface showcases the user's current document version on the left and the changes retrieved from the hosted Git repository on the right. It facilitates Git conflict resolution by displaying the most recent version of the document if new changes exist.

---

[10] https://github.com/wickedest/Mergely.

5.2   Collaboration

This module utilizes Socket.io[11] over WebSockets for real-time editing, cursor tracking, and multi-user communication. Socket.io enables bi-directional, event-driven client-server communication via JavaScript libraries, aimed at simplifying the complexity of editing operations into independent microservices.

Collaboration Service and File Storage in shareDB: To enable real-time collaboration, the file content is stored in our Database. Users initiating edits on the client-side trigger a REST API call to the server, providing Git authentication parameters, file details, and the chosen RDF serialization format. ShareDB is initialized on the server, allowing connections via a separate port. We uniquely identify each file using SHA-1 hashes [3], mirroring how Git stores file information. The hash becomes the document's ID in shareDB. If the document doesn't exist, we create it with an initial version of 0, storing both current data and individual operations.

A unique URL path is generated for each file, containing vital information such as ProjectID, repository details, branch, file name, and RDF serialization format. This link is sent to the client for editing. If the document exists, we send its path to the user, enabling collaboration by sharing the link.

The data is stored in MongoDB for persistence. Locally, ShareDB's in-memory database suffices. Real-time communication begins when a user starts editing, establishing a WebSocket session to the server. Users' names, cursor positions, and document details are maintained on the server and broadcasted to all connected clients. An integrated chat widget allows communication within the system.

Document Updates and Conflict Resolution: ShareDB, as a real-time database, ensures users view the latest document state. Operational Transformation resolves conflicts, managing concurrent editing. CodeMirror's API triggers *onChange* events for any insertion or deletion in the editor.

Insertion operations, demanding an index position and added text, elevate the document's version, while deletions require the index and deletion amount, similarly increasing the version. These actions, applied in ShareDB, are broadcasted to all clients and locally updated. Insert operations synchronize by replacing text using CodeMirror's *replaceRange* function, while deletions are executed by replacing text with an empty string. This process guarantees consistent real-time collaboration across multiple users.

5.3   RDF Validation and Error Notification

OntoEditor utilizes JavaScript parser libraries for real-time validation of various RDF serialization formats (Turtle, RDF/XML, JSON-LD). Users receive instantaneous error messages and can rectify syntax errors seamlessly during collaborative editing of RDF data.

---

[11] https://socket.io.

To validate RDF, established parsers are employed: N3.js[12] for Turtle, RDF/XML streaming parser[13] for RDF/XML, and JSON-LD streaming parser[14] for JSON-LD. These parsers operate streamingly, ensuring efficient handling of large documents with limited memory.

During the editing process, users can select their desired format. The chosen parser is activated accordingly, integrated with an *onChange* function to check syntax while typing automatically. The syntax checker can be toggled on or off, with default activation. The syntax checker identifies the format from the URL path and calls the corresponding parser. Parsing occurs in a streaming manner, providing parsed triples and highlighting any syntax errors. Meaningful error messages are displayed atop the editor for immediate user visibility. Upon error correction, a *Syntax correct, all triples parsed successfully* message is shown (Fig. 5).

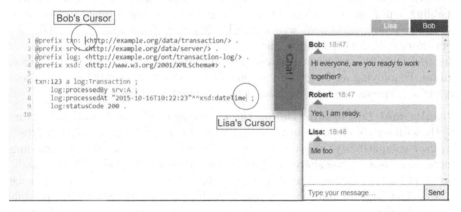

Fig. 5. Real-Time Collaboration Snapshot. Robert, Lisa, and Bob are concurrently editing, each represented by their cursor positions. Additionally, they are engaging in communication through an integrated chat widget.

6   Evaluation

OntoEditor underwent comprehensive evaluation through both functional testing and a user study. The functional tests involved experimenting with collaborative editing via an automated typing script on RDF Turtle documents. In contrast, the user study specifically targeted participants from a computer science background to assess their experiences and feedback.

6.1   Functional Testing

This task assesses collaborative editing performance by multiple automated users on RDF Turtle documents. It tests real-time collaboration with consistency using various browsers and clients.

---

[12] https://github.com/rdfjs/N3.js.

[13] https://github.com/rdfjs/rdfxml-streaming-parser.js.

[14] https://github.com/rubensworks/jsonld-streaming-parser.js.

Objective and Experiment Configuration: The testing aimed to assess real-time collaborative editing under simultaneous input from multiple users subscribed to the same document. Conducted on a Windows 10 machine with a 3rd Gen Intel Core i7-3630 CPU, 2.40 GHz, and 8 GB RAM, the web application was tested across various browsers: Chrome, Firefox, and Opera.

Procedure: Five clients, denoted as tabs A to E, were concurrently opened and tasked with typing different sections of an RDF Turtle file using automated scripts. Initially, an empty document shareable link was generated and accessed by these five clients across separate browser tabs (one client per tab). To assess cross-browser functionality, three clients were opened in Chrome across three tabs, one in Firefox, and one in Opera.

Each client was associated with a distinct segment of the RDF turtle file, enabling simultaneous collaborative editing. To automate typing, a code snippet tailored for each client was utilized. This script, available in our GitHub repository[15], assigned separate instances of the ontology to each client from the shared RDF file. The script prompts for *RDF* input and *Starting time* for execution. Employing an interval function, the script simulates the typing process at variable speeds until completion of the assigned code segment

Results and Discussion: During the experiment, all clients worked concurrently without conflicts, ensuring a consistent document state and robust cross-browser compatibility. Functional tests validated key functionalities, including connection to a unique shareable project link, display of connected users, indication of cursor positions, simultaneous typing by multiple clients, maintenance of a conflict-free document state, and verification of cross-browser compatibility. Client identities from A to E were assigned, and in Fig. 6, client A's browser view exhibits connected clients' names and their cursor positions. The conducted typing script affirmed successful and seamless collaboration, enabling consistent and conflict-free document production among all connected clients.

6.2   User Study

The user study presents a comparative analysis of the user experiences between OntoEditor, TurtleEditor, and WebProtégé. Additionally, it comprehensively outlines the evaluation process steps. To gauge the accessibility of our tool, we employed the Concurrent Think-aloud method. This method involved observing participants closely as they performed tasks, allowing us to capture their real-time thoughts and insights.

Participants & Procedure: Nine participants, varying in expertise from basic to advanced in computer science ontology and modeling, took part in the evaluation. They possessed some familiarity with VCS, particularly Git. The evaluation included comprehensive introductions to OntoEditor, TurtleEditor, and WebProtégé for a comparative analysis. Tasks were assigned across all three tools, with continuous monitoring of participants' approaches to solve each task.

---

[15] https://w3id.org/ontoeditor.

Fig. 6. Snapshot of Functional Testing Result: Client A's browser interface displays real-time status after executing an automated typing script. Visible are the names of the four connected users, along with their respective cursor positions, synchronized within the editor.

Participants could access guidance in the *Help* section as needed. After completing tasks, the focus shifted to evaluating user comments on tool usability rather than just result accuracy. Participants offered suggestions for tool improvement, contributing to future iterations. A survey featuring scaled questions (from 1 to 5) was used to gather detailed user feedback.

Tasks & Questionnaire: The evaluation encompassed nine participants with varying levels of expertise in computer science and Git. They received comprehensive introductions to OntoEditor, TurtleEditor, and WebProtégé for comparative analysis, followed by tasks assigned across all three tools. Continuous monitoring tracked their approaches, with *Help* section guidance available.

Tasks were designed to cover activities from token-based authentication to the final Git commit, executed by three groups. Participants freely chose their preferred GIT VCS platform (GitHub, Bitbucket, or GitLab), completing similar tasks with minor variations in defining properties and instances within different base examples. These base examples in Fig. 7 were extended using TurtleEditor and WebProtégé for comparison.

Post-tasks, participants completed an electronic questionnaire, including the USE Questionnaire[16] utilizing a Likert scale. It assessed usefulness, ease of use, ease of learning, and satisfaction. The second section explored specific areas, gathering insights into individual service importance within OntoEditor. Open-response questions captured participant perceptions of pros and cons, influencing future service integration possibilities. The detailed survey questionnaire is available in our project's repository [6].

---

[16] https://garyperlman.com/quest/quest.cgi?form=USE.

Group 1	Group 2	Group 3
:Person rdf:type owl:Class ; rdfs:comment "Human"@en; rdfs:label "Person"@en .	:Media rdf:type owl:Class ; rdfs:comment "Media"@en; rdfs:label "Media"@en .	:Vehicle rdf:type owl:Class ; rdfs:comment "Transportation Mean"@en ; rdfs:label "Vehicle"@en .
:Actor rdf:type owl:Class ; rdfs:comment "Acting"@en ; rdfs:label "Actor"@en ; rdfs:subClassOf :Person .	:Movie rdf:type owl: Class ; rdfs:comment "Movie"@en ; rdfs:label "Movie"@en ; rdfs:subClassOf :Media .	:Bus rdf:type owl:Class ; rdfs:comment "Bus"@en ; rdfs:label "Bus"@en ; rdfs:subClassOf :Vehicle .

Fig. 7. Participants Group Assignments. Tasks included defining new properties and instances using those base examples.

Results: Participants efficiently completed tasks within 15 to 20 min. Post-study, the USE questionnaire revealed high ratings for OntoEditor: usefulness (4.45), ease of use (3.98), ease of learning (4.29), and satisfaction (4.41), indicating substantial usability favorability. Figure 8 shows participant ratings on OntoEditor, emphasizing high ratings for *Collaboration* and *Syntax Validation*. Key findings highlighted high satisfaction with collaboration for multi-user tasks, positive feedback on syntax checking across RDF formats, and suggestions for integrating a user login system and favoring single sign-on for authentication.

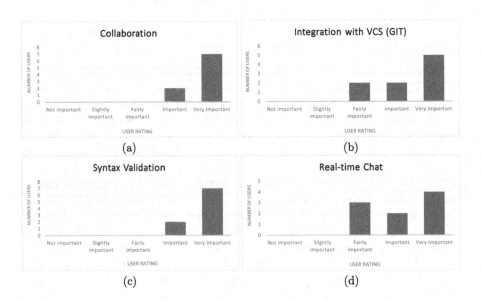

Fig. 8. User ratings. User Satisfaction Assessment of OntoEditor.

Comparing OntoEditor, TurtleEditor, and WebProtégé: OntoEditor scored 95%, outperforming both in collaboration and syntax validation. TurtleEditor integrates with VCS but lacks real-time collaboration, while WebProtégé supports collaboration but lacks synchronization with repositories and textual RDF editing compared to OntoEditor (Fig. 9).

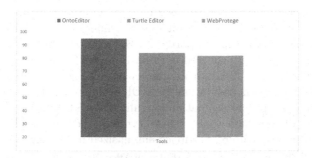

Fig. 9. Comparative study results. Showcasing OntoEditor's higher score of 95%, TurtleEditor's 84%, and WebProtégé's 82% in terms of collaboration and syntax validation.

7   Limitations

Throughout project testing tasks and user reviews, OntoEditor has revealed certain limitations, highlighting potential areas for improvement and future tasks:

1. **Lack of user authentication:** OntoEditor lacks its own authentication and user management system for controlling user access and permissions.
2. **No support for ontologies stored locally on file systems:** The primary focus of OntoEditor was to resolve Git conflicts, thus its utilization for ontology development on local file systems was not incorporated.
3. **Inability to edit multiple files via a single project link:** OntoEditor's current incapacity to simultaneously edit multiple files through a single project link hampers collaborative efficiency, especially when parallel edits across multiple files are necessary.
4. **Limited support for serialization formats:** While OntoEditor robustly supports ontology development, its compatibility remains restricted to specific serialization formats, including Turtle, RDF/XML, and JSON-LD.

8   Adoption

An outstanding attribute of OntoEditor is its seamless integration potential within VoCoREG [8], a comprehensive ontology development environment. VoCoREG augments OntoEditor by offering an array of functionalities such as Ontology Metrics, Evolution details, Query services, and Visualization of Ontologies. The integration promises an efficient platform for real-time collaborative editing of various RDF serialization formats among multiple contributors.

9   Conclusion

This paper introduces OntoEditor, a collaborative ontology development tool leveraging Version Control Systems. Its core modules-Repository Communication, Collaboration, and RDF Validation-are integral to its functionality. The Repository Communication module integrates RESTful APIs from GitHub, GitLab, and Bitbucket, ensuring direct user-repository interaction and conflict prevention. The Collaboration module, a separate microservice, enables real-time collaboration via WebSocket, supported by ShareDB for storage. Unique links allow simultaneous editing, coupled with an in-built chat feature for user communication. The RDF Validation module ensures error-free ontology development by real-time syntax validation for RDF serialization formats, enhancing typing accuracy. Empirical evaluations highlighted OntoEditor's standout features: praised collaboration tools, robust syntax validation, and user productivity, especially for Git users. Feedback emphasized improving user integration and implementing single sign-on. Comparative assessments affirmed OntoEditor's superiority over TurtleEditor and WebProtégé in collaboration and syntax validation. OntoEditor emerges as a versatile tool, poised to revolutionize collaborative ontology editing. Its integration potential with VoCoREG expands collaborative development across RDF serialization formats, significantly contributing to ontology development.

10   Future Work

OntoEditor holds potential for advancement in several crucial areas. First, implementing a user authentication system could significantly enhance collaboration by meticulously tracking individual changes. Moreover, integrating a single sign-on authentication method would simplify access to remote repositories, improving user experience and workflow efficiency. Expanding its support to encompass additional RDF serialization formats, such as RDFa and Notation3, stands as another pivotal area for OntoEditor's evolution. Additionally, enabling the platform to import local files for ontology development, independent of VCS, would mark a substantial stride toward enhanced versatility and accessibility.

Acknowledgement. We express gratitude to the Cognitive Internet Technologies Research Center at Fraunhofer for their vital support, as well as to our colleagues and students for their collaborative efforts. This work has been partially funded by the German Federal Government Commissioner for Culture and the Media (BKM) under grant number 2522DIG012. Special thanks to ChatGPT [9] for enhancing writing quality, optimizing sentence structure, and eliminating errors in this paper. We also acknowledge its use in summarizing initial notes and proofreading the final draft, extending our appreciation to its developers.

References

1. Berners-Lee, T.: Semantic Web Road Map (1998). https://www.w3.org/DesignIssues/Semantic.html. Accessed 01 Dec 2023
2. Cyganiak, R., Wood, D., Stones, R.M.L.: Resource Description Framework (RDF): Concepts and Abstract Syntax. https://www.w3.org/TR/rdf11-concepts/
3. Dang, Q.: Secure Hash Standard (SHS), Federal Information Processing Standard (NIST FIPS). National Institute of Standards and Technology, Gaithersburg, MD (2012). https://doi.org/10.6028/NIST.FIPS.180-4
4. Ellis, C.A., Gibbs, S.J.: Concurrency control in groupware systems. SIGMOD Rec., pp. 399–407 (1989). https://doi.org/10.1145/66926.66963
5. Fette, I., Melnikov, A.: The WebSocket Protocol (2011). https://tools.ietf.org/html/rfc6455. Accessed 11 Nov 2023
6. Hemid, A.: OntoEditor Survey Form. https://github.com/ahemaid/OntoEditor/blob/main/SURVEY_FORM.pdf. Accessed 21 Mar 2024
7. IsSemantic RDF Tools: isSemantic.net: Validate, visualize, generate, and convert structured data. https://issemantic.net/rdf-converter. Accessed 12 Nov 2023
8. Khiat, A., Halilaj, L., Hemid, A., Lohmann, S.: VoColReg: a registry for supporting distributed ontology development using version control systems. In: 2020 IEEE 14th International Conference on Semantic Computing (ICSC), pp. 393–399 (2020). https://doi.org/10.1109/ICSC.2020.00078
9. OpenAI: ChatGPT 3.5 (2023). https://chat.openai.com/, Large Language Model. Accessed 01 Mar 2024
10. Petersen, N., Coskun, G., Lange, C.: TurtleEditor: an ontology-aware web-editor for collaborative ontology development. In: Proceedings of the Tenth IEEE International Conference on Semantic Computing, 3–5 Feb 2016, Laguna Hills, California, USA (2016). https://doi.org/10.5281/zenodo.35499
11. Prud'hommeaux, E.: RDF Validation Service. http://www.w3.org/RDF/Validator/. Accessed 11 Jan 2024
12. Stellato, A., et al.: VocBench: a web application for collaborative development of multilingual thesauri. In: ESWC (2015)
13. Tudorache, T., Vendetti, J., Noy, N.: Web-Protege: a lightweight OWL ontology editor for the web. In: OWLED (2008)

Author Index

Printed in the United States
by Baker & Taylor Publisher Services